GIS *for* Environmental Decision-Making

INNOVATIONS IN GIS

SERIES EDITORS

Jane Drummond
University of Glasgow, Glasgow, Scotland

Bruce Gittings
University of Edinburgh, Edinburgh, Scotland

Elsa João
University of Strathclyde, Glasgow, Scotland

GIS for Environmental Decision-Making
Edited by Andrew Lovett and Katy Appleton

GIS and Evidence-Based Policy Making
Edited by Stephen Wise and Max Craglia

Dynamic and Mobile GIS: Investigating Changes in Space and Time
Edited by Jane Drummond, Roland Billen, Elsa João, and David Forrest

INNOVATIONS IN GIS

GIS for Environmental Decision-Making

Edited by
Andrew Lovett
Katy Appleton

CRC Press
Taylor & Francis Group
Boca Raton London New York

CRC Press is an imprint of the
Taylor & Francis Group, an **informa** business

CRC Press
Taylor & Francis Group
6000 Broken Sound Parkway NW, Suite 300
Boca Raton, FL 33487-2742

Printed in the United States of America on acid-free paper
10 9 8 7 6 5 4 3 2 1

International Standard Book Number-13: 978-0-8493-7423-4 (Hardcover)

Library of Congress Cataloging-in-Publication Data

National Conference on GIS Research UK (12th : 2004 : Norwich, England)
GIS for environmental decision-making / editors, Andrew Lovett and Katy Appleton.
p. cm. -- (Innovation in GIS)
Papers from the GIS Research UK (GISRUK) 12th Annual Conference held at the University of East Anglia, Norwich; April 28-30, 2004.
Includes bibliographical references and index.
ISBN 978-0-8493-7423-4 (alk. paper)
1. Environmental policy--Decision making--Equipment and supplies--Congresses. 2. Environmental sciences--Information technology--Congresses. 3. Geographic information systems--Congresses. 4. Environmental monitoring--Data processing--Congresses. I. Lovett, Andrew A. II. Appleton, Katy, 1975- III. University of East Anglia. IV. Title. V. Series.

GE170.N358 2008
628--dc22 2007034392

Visit the Taylor & Francis Web site at
http://www.taylorandfrancis.com

and the CRC Press Web site at
http://www.crcpress.com

Table of Contents

Preface

This book stems from the GIS Research UK (GISRUK) 12th Annual Conference which was held at the University of East Anglia in Norwich from 28–30th April 2004. The conference was attended by over 190 participants from more than 20 countries, reflecting the way in which the GISRUK series has evolved into a prominent international event for those interested in research on Geographical Information Science and Systems. Details of future GISRUK conferences are available at http://www.geo.ed.ac.uk/gisruk/gisruk.html.

The conference at UEA was hosted by staff from the School of Environmental Sciences and several of the sessions stemmed from the research interests of the local organizing committee. These included coastal and health applications of GIS, decision-making, environmental hazards and landscape visualization. With the assistance of the National Institute for Environmental eScience at the University of Cambridge (http://www.niess.ac.uk) it was also possible to organize a session on the implications of developments in Grid computing for GIS. Plenary presentations relating to several of these topics were given by Ian Bishop (University of Melbourne) and Mark Gahegan (Pennsylvania State University). We would like to thank all the conference sponsors whose financial input supported the participation of the plenary speakers, two evening events, speaker prizes and the provision of a number of student bursaries. Considerable thanks are also due to the other members of the local organizing committee (Iain Brown, Ollie Bennett, Trudie Dockerty, Iain Lake, Andy Jones, Simon Jude and Paulette Posen) for all their hard work. We would also like to acknowledge the contribution from members of the GISRUK National Steering Committee in reviewing abstracts submitted to the conference. Particular thanks are due to Bruce Gittings (Chair of the National Steering Committee) and Peter Halls (organizer of the Young Researchers Forum) for their input to the conference and valuable advice during the planning stages.

A set of papers from the conference on the theme of 'Extracting Information from Spatial Datasets' were published in January 2007 as a special issue of the journal *Computers, Environment and Urban Systems* (Volume 31/1). This book brings together other papers from the meeting around the theme of environmental decision-making. It includes a contribution based on the plenary presentation by Ian Bishop and two others that won prizes for being the best presentations during the conference (those by Foyfa Shutidamrong and Christian Castle). All of the original conference papers have been reviewed and revised for this book and two other contributions were invited to extend the coverage of particular topics. Our aim in compiling the book was to illustrate the current 'state of the art' in the use of GIS for environmental decision-making, and we would like to think that the contents provide a sense of some interesting and important research advances.

Many people have helped during the production of this book. We would like to thank many of the authors for their patience and prompt responses to our queries or requests. Jill Jurgensen and Tai Soda at Taylor & Francis have also provided invaluable guidance and support, particularly in the later stages of manuscript preparation. A significant amount of the final editing was undertaken while Andrew Lovett was on study leave at the Institüt fur Umweltplanung, Leibniz Universität Hannover and we would like to thank Professor Christina von Haaren for her assistance in making departmental facilities available. Research support from the ESRC to the Centre for Social and Economic Research on the Global Environment (CSERGE) Programme on Environmental Decision-making is also acknowledged. In the final stages of manuscript preparation Trudie Dockerty and Paulette Posen made some very timely contributions to the compilation of the book preliminaries and index. To all who have assisted in some way our very real thanks.

Andrew Lovett and Katy Appleton
Norwich, June 2007

List of Contributors

Katy Appleton is the Research Officer for the Social Science for the Environment, Virtual Reality and Experimental Laboratories, within the Zuckerman Institute for Connective Environmental Research, in the School of Environmental Sciences at the University of East Anglia. Her interests are GIS-based landscape visualization, particularly for communication and participation purposes relating to environmental management and decision-making; perceptions of different visualization content and presentation methods.

Zuckerman Institute, School of Environmental Sciences, University of East Anglia, Norwich NR4 7TJ, UK; Email: k.appleton@uea.ac.uk

Peter Bebi is leader of the Forest and Treeline Ecotone Group in the Ecosystem Boundaries research unit of the Swiss Federal Institute for Snow and Avalanche Research (WSL/SLF). His research interests include analyzing the dynamics and pattern of forest and treeline ecotones in response to land-use change, climatic changes and natural disturbances; evaluating changes in ecosystem services in relation to changes in forest and treeline ecotones and studying interrelationships between subalpine forests and snow avalanches and proving practical recommendations in questions related to avalanche protection forests and risk-based management of mountain forests.

WSL-Swiss Federal Institute for Snow and Avalanche Research SLF, Fluelastrasse 11, CH-7260, Davos, Switzerland; Email: bebi@slf.ch

Ian Bishop is a Professor in the Department of Geomatics, University of Melbourne. He is Director of the Centre for Geographic Information Systems and Modelling (CGISM). His research interests include GIS applications, visualization and visual analysis, particularly using interactive visual displays of planning data and options to assist design and decision processes; generation of realistic visual imagery for communication of options to the public; research into public perceptions and choices using virtual environments and use of visualization tools to predict the effect of environmental changes on people's attitudes.

Department of Geomatics, Engineering Building B, University of Melbourne, 3010 Australia; Email: i.bishop@unimelb.edu.au

Iain Brown is co-ordinator of the thematic research program on Landscapes and Rural Communities at the Macaulay Institute. His research focuses on developing integrated responses to climate change, including links with soils, ecology, freshwater and coastal systems. He is also interested in examining people's perception of, and responses to, climate change, and therefore how this can be better connected with the emerging science. A key technique for this is the modeling and analysis of future scenarios, together with the application of Geographic Information Systems (GIS) and 3D landscape visualization.

Macaulay Institute, Craigiebuckler, Aberdeen, AB15 8QH, UK; Email: i.brown@macaulay.ac.uk

Christian Castle is a postgraduate researcher working on an ESRC CASE studentship entitled 'A GIS-Based Spatial Decision Support System (SDSS) for Emergency Services: London's King's Cross Redevelopment'. Current interests relate to emergency planning, SDSS, pedestrian simulation and network analysis, and public participation in GIS.

Centre for Advanced Spatial Analysis, University College London, 1-19 Torrington Place, London, WC1E 7HB; Email: c.castle@ucl.ac.uk

Alastor Coleby was previously a PhD student at Heriot Watt University and the Macaulay Institute, researching public attitudes towards changes in the landscape, with specific reference to wind turbines. Following completion of his PhD he is now working at Guangdong University in China.

Email: alastorcoleby@yahoo.co.uk

Mark Dickson is a geomorphologist specializing in coastal erosion, particularly the use and development of numerical models to study shoreline change over the scale of decades to millennia. At present he is working on alluvial fan coasts in South Island, New Zealand, with previous work on soft-rock glacial-till shores in East Anglia and hard-rock basaltic shores around Lord Howe Island, southwest Pacific.

National Institute of Water and Atmospheric Research (NIWA), Christchurch New Zealand. E-mail: m.dickson@niwa.co.nz

Helen Dunsford is a member of the Division of Geography and Environmental Management, School of Applied Sciences, Northumbria University. She is particularly involved in GIS applications as part of the Centre for Environmental and Spatial Analysis (CESA).

School of Applied Sciences, Northumbria University, Newcastle upon Tyne, NE1 8ST; Email: helen.dunsford@northumbria.ac.uk

Sarah Evers is a Hydrological Systems Senior Scientist working in Earth Sciences within the Environment Agency's Science Group. She is involved in national-scale GIS groundwater modelling studies and is currently working on a methodology for the assessment of significant damage to wetlands for the Water Framework Directive and in developing a framework for ecological modelling in the Agency.

Environment Agency, Olton Court, 10 Warwick Road, Olton, Solihull, B92 7HX, UK; Email: sarah.evers@environment-agency.gov.uk

Ángel Felicísimo is a Professor at the University of Extremadura. His work focuses on the development and application of environmental predictive models.

Universidad de Extremadura, Escuela Politécnica, Adva. de la Universidad s/n 10071 Cáceres, Spain; Email: amfeli@unex.es

Alicia Gómez-Muñoz is a doctoral student at the University of Extremadura, working with applications of GIS and remote sensing to ecological models.

Universidad de Extremadura, Escuela Politécnica, Adva. de la Universidad s/n 10071 Cáceres, Spain; Email: aligm@unex.es

Simon Gomm has worked at the Ordnance Survey of Great Britain in a variety of roles ranging from geodetic surveying to data quality assurance and, more recently, research, particularly related to data enhancement and data integration. He is currently Senior Technical Product Manager responsible for technical issues related to large-scale topographic products.

Ordnance Survey. Romsey Road, Southampton, United Kingdom, SO16 4GU

Sylvia Herrmann is a researcher in the Institute for Environmental Planning, Leibniz University, Hannover. Her main research interests are the integrated planning and development of rural areas under the conditions of global change, including interactions between economic, environmental and social problems; scenario techniques (based on interdisciplinary modelling systems) to develop integrated strategies and to derive information for political decision support; and the environmental effects of different land-use types in rural areas.

Institute for Environmental Planning, Leibniz University Hannover, Herrenhäuserstrasse 2, D- 30419 Hannover, Germany; Email: herrmann@umwelt.uni-hannover.de

Kevin Hiscock is a hydrogeologist and Senior Lecturer in the School of Environmental Sciences at the University of East Anglia. His interests include the application of stable isotope methods and dissolved gases in hydrogeological investigations and understanding groundwater recharge and flow processes. He is currently researching the impacts of land management practices and climate change on groundwater resources through the application of groundwater models.

School of Environmental Sciences, University of East Anglia, Norwich NR4 7TJ, UK; Email; k.hiscock@uea.ac.uk

Claire Jarvis is a Senior Lecturer in Geographical Information at the University of Leicester. She is currently exploring the emerging GIScience implications of the connections between input data quality/quantity with the quality and delivery speed of modelled outputs and how visualization and integration methods impact upon the transfer, perception and usability of the results. Associated work has included the development of spatial analysis methods for projecting climate surfaces across the landscape, using both geostatistical methods and techniques from artificial intelligence. Of late Claire has become particularly interested in developing collaborative geographical modelling environments using web service and emergent Grid technologies.

Department of Geography, University of Leicester, Bennett Building, University Road, Leicester LE1 7RH; Email: chj2@le.ac.uk

Simon Jude is a Senior Research Associate at the Tyndall Centre for Climate Change Research at the School of Environmental Sciences, University of East Anglia UK. His research involves developing the use of GIS, virtual reality, and visualization techniques for coastal decision-making.

Tyndall Centre for Climate Change Research, School of Environmental Sciences, University of East Anglia, Norwich, Norfolk, NR4 7TJ, E-mail s.jude@uea.ac.uk

Zarine Kemp is an Honorary Senior Lecturer in Computer Science at the University of Kent. Her research interests include interoperable spatiotemporal information systems, knowledge representation and multi-dimensional reasoning for environmental decision support. She has been actively involved in developing systems for fisheries and marine environmental research as a senior partner in EU-funded projects.

Computing Laboratory, University of Kent, Canterbury, CT2 7NF, UK; Email: z.kemp@kent.ac.uk

Sotirios Koukoulas received his BSc in Environmental Studies from the University of the Aegean (Greece) and a PhD degree in Geography from King's College London (UK) in 1994 and 2001, respectively. He is currently an Assistant Professor in Geographical Information Systems at the University of the Aegean (Greece) in the Department of Geography.

Department of Geography, University of the Aegean, University Hill, Mytilene 81100, Greece. E-mail skouk@geo.aegean.gr

Paul Longley, Professor of Geographic Information Science (Department of Geography, UCL), acts as Deputy Director of CASA. He was previously Professor of Geography at the University of Bristol and has also worked in the universities of Cardiff, Reading and Karlsruhe. His research is focused around the principles and techniques of geographic information science (GISc), with research applications in geodemographics, public service provision and urban morphological analysis.

Centre for Advanced Spatial Analysis, University College London, 1-19 Torrington Place, London, WC1E 7HB; Email: p.longley@ucl.ac.uk

Andrew Lovett is a Professor in Environmental Sciences at the University of East Anglia, Norwich. His background is in human geography, specializing in applications of GIS, statistical and virtual reality techniques. Recent research has included work on modelling accessibility to health services and the assessment of measures for reducing diffuse agricultural pollution and flooding risks. At present, he has a particular interest in issues of future rural land-use change, including involvement in projects that make use of the specialist visualization equipment within the School of Environmental Sciences.

Zuckerman Institute, School of Environmental Sciences, University of East Anglia, Norwich NR4 7TJ, UK; Email: a.lovett@uea.ac.uk

Robert MacFarlane is Assistant Director, Training and Doctrine, at the Cabinet Office Emergency Planning College. Prior to this, at Northumbria University, he conducted research and consultancy across a wide range of areas, including land use and energy planning, crime analysis, aircraft birdstrike hazard assessment, post-conflict reconstruction and emergency/disaster preparedness and response in the developed and developing world. A particular area of specialty is the application of

GIS in Integrated Emergency Management; he currently chairs the Association for Geographical Information (AGI) Emergency Planning Group.

Cabinet Office Emergency Planning College, The Hawkhills, Easingwold, York YO61 3EG; Email: robert.macfarlane@cabinet-office.x.gsi.gov.uk

Pernette Messager is a researcher at the Macaulay Institute. Her research interests include virtual environment rendition types and understanding of information in a rural landscape context, along with evaluation of visual impacts of rural development measures on rural landscape components.

Macaulay Institute, Craigiebuckler, Aberdeen, AB15 8QH, UK; Email: pe.messager@macaulay.ac.uk

David Miller is a researcher into issues of landscape change, including quantitative and qualitative assessments of the effects of drivers of change on the visual landscape. He also oversees the Macaulay Institute's Virtual Landscape Theatre.

Macaulay Institute, Craigiebuckler, Aberdeen, AB15 8QH, UK; Email: d.miller@macaulay.ac.uk

Jane Morrice is a researcher on issues relating to landscape change at the Macaulay Institute.

Macaulay Institute, Craigiebuckler, Aberdeen, AB15 8QH, UK; Email: j.morrice@macaulay.ac.uk

Dean Morton is a graduate of the University of Newcastle upon Tyne with a BA Hons in Geography. He joined Ordnance Survey in 2001 after some 20 years work in property development consultancy in the leisure, retail and residential sectors. In his first three years he represented Ordnance Survey in inter-government department committees developing the scope and user needs for a land-use map of Great Britain and has gone on to manage and represent them in several pan-European data projects.

Ordnance Survey. Romsey Road, Southampton, United Kingdom, SO16 4GU

Stefan Neumeier is a researcher in the Institute of Rural Studies at the Federal Agricultural Research Centre. His research interests include economic and living conditions in rural areas.

Federal Agricultural Research Centre, Institute of Rural Studies, Bundesallee 50, 38116 Braunschweig, Germany; Email: stefan.neumeier@fal.de

Robert Nicholls is Professor of Coastal Engineering at the University of Southampton, and Programme Leader for Coasts in the Tyndall Centre for Climate Change Research. His research focuses on understanding and managing long-term coastal change, especially impacts and adaptation to climate change and sea-level rise in coastal zones, including methodological developments, case studies and regional and global analyses. He is co-ordinating lead author of the 'Coastal Systems and Low-Lying Areas' chapter in the IPCC fourth assessment report.

School of Civil Engineering and the Environment and the Tyndall Centre for Climate Change Research, University of Southampton, Southampton, SO17 1BJ, UK. E-mail r.j.nicholls@soton.ac.uk

Chris Parker is Head of the Research Labs at Ordnance Survey.

Research and Innovation, Ordnance Survey. Romsey Road, Southampton, United Kingdom, SO16 4GU; Email: chris.parker@ordnancesurvey.co.uk

Paulette Posen is Senior Research Associate in the Zuckerman Institute for Connective Environmental Research, School of Environmental Sciences at the University of East Anglia. Her interests are Geographical Information Science; relationships between land use and water quality; integrated modelling of EU Water Framework Directive impacts upon rural land use and farm incomes.

Zuckerman Institute, School of Environmental Sciences, University of East Anglia, Norwich NR4 7TJ, UK; Email: p.posen@uea.ac.uk

Ross Purves is a Lecturer and Project Coordinator of the GIS Division, Department of Geography, University of Zurich-Irchel. His research focuses on two areas – environmental modelling and Geographic Information Retrieval. He has been involved in projects concerning the influence of terrain representation on large-scale environmental modelling and its contribution to uncertainty in model results and additionally in research using ice-sheet models to examine potential climatic scenarios. Currently he is involved in research on the automatic captioning of images, based on their locations.

Department of Geography, University of Zurich-Irchel, Winterhurerstrasse 190, Zurich, Switzerland; Email: rsp@geo.unizh.ch

Brian Reid is a Senior Lecturer in the School of Environmental Sciences at the University of East Anglia. His areas of interest include environmental chemistry and soil science; fate and behavior of persistent organic pollutants (POPs), pesticides and antibiotics in soil; contaminated land, bioremediation and the legislative position regarding contaminant availability; ecological impacts of contamination; the formation and stability of bound residues; the potential for groundwater contamination and natural attenuation.

School of Environmental Sciences, University of East Anglia, Norwich NR4 7TJ, UK; Email: b.reid@uea.ac.uk

Elizabeth Seaman has BSc (Hons) in Geography from the University of Plymouth and a Master's Degree in Applied Remote Sensing from Cranfield University. She has more than 12 years' experience in the use of spatial data and has worked in both the public and private sector. While at Ordnance Survey, she was involved in its research department investigating the use of remotely sensed data for feature extraction, and then working as a GI Consultant assisting customers in their use of geographic information to support business processes. Elizabeth is currently Geography Project Manager at Communities and Local Government, delivering access to geographic data within central government and promoting best practice.

Communities and Local Government, Eland House, Bressenden Place, London, SW1E 5DU

Foyfa Shutidamrong was formerly a PhD student in the School of Environmental Sciences at the University of East Anglia and is now Project Officer with The Royal Chitralada Projects in Bangkok. Her interests include evaluating techniques for stakeholder participation in land-use planning.

The Royal Chitralada Projects, Bureau of Royal Household, Chitralada Villa, Dusit, Bangkok 10303, Thailand; Email: zahsamerr@yahoo.com

Aidan Slingsby is a post-graduate researcher with the Centre for Advanced Spatial Analysis at University College London and is working on a three-dimensional digital mapping project with the Ordnance Survey. He is interested in methods and techniques for modelling and representing space. His MSc project, ‘An object-orientated approach to hydrological modelling based on a triangular irregular network’, designed, evaluated and implemented an alternative approach for hydrological modelling, compared to traditional raster-based methods.

Centre for Advanced Spatial Analysis, University College London, 1-19 Torrington Place, London, WC1E 7HB; Email: a.slingsby@ucl.ac.uk

William Tompkinson joined Ordnance Survey in 2001 as a Research Scientist. Here, he pursued his interests in remote sensing and landscape modelling in a range of research themes that included land-use classification. Since 2005, he has worked in Ordnance Survey’s consultancy team advising customers on the use of geographic information. In January 2007, William was appointed as an Industrial

Fellow in the School of Geography at the University of Nottingham.

Ordnance Survey. Romsey Road, Southampton, United Kingdom, SO16 4GU; Email: william.tompkinson@ordnancesurvey.co.uk

Mike Walkden works for the Tyndall Centre at the University of Newcastle where he develops large-scale numerical models of coastal morphology under climate change. He has also worked in coastal research at the universities of Plymouth, Belfast, Edinburgh, Bristol and at Halcrow.

School of Civil Engineering and Geosciences and the Tyndall Centre for Climate Change Research, University of Newcastle upon Tyne, Newcastle upon Tyne, NE1 7RU, UK; E-mail: mike.walkden@newcastle.ac.uk

Ariane Walz has research interests that include land-use modelling, participatory modelling, scenario analysis, integrated regional modelling and natural hazards and risk management. She has worked on simulation of land-use changes in the Swiss mountain area (LUCALP) as part of the ALPSCAPE Integrated Regional Modelling project.

WSL-Swiss Federal Institute for Snow and Avalanche Research SLF, Fluelastrasse 11, CH-7260, Davos, Switzerland; Email: walz@slf.ch

Rob Ward is a Technical Advisor in the Geoscience Process Team and manages the Environment Agency's national groundwater quality monitoring program. He is also involved in implementing the Water Framework Directive (WFD), acting as an expert advisor to the European Commission and co-leading drafting groups producing WFD guidance documents on monitoring, threshold setting, status classification and trend assessment.

Environment Agency, Olton Court, 10 Warwick Road, Olton, Solihull, B92 7HX, UK; Email: rob.ward@environment-agency.gov.uk

Jacqueline Whalley is a Senior Lecturer in the School of Computer and Information Sciences, Auckland University of Technology.

School of Computer and Information Sciences, Auckland University of Technology, Auckland, New Zealand; Email: jacqueline.whalley@aut.ac.nz

CHAPTER 1

Developments in GIS for Environmental Decision-Making

A. Lovett and K. Appleton

1.1 INTRODUCTION

Decision-making on environmental issues is rarely straightforward. In part, this reflects several distinctive features of environmental systems, namely their dynamic nature, the complexity of interactions involving physical, chemical or biological processes, and uncertainty regarding the functioning of such processes[1]. There is also increasing recognition that longer-term sustainability in environmental management often requires consideration of socio-economic issues as well, including the development of methods to assess different options and facilitate stakeholder participation in decision-making processes[2,3]. This, in turn, reinforces the need for multi-disciplinary or interdisciplinary perspectives[4].

Computer-based decision support systems (DSS) have been one, increasingly widespread, response to the challenges of improving environmental management and planning[5,6]. An environmental DSS can be defined and constructed in many different ways, but typically involves capabilities to acquire and structure data, model processes, and provide guidance to the user on different courses of action[7,8]. Geographical information systems (GIS) are seen as a key component of environmental DSS, not least because there is almost always a spatial element to the decisions to be made and the data used to make them. However, the use of GIS also has a role in environmental decision-making beyond the implementation of DSS. This book takes such a broader perspective and seeks to illustrate the current 'state of the art' across a range of conceptual, technical and application issues. To set the individual contributions in context the following section provides a brief overview of how the nature and use of GIS technology has evolved.

1.2 THE ROLE OF GIS

Environmental applications have long been a core use of GIS. Many of the earliest applications were primarily concerned with matters of inventory and measurement, but from the mid 1980s a much greater emphasis on statistical analysis and modelling was apparent. Much of this work involved coupling GIS with other software tools and has gradually become more sophisticated in terms of numerical complexity and the representation of change over time[9,10].

In the past decade, innovations such as the use of internet and wireless communications to support Distributed GIS and Location-Based Services (LBS)[11,12], the availability of higher-resolution GPS and satellite data, improvements in computer processing power and interoperability and the wider adoption of object-orientated software concepts have opened up new opportunities for the dissemination, analysis and display of spatial data[13]. These have included individual level (agent-based) modelling of processes[14] and much more detailed forms of 3D geovisualization[15,16].

Alongside these technical developments, the years since the mid 1990s have also seen a greater emphasis on the societal implications of GIS use. One aspect of this has been a growth in research on Public Participation GIS (PPGIS), something which can be seen as consistent with a broader emphasis on communication and collective design in planning[17,18]. Recent reviews recognize that the use of PPGIS has brought some improvements to decision-making processes, but also acknowledge that there are still limitations with respect to such issues as access (e.g., to the internet), digital representation of knowledge and capacity for conflict resolution[19-21]. Other 'grand challenges' relating to data availability, the representation of geographical phenomena, modelling capabilities and user-empowerment have been highlighted for GIS as a whole[22,23].

1.3 THE STRUCTURE OF THE BOOK

This book reflects the above developments and challenges by being divided into three parts. These relate to the data required, tools being developed, and aspects of participation; three key aspects of the decision-making process as supported by GIS. Furthermore, as each of these elements improves and increases in complexity, the others are enabled to do so too – there is a cycle of development. Data must be of high quality to support robust decision-making, and procedures to integrate different sources or automate the extraction of features add further value. These data feed in to GIS-based modelling of environmental features and processes, which is fundamental to the decision-making process; the possibilities presented by ever-improving network technology (internet or wireless) considerably widen the scope for this type of activity. Increased network connectivity among all sections of society also increases the scope for new methods of presenting information and opens up the decision-making process to a wider audience. Such participation also has the potential to improve the underlying information base, particularly through providing data that would otherwise be hard to collect.

1.3.1 Data for Decision-Making

The first section of the book deals with the fundamental requirement for all GIS operations: data. An ability to integrate data from different sources has long been regarded as one of the defining characteristics of a GIS and Tompkinson et al.

provide an illustration of the benefits that this can bring in the context of deriving land-use information. They present an approach for automatically deriving such information from a range of sources, particularly Ordnance Survey MasterMap®, and demonstrate that the proposed combination of techniques has the potential to provide an important base resource for many aspects of environmental decision-making.

Issues of improved feature representation are frequently mentioned in GIS research agendas. In their chapter, Slingsby et al. focus on the development of a 3D feature framework for urban environments. They begin by reviewing existing frameworks in Great Britain and then discuss how these would need to be extended to provide a 3D data model that could be implemented on a national scale. As they note, such a development would provide many benefits for urban management, particularly in respect to issues of access and to support work on virtual cities[24,25].

Matters of representation and data integration are also discussed by Whalley and Kemp in their chapter on the development of the FISHCAM GIS. This modular system was designed to support implementation of the EU Common Fisheries Policy. It includes facilities to integrate data from a range of sources (including real-time GPS), as well as capabilities to query and visualize the resulting information in a very flexible manner. On a broader level, the chapter also highlights the wide variety of data typically involved in environmental decision-making, and the challenges associated with structuring and combining them.

1.3.2 Tools to Support Decision-Making

Analysis and modelling techniques are at the heart of many GIS applications. MacFarlane and Dunsford begin this section by discussing a range of issues concerning the use of GIS for suitability mapping and strategic planning exercises. Through a series of examples they illustrate some of the problems and dilemmas that can arise in such work, including the judgments that are often involved in translating criteria into map layers or implementing particular techniques. Furthermore, they highlight the importance of the political context in which such works takes place, especially when it is a matter of translating strategic plans into actual siting decisions.

The next three chapters are concerned with the use of different statistical techniques within a GIS context. Posen et al. examine the use of different spatial interpolation methods to generate surfaces of depth to water table, a key variable in groundwater vulnerability assessment. They conclude that an approach based on kriging produces the most robust results, though issues related to outlier values and edge effects still require careful consideration.

Felicísimo and Gómez-Muñoz compare three regression methods to model tree species distributions in the Extremadura region of Spain. Their analysis suggests that it is important to allow for non-linearity in such models and that multiple adaptive regression splines produce the most appropriate potential vegetation maps for forest planning. Walz et al. also use regression techniques and compare them with transition probability matrices as a means of simulating land-use changes in the Swiss Mountain Areas. Both methods are found to have merits, though the simpler input requirements of a regression approach mean that it is easier to implement on a regional scale and to assess future scenarios. These two chapters also raise a number of issues concerning the linkage of statistical techniques and GIS. Recent years have seen considerable improvements in this type of integration, but as Anselin[26, p.106] notes *'much remains to be done'*.

Multi-criteria evaluation (MCE) techniques provide a means of incorporating stakeholder preferences into decision processes and have been used in a wide variety of environmental contexts[27]. Shutidamrong and Lovett present an application of such methods to land-use planning for a watershed in northern Thailand. Their research does not succeed in identifying a single compromise scenario for land-use zoning, but does demonstrate how stakeholder involvement can be facilitated and help identify both the extent of current consensus and the problems that still require attention. They also comment on the practical issues involved in implementing MCE techniques and caution that the data and personal effort required can be considerable.

Developments in e-science are now starting to offer new opportunities for GIS analysis, particularly in terms of data access, shared computing resources and collaborative working. The possibilities and challenges offered by Grid technology are reviewed by Jarvis in her chapter. In particular, she highlights the need for work on research infrastructures, ontologies and metadata, as well as the adjustment in attitudes that is often required to support successful 'virtual organizations' and interdisciplinary investigations.

1.3.3 Participation in Decision-Making

The final section of the book focuses on the use of GIS-based techniques to facilitate public participation in decision-making processes. In the opening chapter Bishop provides an overview of developments in this area, concentrating particularly on how GIS, modelling and 3D landscape visualization techniques have gradually achieved closer integration. Ongoing work described in this chapter illustrates the possibilities of much more interactive group decision-making experiences, with 'what if' capabilities that extend substantially beyond traditional GIS outputs.

Integration of GIS, simulation modelling and visualization tools is also a feature of the research by Brown et al. on coastal erosion scenarios. They discuss the data processing required to produce real-time visualizations of cliff recession scenarios and the value of such displays when engaging with coastal managers or other stakeholders. However, they also highlight the manner in which model outputs had to be extrapolated to provide a sufficiently detailed basis for the visualizations and the need for future research on the quantification, representation and communication of uncertainties in the scenarios. Such issues related to uncertainty in geographical data are widely recognized as a priority for GIS research[28].

In addition to developing techniques to facilitate stakeholder engagement and participation it is also important to evaluate the effectiveness of such tools. The next two chapters both address this topic. Castle and Jarvis discuss the creation and use of a web-based PPGIS to rate the accessibility of buildings to mobility impaired and able bodied users on the University of Leicester campus. The results indicate the value of such an approach in compiling an evidence base for collective action and also in providing a sense of empowerment for the mobility impaired. It is emphasized, however, that further initiatives are required to translate the information and experiences associated with the PPGIS into emancipatory outcomes.

Miller et al. present the findings of an experiment using a portable virtual reality theater to assess public attitudes to the design and layout of wind turbine developments. Participants in their study were shown a sequence of real-time landscape visualizations and asked to vote (via handsets) on which turbine, if any, should be removed from the display. Evaluation of the outcomes suggests that such an approach to stakeholder engagement was received positively, but also highlights the importance of issues such as realism in the visualizations and the need to explore the reasons behind the choices made.

The final chapter by Herrmann and Neumeier provides further examples of web-based PPGIS, but is less concerned with technical aspects than the social implications of creating such tools. In particular, using examples from Bavaria and applying concepts from Actor-Network Theory, they discuss how the planning and introduction of web-GIS services for tourists also facilitated the creation of new stakeholder communities concerned with regional development goals. The ability of computer technology to provide a nexus around which multi-disciplinary collaboration can take place has also been noted for GIS as a whole[29]. Indeed, given the nature of environmental problems noted earlier this may be one of the most valuable indirect contributions of GIS to improvements in decision-making.

1.4 WHERE NEXT?

The chapters in this book provide a sense of the current state of GIS for environmental decision-making. Together, they emphasize the importance of matters related to data, analysis and modelling tools, and stakeholder participation. Further advances in all of these aspects can be expected in the next few years, driven by developments such as the wider shift to Distributed GIS and Location-Based Services, which will enhance access to data and processing resources. There is also a continuing political emphasis on increasing public participation in decision-making, e.g., the implementation of the Aarhus Convention[30] in Europe. Moreover, the recent introduction of geobrowsers such as Google Earth[31], NASA

World Wind[32] and ArcGIS Explorer[33] is already having major impacts on public familiarity with, and ability to interact with, geographical information, and could even be considered to be driving an expectation of such information being provided. Through the capacity to 'mash' such tools with other software and sources of data the manner in which many environmental issues are discussed and decisions taken could also undergo substantial changes.

Beyond these technical innovations, however, a theme that runs through many of the chapters in this book is that the effectiveness of GIS-based methods depends on the decision-making frameworks and contexts within which they are employed; careful thought must be put into the design of decision-making processes to allow spatial information to be used effectively and properly understood by all participants. By itself GIS is not, and never will be, a universal panacea for environmental decision-making. However, its capacities to integrate, analyze and display data are clearly extremely valuable in drawing together the ever more varied pieces of information on which decisions need to be based, implying that greater involvement of GIS specialists would be highly beneficial in future interdisciplinary or transdisciplinary[34] initiatives to address environmental problems. The evidence from this book suggests that the use of GIS can make a significant contribution to such collaborations, which are becoming increasingly necessary given the complex challenges presented by global environmental change.

1.5 REFERENCES

1. Guariso, G. and Werthener, H, *Environmental Decision Support Systems*, Wiley, New York, 1989.

2. Hemmati, M., Dodds, F., Enayati, J., and McHarry, J., *Multi-Stakeholder Processes for Governance and Sustainability: Beyond Deadlock and Conflict*, Earthscan Publications, London, 2002.

3. Gibson, R.B., Hassan, S., Holtz, S., Tansey, J., and Whitelaw, G., *Sustainability Assessment: Criteria, Processes and Applications*, Earthscan Publications, London, 2005.

4. O'Riordan, T., Environmental science on the move, in *Environmental Science for Environmental Management*, 2nd Edition, O'Riordan, T., Ed., Prentice Hall, Harlow, 2000, 1-27.

5. Geertman, S. and Stillwell, J., Planning support systems: an introduction, in *Planning Support Systems in Practice*, Geertman, S. and Stillwell, J., Eds., Springer, Berlin, 2003, 3-22.

6. Matthies, M., Giupponi, C., and Ostendorf, B., Environmental decision support systems, current issues, methods and tools, *Environmental Modelling and Software*, 22, 123-127, 2007.

7. Rizzoli, A.E. and Young, W.J., Delivering environmental decision support systems: software tools and techniques, *Environmental Modelling and Software*, 12, 237-249, 1997.

8. Poch, M., Comas, J., Rodriguez-Roda, I., Sànchez-Marrè, M. and Cortés, U., Designing and building real environmental decision support systems, *Environmental Modelling and Software*, 19, 857-873, 2004.

9. Fedra, K., GIS and environmental modeling, in Goodchild, M.F., Parks, B.O., and Steyaert, L.T., Eds., *Environmental Modeling with GIS*, Oxford University Press, Oxford, 1993, 35-50.

[10.] Skidmore, A., Ed., *Environmental Modelling with GIS and Remote Sensing*, Taylor & Francis, London, 2002.

[11.] Peng, Z-R. and Tsou, M-H., *Internet GIS: Distributed Geographic Information Services for the Internet and Wireless Networks*, Wiley, Chichester, 2003.

[12.] Jiang, B. and Yao, X., Location-based services and GIS in perspective, *Computers, Environment and Urban Systems*, 30, 712-725, 2006.

[13.] Maguire, D.J., Batty, M., and Goodchild, M.F., Eds., *GIS, Spatial Analysis and Modeling*, ESRI Press, Redlands, California, 2005.

[14.] Parker, D.C., Manson, S.M., Janssen, M.A., Hoffmann, M.J., and Deadman, P., Multi-agent systems for the simulation of land-use and land-cover change: a review, *Annals of the Association of American Geographers*, 93, 314-337, 2003.

[15.] Dykes, J., MacEachren, A.M. and Kraak, M-J., Eds., *Exploring Geovisualization*, Elsevier, Amsterdam, 2005.

[16.] Bishop, I. and Lange, E., Eds., *Visualization in Landscape and Environmental Planning*, Taylor & Francis, Abingdon, 2005.

[17.] Klosterman, R.E., Planning support systems: a new perspective on computer-aided planning, in *Planning Support Systems: Integrating Geographic Information Systems, Models, and Visualization Tools*, Brail, R.K. and Klosterman, R.E., Eds., ESRI Press, Redlands, California, 2001, 1-23.

[18.] Malczewski, J., GIS-based land-use suitability analysis: a critical overview, *Progress in Planning*, 62, 3-65.

[19.] Carver, S., The future of participatory approaches using geographic information: developing a research agenda for the 21st Century, *URISA Journal*, 15, Access and Participatory Approaches I, 63-71, available at http://www.urisa.org/journal-archives, 2003.

[20.] Kyem, P.A.K., Of intractable conflicts and participatory GIS applications: the search for consensus amidst competing claims and institutional demands, *Annals of the Association of American Geographers*, 94, 37-57, 2004.

[21.] Elwood, S., Critical issues in participatory GIS: deconstructions, reconstructions and new research directions, *Transactions in GIS*, 10, 693-708, 2006.

[22.] Longley, P.A., Goodchild, M.F., Maguire, D.J., and Rhind, D.W., *Geographic Information Systems and Science*, Second Edition, Wiley, Chichester, 2005, 478-485.

[23.] Longley, P.A., Grand challenges, environment and urban systems, *Computers, Environment and Urban Systems*, 30, 1-9, 2006.

[24.] Hudson-Smith, A. and Evans, S., Virtual cities: from CAD to 3-D GIS, in *Advanced Spatial Analysis: The CASA Book of GIS*, Longley, P.A. and Batty, M., Eds., ESRI Press, Redlands, California, 2003, 41-60.

[25.] Centre for Advanced Spatial Analysis, http://www.casa.ucl.ac.uk/index.asp, 2007.

[26.] Anselin, L., Spatial statistical modeling in a GIS environment, in *GIS, Spatial Analysis and Modeling*, Maguire, D.J., Batty, M., and Goodchild, M.F., Eds., ESRI Press, Redlands, California, 2005, 93-111.

[27.] Malczewski, J., GIS-based multicriteria decision analysis: a survey of the literature, *International Journal of Geographical Information Science*, 20, 703-726, 2006.

28. Zhu, A-X, Research issues on uncertainty in geographic data and GIS-based analysis, in *A Research Agenda for Geographical Information Science*, McMaster, R.B. and Usery, E.L., Eds., CRC Press, Boca Raton, Florida 2005,197-223.

29. DiBase, D.W., Is GIS a Wampeter?, *Transactions in GIS*, 11, 1-8, 2007.

30. United Nations Economic Commission for Europe, Introducing the Aarhus Convention, http://www.unece.org/env/pp/welcome.html, 2007.

31. Google, Google Earth, http://earth.google.com, 2007.

32. NASA, World Wind, http://worldwind.arc.nasa.gov/, 2007.

33. ESRI, ArcGIS Explorer, http://www.esri.com/software/arcgis/explorer/index.html, 2007.

34. Tress, B., Tress, G., and Fry, G., Integrative studies on rural landscapes: policy implications and research practice, *Landscape and Urban Planning*, 70, 177-191, 2005.

Part I

Data for Decision-Making

CHAPTER 2

An Optimized Semi-Automated Methodology for Populating a National Land-Use Dataset

W. Tompkinson, D. Morton, S. Gomm and E. Seaman

2.1 INTRODUCTION

Land use can be defined as the *'activity or socio-economic function for which land is used'*[1, p.2]. Barnsley et al.[2] note that the terms 'land cover' and 'land use' are often used interchangeably within a single classification scheme and provide the distinct definitions of land cover referring to the *'physical materials on the surface of a given parcel of land'* and land use being the *'human activity that takes place upon it'*. Although the increased availability of high-resolution remotely-sensed imagery offers the potential for the regular update of land cover data, land use information cannot be derived solely from identification of land surface characteristics. This is because land use is defined in terms of function rather than physical form[3], and consequently has a high dependency on *in situ* manual survey.

With the release of the Ordnance Survey (OS) MasterMap® Topography Layer[4], a national feature-based topographic dataset for Great Britain, there are opportunities to integrate and associate a diverse range of data sources, and also to infer the socio-economic function of topographic objects based upon their context and relationships to other objects. This chapter outlines a methodology for populating a land-use dataset based on the OS MasterMap® Topography Layer, using object-based analysis techniques and information derived from existing Ordnance Survey and third-party datasets. The aim of the research was to further classify the land use of topographic objects in terms of their morphology and spatial relationships using an object-based approach.

2.2 DATA

The research in this paper is based upon Great Britain's large-scale geographic information. Responsible for Great Britain's National Geographic Database, Ordnance Survey maintains digital topographic datasets that are surveyed at the basic scales of 1:1250 in urban areas, 1:2500 in rural areas and 1:10000 in mountain and moor land environments. In 1991, Ordnance Survey released Land-Line®, a digital vector topographic dataset (composed of a 'spaghetti' vector data structure, i.e., only points and lines) produced at these scales[5].

A successor to Land-Line®, the OS MasterMap® Topography Layer, was introduced in 2001. This product comprises a seamless database of over 400 million objects that depict detailed vector topographic features such as houses, land parcels and pavements[6] and constitutes a major change in how Great Britain's topographic data are represented and manipulated within geographic information systems and databases.

Feature-based datasets that are similar to OS MasterMap® have been produced by several other national mapping agencies, especially within Europe. Examples include the Vector25 digital landscape model of Switzerland[7] and the Dutch Top10vector product[8]. Differences between these products and the OS MasterMap® Topography Layer that would have implications for the modelling processes described here include:

- *The scale at which the products are delivered*: 1:25000 and 1:10000, for the Swiss and Dutch products respectively.
- *The associated simplification of their data models.* For example, both the Swiss and Dutch products contain a less detailed range of topographic feature types relative to the OS MasterMap® Topography Layer.

It should be noted that the OS MasterMap® Topography Layer also differs from the UK Land Cover Map 2000 developed by the Centre for Ecology and Hydrology[9]. This is a parcel-based vector dataset derived through the processing of satellite imagery that depicts land cover classes and was designed to provide a census of broad habitat types in the United Kingdom[10].

Additional Ordnance Survey data used in this work includes ADDRESS-POINT®, a point-based dataset that identifies the precise geographical location of residential, business and public postal addresses[11], and OSCAR®, a national large-scale vector dataset depicting road information[12].

2.3 PREVIOUS APPROACHES TO LAND-USE CLASSIFICATION

As described by Wyatt[13], between the mid 1980s and early 1990s, the (then) Department of the Environment commissioned a series of studies[14,15] to assess the feasibility of a national land-use stock survey (interests in land use information are now being taken forward by Communities and Local Government[16]). The main conclusion from these studies was that land use should be collected and maintained in collaboration with Ordnance Survey's large-scale digital mapping. Following the further recommendation by Dunn and Harrison[17] that a national land-use stock system should be created, additional evaluation by the Ordnance Survey[18] in 1996 concluded that a desk-based method that used surveyors' local knowledge, together with large-scale Ordnance Survey data would be the most appropriate approach in urban areas. For rural environments, stereo interpretation of color aerial photography was the preferred method.

It has long been recognized that physically visiting land parcels to assess their use is a most inefficient approach to developing and maintaining extensive land-use maps, and that manual interpretation of high-resolution remotely-sensed images offers a more effective and efficient procedure. Such an approach has the advantage of utilizing human intelligence to understand the function of land parcels. The operator intuitively studies the spatial context of a topographical feature within its overall environment inferring, for example, that a building next to a garden is probably residential, and one peripheral to a residential area and surrounded by a car park might be retail. The obvious disadvantage to this approach is that it is very time consuming when applied to anything beyond a local scale. Consequently, there has been a history of research into increasingly automated methods of land-use classification, both at Ordnance Survey and in the wider research community.

Given the difficulties in applying the intuitive rules of visual interpretation to imagery in an automated manner, initial research at Ordnance Survey sought to populate a land-use dataset using attributes drawn from a polygonized version of Land-Line®, ADDRESS-POINT® and OSCAR®, in combination with intelligence from third party sources[19]. The datasets that were proposed for use in future incarnations of this methodology included: a remotely-sensed land cover data source to populate land cover attributes, a range of commercial business directories and the Valuation Office's National Non-Domestic Rating List[20] and Council Tax Valuation List[21] (these are databases that relate to local taxation and can be used to aid differentiation between commercial and residential properties). The potential for using remotely-sensed satellite data was also explored[22]. Remotely sensed satellite data have the advantages over aerial photography of larger areal coverage and higher revisit rates, but their contribution was limited at this time (1997) by the non-availability of very high-resolution satellite imagery.

The data integration method proposed in the Ordnance Survey reports[19,22,23] was developed further through a research contract undertaken on behalf of the National Land Use Database (NLUD®) Partnership[24] that was placed by the then Department for Transport, Local Government and the Regions (now Communities and Local Government) with Infoterra Ltd. This study combined a limited number of national coverage datasets, including those trialled in earlier Ordnance Survey work[19], to form a prior information dataset. The integration was performed using rule-based searches and was known as the Semi-Automated Data Driven Analysis (SADDA) technique[1]. This approach had two main elements, the first consisting of data driven methods to directly classify polygons, and the second involving rules to assign labels to unclassified polygons[25]. The latter inferences were based upon:

- *Clump analysis*: following the classification of a group of buildings using data-driven methods, any remaining adjacent polygons within the same 'clump' were assigned a similar use.
- *Block analysis*: Ordnance Survey OSCAR data were used to define blocks of one land use and all unclassified polygons in such blocks were assigned that use.
- *Residential adjacency*: polygons with an OS MasterMap® Topography Layer 'Make' attribute (identifying features as man-made, natural, etc.) of 'Multiple' were classified as residential if they were next to other residential polygons.

It was recognized, however, that these were relatively primitive methods and Infoterra[25] concluded that there was scope for more sophisticated contextual analysis that was beyond the remit of their study.

A fundamental weakness in only using a 'data integration' approach to land-use classification is that features are treated in a very individualistic manner, i.e., polygons will be classified only if a point from another data source falls within them. Curtilage (i.e., land surrounding a building[26]) or other buildings on a large site such as a hospital complex may be left unclassified. As a result, a number of approaches have been adopted in remote sensing research to mimic the rules used by a manual photo-interpreter to infer land use based upon relationships between objects in the landscape.

At their simplest level, these image-based approaches have involved focusing upon the immediate relationships between pixel values in an image. These 'kernel-based' approaches involve studying pixel values within a convolution kernel and investigating the spatial pattern of pixel values in terms of edge and vertex adjacency events[2]. If pixel values are classified to represent land cover or material types, the land use is then inferred from the spatial relationships between those surface categories. However, capturing the spatial composition of high-resolution imagery (< 5 m) using this method of analysis appears to be problematic. Large kernels are required to better capture the spatial structure within such imagery, but their application introduces the unwanted effects of blurring and smoothing and so constrains the use of such techniques. A refinement, implemented upon medium-resolution SPOT-HRV and Landsat-TM satellite data by Barnsley et al.[2], is a 'higher-level' approach that takes into account better functional relationships between pixels, combined with image texture models.

Increasingly sophisticated approaches have been used to identify contiguous objects in imagery and then assess the relationships between them (as opposed to pixel-by-pixel analyses). Initial attempts to analyze the contextual relationships between objects included the application of graph-based structural pattern recognition techniques. For instance, Barnsley and Barr[3] developed the eXtended Relational Attribute Graph (XRAG) that analyzed structural properties and the relationships between parcels in rasterized digital maps, with the objective that the methodology could also be applied to parcels in high-resolution satellite imagery.

Further research at Ordnance Survey[27,28] studied the relationships between land use and the morphological characteristics of objects in a topographic database. Rules were constructed that examined both the shape of objects and the patterns in the immediate vicinity of the objects. A qualitative assessment of the results indicated that when discriminating between uses such as residential, roads, fields of crops and field margins, a 60-70% attribution could be achieved[28]. However, the rule base used in this case was considered inappropriate for classifying all land uses since it concentrated upon the geometrical properties of individual polygons, rather than also considering the spatial pattern of those features around the entity of interest. It was concluded from this study, as well as the experience of being involved in research such as the CLEVER-mapping project[19,29], that the integration of raster and vector data and the attributes of third-party datasets all show potential for classifying land use.

In recent years commercially available object-oriented image analysis software such as eCognition[30] has enabled imagery to be easily segmented and facilitated the implementation of more complex classification rules based upon a range of morphological and relational characteristics. This type of methodology has been applied to land-use classification (e.g., Bauer and Steinnocher[31]). The main difficulty in applying a segmentation-based technique in a mapping context is the likelihood of incompatibilities between objects in the imagery and those in existing topographic data. These limitations in image-led object-based methodologies primarily result from the characteristics of image-dependent segmentation algorithms. By their very nature, segmentation routines are usually based upon low-level image analysis techniques designed to be an initial step in object recognition[32]. As such, due to the inherent heterogeneity of remotely-sensed data, factors such as differences in illumination (e.g., across a roof apex) will result in segments that do not fully match the desired topographic features. Indeed, with respect to applying segmentation routines for large-scale mapping purposes, although there has been over two decades of research in this area, there is no universal routine available that can extract the geometries of specific types of topographic feature in a variety of contexts[33]. That said, methods for deriving information from imagery for mapping purposes are increasing in their sophistication, and in their use of additional knowledge (such as other spatial information), to better target processing routines[34]. This is especially true if these routines are applied to the revision of existing information. Nevertheless, despite these advances, drawing upon initial experimentation in work that used the outline of topographic features in the OS MasterMap® Topography Layer to segment recently flown aerial photography, it became apparent that there were too many factors affecting the integrity of the resulting land-use dataset. As a consequence, the methodology outlined below does not depend upon remotely-sensed data.

2.4 ORDNANCE SURVEY CLASSIFICATION METHODOLOGY

The research at Ordnance Survey sought to produce an automated methodology that would improve the quality of a land-use dataset initially populated through data integration techniques by using morphological and relational rules to increase both completeness and usability. To help avoid confusion between techniques the latter rule-based element has been given the acronym OOLUC (object-oriented land-use classification), with the initial development utilizing OS MasterMap® Topography Layer and ADDRESS-POINT® data[35]. OOLUC is essentially a fully-automated rule base that applies rules in a cyclical manner, classifying objects into functional groups, and then reclassifying objects, labelled with functional classifications, in accordance with the NLUD classification scheme. For example, a school building and a playground would be classified separately (with a playground being classified using its land cover and proximity to a school), before both objects become reclassified as 'Education'.

Rule-based procedures require many assumptions. For instance, in Cassettari[36] retail areas are assumed to take precedence over other types of town center land use. In cases of low classification certainty, OOLUC incorporates similar rules that assess the relative area of different classifications surroundings a polygon. For example, if an unclassified polygon has a high proportion of properties labelled as 'Retailing' nearby, then it will be assigned a similar status. However close inspection of the results produced by OOLUC indicates that although these rules lead to an increase in the completeness of the classification, there are high numbers of misclassifications between classes such as 'Retailing', 'Offices' and 'Industrial'. In addition, the presence of some imprecisely defined land cover classes (for instance water ditches in urban areas) results in misclassification by relational rules. Consequently, when only using OOLUC there is a lower assurance that polygons are classified correctly.

SADDA (see Section 2.3) was developed from a body of research that took place at Ordnance Survey through the 1990s. The overall Ordnance Survey land-use classification methodology (OSLUM) consists of applying a SADDA-like technique followed by OOLUC. Such an approach acknowledges that an expanded range of third-party datasets are required to differentiate between classes that otherwise have a low confidence of classification, and that an optimal methodology should consist of both data integration techniques (e.g., those in SADDA) and a rule base (e.g., OOLUC). It should be noted that to avoid any inconsistencies that might arise after applying SADDA's inferred methods, only the data-driven part of SADDA (which employs the integration techniques) is used within OSLUM.

Figure 2.1 illustrates the relationship between the SADDA and OOLUC components of the OSLUM methodology. Within the OSLUM flowline, OOLUC takes, as input data, OS MasterMap® Topography Layer with attributes for land cover and land use that have been assigned during the SADDA procedure. Both

SADDA and OOLUC have been designed to populate the OS MasterMap® Topography Layer in accordance with the NLUD classification scheme from version 4.1 onwards[1]. In these versions of the NLUD classification, a polygon should be assigned both a land use and land cover attribute. However, OOLUC is only designed to provide land use attributes and so it is assumed that SADDA, on its own, is sufficient to generate land cover information. Furthermore, polygons with a land use attribute from SADDA do not have this information changed by OOLUC, since priority is given to the direct, rather than inferred, method.

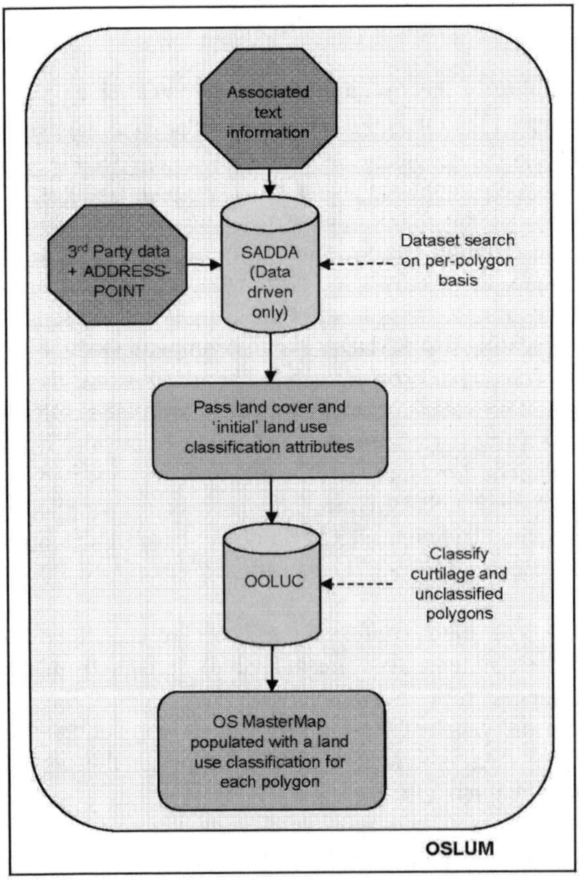

Figure 2.1 The structure of the Ordnance Survey land-use classification methodology (OSLUM).

Within the OOLUC component, polygons that have not been classified using SADDA are populated according to the following sequence of rules:

1. Direct attribution.
2. Adjacent to directly-attributed object and fulfills certain morphological criteria.
3. Within a defined distance of a directly-attributed object and fulfills certain morphological criteria.
4. Fulfills defined morphological criteria and adjacent to another object of defined morphological criteria.
5. Fulfills defined morphological criteria and within a defined distance of another object of defined morphological criteria.
6. Fulfills defined morphological criteria.

To ensure stability in the resultant classification, after empirical investigation during development of the methodology, it was concluded that the rule base should be applied to each polygon four times. The order of precedence during each of the cycles is recorded as an extra attribute for each polygon. Using this information, the operator is able to obtain a confidence measure for the classification of each polygon. After the program has finished applying the rule-base cycles, a further rule is implemented that deals with curtilage and any remaining unclassified polygons to ensure completeness in the final dataset. Example land use maps that result from implementing just SADDA and the complete OSLUM methodology are shown in Figures 2.2 and 2.3 respectively. The illustrations also demonstrate that the OSLUM method provides a greater completeness in the resulting dataset

To summarize, Table 2.1 compares the characteristics of OSLUM with those of the alternative methods for land-use classification described in Section 2.3. This evaluation suggests that OSLUM incorporates many favorable characteristics of previously documented techniques (e.g., an object-based focus and no requirement for direct use of imagery). Overall, the key advantages of OSLUM are that it:

- Preserves the high confidence of correct classification of individual polygons (if suitable data are available) derived from a data integration (data-driven) component, and
- Increases the value of the final dataset by populating objects that do not have other data associated with them through the use of relational and morphological modelling.

2.5 ACCURACY ASSESSMENT OF SADDA AND OSLUM

A quantitative comparison was conducted to compare the SADDA and OSLUM methodologies described in the previous section. Reference data were collected for nine study sites (each between 1–2 km^2 in area) in both urban and rural environments. These data were used to validate results from SADDA by Infoterra[25]. Analysis of the OSLUM results was performed on the same sites.

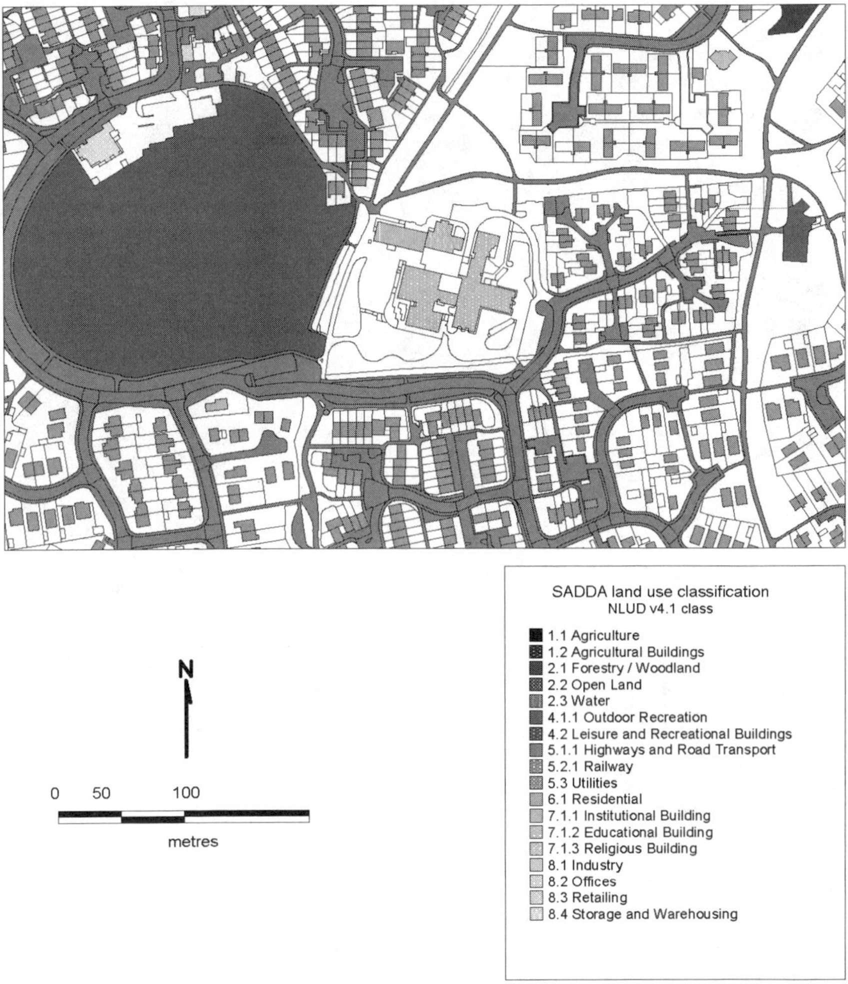

Figure 2.2 A land-use classification of central Sheffield using the SADDA methodology. Polygons depicted in white are unclassified. Ordnance Survey data © Crown Copyright. All rights reserved.

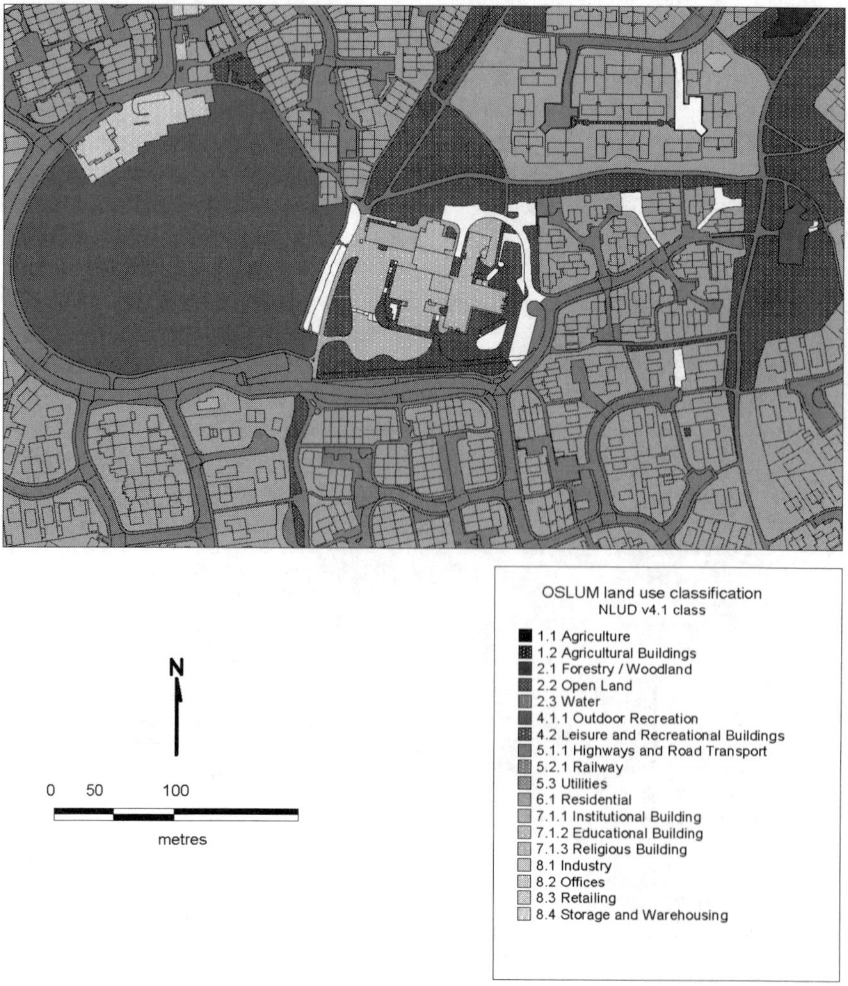

Figure 2.3 A land-use classification of central Sheffield using the OSLUM methodology. Polygons depicted in white are unclassified. Ordnance Survey data © Crown Copyright. All rights reserved.

Table 2.1 A comparison of methodologies for land-use classification

Land-Use Classification Methodology	Input Data	Pixel or Object-Based	Rules to Take Account of Context
Manual interpretation	Remotely-sensed imagery	Intuitively object-based	Intuitive assumptions
Data integration	Feature-based vector data, raster land cover data, and a range of point-based spatial datasets	Object-based	None
Kernel-based	Remotely-sensed imagery	Pixel-based	Vertex and edge adjacency events
Graph-based	Rasterized digital map imagery	Object-based	Distance and direction relationships
Early morphological techniques	Polygonized vector data	Object-based	Immediate adjacency counts
Image-based object-oriented methodologies	Remotely-sensed imagery	Object-based	Morphological and spatial relationship rules between objects derived solely from imagery
Ordnance Survey land-use classification methodology (OSLUM)	Feature-based vector data, raster land cover data, and a range of point-based spatial datasets	Object-based	Morphological and spatial relationship rules between objects derived from object-based data. Non-visual prior information from point-based data

On a per-class basis, Table 2.2 (urban sites) and Table 2.3 (rural sites) summarize accuracy values for both the SADDA (using only direct attribution and then this followed by inferred methods) and the OSLUM methodologies. The accuracy values listed represent percentages of total areas. Based upon the original SADDA results from Infoterra[25], there are two types of accuracy value provided:

- *Confidence* (also known as user's accuracy): The probability that an area classified with a given class actually represents that class on the ground.
- *Class Completeness* (also known as producer's accuracy): The proportion of total area in each class of reference data that is classified correctly.

Table 2.2 Total per-class accuracy figures for urban reference sites

	Confidence (% Area)			**Class Completeness (% Area)**		
NLUD Classification (Version 4.1)	**SADDA (data - driven only)**	**SADDA (including inference techniques)**	**OSLUM**	**SADDA (data-driven only)**	**SADDA (including inference techniques)**	**OSLUM**
1.1 Agriculture	52.78	52.77	52.78	91.21	91.21	91.21
1.2 Agric. bldg.	0.00		8.94	0.00	0.00	18.92
2.1 Forest/wood	12.45	12.45	12.45	29.33	29.33	29.33
2.2 Open land	44.51	0.00	28.87	14.79	0.00	28.28
2.3 Water	88.71	88.71	88.71	99.27	99.27	99.27
3.1 Mineral/quarry						
3.2 Landfill site						
4.1.1 Recreation	75.69	75.14	74.69	43.40	44.44	43.39
4.1.2 Allotment	100.00	100.00	100.00	85.60	85.60	85.60
4.2 Leisure bldg.	75.28	74.28	76.66	33.50	33.70	37.48
5.1.1 Highway	89.77	88.77	88.77	80.10	80.11	80.11
5.1.2 Car park	28.17	28.17	28.42	7.68	7.68	7.93
5.2.1 Railway	88.69	88.69	88.69	94.11	94.11	94.11
5.2.2 Airport						
5.2.3 Dock		0.00	0.00			
5.3 Utilities	0.00	14.27	15.84	18.75	16.68	20.89
6.1 Residential	90.44	95.38	86.84	39.30	86.96	86.30
6.2 Institutional	7.18	29.06	7.18	1.18	0.00	1.18
7.1.1 Inst. bldg.	85.35	83.92	81.42	23.67	18.95	27.35
7.1.2 Educat. bldg.	100.00	100.00	100.00	23.88	8.50	23.88
7.1.3 Relig. bldg.	100.00	100.00	100.00	4.26	4.26	7.53
8.1 Industry	31.78	27.69	28.03	10.70	11.71	16.49
8.2 Offices	48,56	48.63	40.12	30.47	31.22	31.36
8.3 Retailing	64.31	83.41	43.15	43.72	28.98	68.49
8.4 Warehousing	25.84	26.4	14.11	15.77	16.11	22.68
9.1.1 Vacant land	0.00			0.00	0.00	0.00
9.1.2 Vacant bldg.	0.00			0.00	0.00	0.00
9.2 Derelict land				0.00	0.00	0.00
10.1 Defense						

The land-use classification scheme employed in the assessment was the NLUD Classification v4.1 (see Harrison[1] for further information). In Tables 2.2 and 2.3, if no accuracy value is shown then there was no population of the class in question, while if a value of 0.00 is listed then that class was populated with entirely incorrect classifications.

Table 2.3 Total per-class accuracy figures for rural reference sites

NLUD Classification (Version 4.1)	Confidence (% Area)			Class Completeness (% Area)		
	SADDA (data - driven only)	SADDA (including inference techniques)	OSLUM	SADDA (data-driven only)	SADDA (including inference techniques)	OSLUM
1.1 Agriculture	98.13	98.14	98.14	92.95	92.95	92.95
1.2 Agric. bldg.	82.50	82.5	74.96	0.18	0.18	56.25
2.1 Forest/wood	64.81	64.81	64.81	86.91	86.94	86.92
2.2 Open land	77.46	77.46	77.40	86.79	86.79	86.79
2.3 Water	20.10	20.11	20.11	94.38	94.38	94.38
3.1 Mineral/quarry	0.00	0.00	0.00			
3.2 Landfill site						
4.1.1 Recreation				0.00	0.00	0.00
4.1.2 Allotment						
4.2 Leisure bldg.	30.03	22.96	20.51	97.60	97.60	97.60
5.1.1 Highway	79.80	79.80	79.80	84.38	84.38	84.38
5.1.2 Car park				0.00	0.00	0.00
5.2.1 Railway						
5.2.2 Airport						
5.2.3 Dock						
5.3 Utilities	68.45	68.45	93.55	0.02	0.02	0.15
6.1 Residential	63.13	89.72	84.01	4.13	38.91	52.46
6.2 Institutional	0.00	0.00	0.00	0.00	0.00	0.00
7.1.1 Inst. bldg.	61.73	54.50	73.19	16.05	16.05	57.38
7.1.2 Educat. bldg.	100.00	100.00	100.00	6.55	6.61	6.61
7.1.3 Relig. bldg.	100.00	100.00	100.00	8.51	8.59	9.24
8.1 Industry	100.00	100.00	100.00	1.55	1.55	2.93
8.2 Offices	0.00	0.00	0.00			
8.3 Retailing	59.28	44.99	20.38	12.62	13.29	56.19
8.4 Warehousing			0.00			
9.1.1 Vacant land			0.00	0.00	0.00	0.00
9.1.2 Vacant bldg.				0.00	0.00	0.00
9.2 Derelict land	0.00					
10.1 Defense						

The results for urban areas in Table 2.2 indicate that including inferred methods improved the Class Completeness for several NLUD categories characterized by relatively large 'blocks' with multiple buildings and surrounding land. For example, using both the 'full' SADDA approach and the OSLUM methodology dramatically increased the Class Completeness of the 6.1 Residential category when compared to the solely data-driven method. This trend is further demonstrated in the case of OSLUM with higher completeness values for 7.1.1 Institutional Building, 7.1.2 Educational Building and 7.1.3 Religious Building, along with commercial classes such as 8.3 Retailing and 8.4 Storage and Warehousing. For classes 8.1 Industry, 8.3 Retailing and 8.4 Storage and Warehousing, the OSLUM confidence value is lower than those produced by both SADDA methods. Even when compared to the full SADDA method, this is explained by even more inferred rules being used in OSLUM to associate topographic objects with others in order to assign a classification in situations where associated data are sparse.

In the rural environment, Table 2.3 shows that the Class Completeness for 6.1 Residential and 8.3 Retailing was improved through the association of polygons with surrounding land uses that had already been identified. A particular increase in Class Completeness is also apparent for 1.2 Agricultural Buildings using OSLUM. The difference between the full SADDA and OSLUM results for this category reflects a better representation of the context within which such buildings are set in the rules of the latter. In OSLUM, an agricultural building is differentiated from a residential one by stipulating that it should not be adjacent to a garden and that it should be surrounded by a high proportion of natural land cover.

The rules that contribute to increased Class Completeness within the rural setting also tend to produce a slight fall in the corresponding Confidence levels (e.g., see the results for 8.3 Retailing). One exception is the 5.3 Utilities class where the Confidence level actually increases when using OSLUM. This could be due to improved contextual rules (e.g., land owned by a water company being close to a reservoir). However mistakes in SADDA, such as classifying reservoirs into category 2.3 Water (derived directly from the land-cover attribute) and not 5.3 Utilities, are also propagated into OSLUM. This helps to explain the extremely low values of Class Completeness for the 5.3 Utilities category in all methodologies. In addition, it suggests that a future version of the rule base may need to include an extra condition that classifies 2.3 Water as 5.3 Utilities if it is adjacent to a 5.3 Utilities object in a previous classification cycle.

Other features of the results generated by OSLUM are as follows:

- A vacant building in a rural environment was usually put into class 1.2 Agricultural Building by OSLUM. In an urban area it was usually classified as 8.1 Industry or 8.4 Storage and Warehousing.

- If an 8.1 Industry object was incorrectly classified, it was usually defined as 8.4 Storage and Warehousing.
- Car parks in reference data that are associated with sites such as retail parks were sometimes classified as 5.1.2 Car Park and not as 8.3 Retailing.

Table 2.4 presents the overall accuracies of the methods tested in this study. The higher accuracy values in both urban and rural environments indicate that applying the OOLUC rule base after the data-driven component of SADDA really does add extra value to a method that only directly populates polygons with existing data. In addition, the results illustrate the effect of the more sophisticated contextual inference rules in OSLUM, compared to those in the full version of SADDA. For example, OSLUM leaves fewer polygons unclassified and the better accounting for curtilage contributes to the higher overall accuracy of the method in urban areas. The very small differences between the results in rural areas can be attributed to the presence of a higher proportion of classes that can be more reliably populated using existing land-cover classifications (e.g., 1.1 Agriculture).

Table 2.4 Overall accuracies of the SADDA and OSLUM classification methodologies

	SADDA (data-driven approach only)	**SADDA (including inference techniques)**	**OSLUM**
Urban	46.87%	53.25%	59.87%
Rural	87.55%	88.04%	88.34%

2.6 CONCLUSIONS

This chapter has outlined and evaluated a semi-automatic approach to populating an initial baseline land-use dataset. The aim was to advance a methodology that couples techniques for deriving land-use function from morphological and spatial rules, with information from existing geographic datasets utilizing solely semi-automated techniques. OSLUM fulfills these criteria, and enhances previous approaches to classifying land-use in a semi-automated manner.

The results from evaluating OSLUM suggest that for many land-use classes it provides higher levels of per-class completeness and levels of total population (especially within urban areas) than when using the SADDA method on its own. The main advantage of OSLUM compared to other methods is that the functional characteristics of land use are assessed in terms of both visual context (such as in the pattern of topographical objects on a map) and non-visual information (such as address, or business-directory related information). Since these contextual rules are only applied to polygons that are unclassified after applying a direct data-driven approach, higher classification accuracies are bound to arise. In addition, unlike the inference rules employed in the full version of SADDA, the contextual rules in OSLUM are applied in a fully automatic manner in all types of environment.

OSLUM's shortcomings include the expense and availability of data required in the data-driven component and its dependence upon the detail that is present in large-scale topographic data for morphological modelling. Although, in theory, a similar methodology could be applied to other countries that possess a national feature-based topographic dataset, it is likely that the use of a generalized base dataset could have a detrimental effect upon the accuracy and completeness of the results. For some users, the Class Completeness and Confidence values of some individual classes in OSLUM might be too low and this may present a further disadvantage of the methodology. However, this could be improved with subsequent manual interpretation and intervention. The value of OSLUM lies in maximizing the accuracy of a land-use dataset that has been populated using automated methods, thereby minimizing cost in any subsequent intervention or maintenance processes.

The analysis of results produced by OSLUM indicates that the method provides improvements in the overall accuracy and completeness of a land-use dataset produced using close to fully automatic methods, most notably within the urban environment. OSLUM therefore offers one way forward to solve the inherently multi-faceted problem of effective and complete population of a land-use dataset.

2.7 ACKNOWLEDGMENTS

The authors are grateful to John Goodwin for his work on the initial data integration and morphological methodologies developed at Ordnance Survey, and to Tristan Wright, Nicholas Regnauld and David Russell for their assistance in the development and analysis of OOLUC.

2.8 REFERENCES

1. Harrison, A., Extending the dimensionality of OS MasterMap™: land use and land cover, Presented at the AGI Conference at GIS 2002, Available at http://www.nlud.org.uk/draft_one/key_docs/pdf/Harrison_AGI02.pdf, 2002.

2. Barnsley, M.J., Moller-Jensen, L., and Barr, S.L., Inferring urban land-use by spatial and structural pattern recognition, in *Remote Sensing and Urban Analysis*, Donnay, J-P., Barnsley, M.J., and Longley, P.A., Eds., Taylor and Francis, London, 2000, 115-144.

[3.] Barnsley, M.J. and Barr, S.L., Distinguishing urban land-use categories in fine spatial resolution land-cover data using a graph-based, structural pattern recognition system, *Computers, Environment and Urban Systems*, 21, 209-225, 1997.

[4.] Ordnance Survey, Welcome to OS MasterMap, http://www.ordnancesurvey.co.uk/oswebsite/products/osmastermap, 2006.

[5.] Ordnance Survey, *Land-Line® User Guide*, Ordnance Survey, Southampton, 2002.

[6.] Murray, K., A new geo-information framework for Great Britain, in Proceedings of FIG XXII International Congress, Washington DC, 2002, 1/13.

[7.] Swisstopo, Vector25, http://www.swisstopo.ch/en/products/digital/landscape/vec25/, 2006.

[8.] Kadaster, Top10vector, http://www.kadaster.nl/zakelijk/producten/topografische_dienst_top10vector.html, 2003.

[9.] Centre for Ecology and Hydrology, Land Cover Map, http://www.ceh.ac.uk/sections/seo/lcm2000_home.html, 2007.

[10.] Fuller, R.M., Smith, G.M., Sanderson, J.M., Hill, R.A., and Thomson, A.G., The UK Land Cover Map 2000: Construction of a parcel-based vector map from satellite images, *Cartographic Journal*, 39, 15-25, 2002.

[11.] Ordnance Survey, ADDRESS-POINT®, http://www.ordnancesurvey.co.uk/oswebsite/products/addresspoint, 2006.

[12.] Ordnance Survey, *OSCAR® User Guide*, Ordnance Survey, Southampton, 2005.

[13.] Wyatt, P., *Creation of an Urban Land Use Database*, RICS Education Trust, Bristol, 2002.

[14.] Roger Tym and Partners, National Land Use Stock Survey: A Feasibility Study for the Department of the Environment, Roger Tym and Partners, London, 1985.

[15.] Dunn, R. and Harrison, A., A feasibility study for a national land use stock survey, in Proceedings of the AGI Conference 1992, Association for Geographic Information, London, 1992, 1.13.1-1.13.4.

[16.] Communities and Local Government, Planning, building and the environment, http://www.communities.gov.uk/index.asp?id=1503250, 2007.

[17.] Dunn, R. and Harrison, A., Working towards a national land use stock system, in Prooceedings of the AGI Conference 1994, Association for Geographic Information, London, 1994, 8.1.1-8.1.5.

[18.] Ordnance Survey, Research Trial for a National Land Use Stock System, Report to the Department of the Environment, Ordnance Survey, Southampton, 1996.

[19.] Ordnance Survey, Population & Maintenance of a Land Use/Cover Database of England (Stage 1), Internal Report, Ordnance Survey, Southampton, 1997.

[20.] Valuation Office Agency, Business Rates - general information, http://www.voa.gov.uk/business_rates / index.htm, 2005.

[21.] Valuation Office Agency, Council Tax Valuation List – glossary, http://www.voa.gov.uk/cti/MGlossary.asp, 2005.

[22.] Ordnance Survey, Population & Maintenance of a Land Use/Cover Database of England (Stage 2), Internal Report, Ordnance Survey, Southampton, 1997.

[23.] Ordnance Survey, Population & Maintenance of a Land Use/Cover Database of England (Stage 3), Internal Report, Ordnance Survey, Southampton, 1997.

[24.] National Land Use Database, http://www.nlud.org.uk/, 2006.

[25.] Infoterra, Final Report for Research Contract CP01004: NLUD County Demonstrator, Infoterra Limited, Farnborough, 2002.

[26.] ACSM, *Glossary of Mapping Sciences*, American Congress on Surveying and Mapping, American Society for Photogrammetry and Remote Sensing, and American Society of Civil Engineers, Maryland and New York, 1994, 133.

[27.] Ordnance Survey, Land Use/Cover Class Signatures, Internal Report, Ordnance Survey, Southampton, 1998.

[28.] Ordnance Survey, NLUD by Integrated Analysis of Datasets, Internal Report, Ordnance Survey, Southampton, 1999.

[29.] Smith, G.M., Fuller, R.M., Amable, G., Costa, C., and Devereux, B.J., CLEVER-mapping: An implementation of a per-parcel classification procedure within an integrated GIS environment, Proceedings of the Remote Sensing Society conference, Observations and Interactions: RSS97, Remote Sensing Society, Nottingham, 1997, 21-26.

[30.] Baatz, M., Heynen, M., Hofmann, I., Lingenfelder, M., Mimler, A., Schape, M., Weber, M., and Willhauck, G., *eCognition User Guide*, Definiens AG, Munich, 2000.

[31.] Bauer, T. and Steinnocher, K., Per-parcel land-use classification in urban areas applying a rule-based technique, *GIS*, 6, 25-27, 2001.

[32.] Pal, N.R. and Pal, S.K., A review on image segmentation techniques, *Pattern Recognition*, 26, 1277-1294, 1993.

[33.] Agouris, P., Mountrakis, G., and Stefanidis, A., Automated spatiotemporal change detection in digital aerial imagery, in Proceedings of Aerosense 2000, SPIE Proceedings Vol. 4054, Orlando, Florida, 2000, 2-12.

[34.] Baltsavias, E. P., Object extraction and revision by image analysis using existing geodata and knowledge: current status and steps towards operational systems, in Proceedings of ISPRS Commission II Symposium, Xian, China, 2002, 13.

[35.] Ordnance Survey, Land Use Classification of OS MasterMap, Internal Report, Ordnance Survey, Southampton, 2002.

[36.] Cassettari, S., Land use mapping – the GeoInformation Group's approach, *Geomatics World*, 6, 40-41, 2002.

CHAPTER 3

A New Framework for Feature-Based Digital Mapping in Three-Dimensional Space

A. Slingsby, P. Longley and C. Parker

3.1 INTRODUCTION

National mapping agencies have produced definitive maps on regional and national scales for over two centuries. An important principle of these maps is their basis on a standard georeferencing framework and their adherence to a set of standard capture specifications ensuring that elements of different geographical regions within the same framework are comparable to each other. Maps are diagrammatic in style; an object of interest is depicted by a plan geometry (often a simplified footprint) or a symbol, whose color, size and labelling indicates its identity, its classification and other related information.

Within the past decade, many mapping agencies have digitized their original paper-based information. The results are placed into a data management system and they become the definitive version, from which subsequent paper and digital mapping products are produced. Many mapping agencies now consider themselves as data rather than map providers (digital data products now form the largest market for Britain's national mapping agency). We use the term 'framework' to refer to the database that supports the data management system. Its design (the subject of this chapter) has major implications for the type and range of applications that can be supported. In common with most programs involving the wholesale digitization of information, initial efforts have been little more than digital versions of the original paper versions.

The geometry of the urban environment can be complex with parts of multiple-story buildings, bridges, underpasses and tunnels juxtaposed in various ways. Increasingly, a 2D description of the world is becoming inadequate for applications such as recording the ownership of flats on multiple stories or evacuation planning in a multi-tiered urban environment. A challenge for national mapping agencies is to be able to model 3D geometries and relationships without compromising the traditional agenda of providing national coverage in a useable and common framework.

This chapter discusses issues and applications appropriate for the design of a new framework for the storage of digital national mapping data. The framework design is guided by the following design principles and capabilities, whose rationale is described and discussed in later sections.

- The ability to accommodate alternative conceptualizations of real-world features.
- To be a data repository: a unified database storing a rich set of information about the urban environment which is compact and is structured in such a way that a wide range of views (maps) can be extracted according to various criteria sets.
- The ability to construct three-dimensional geometries.
- The ability to deal with space both interior and exterior to buildings in a seamless fashion: e.g., allowing a courtyard space to be retrieved in the same way as an entrance lobby space.
- To be incrementally updatable: the ability to populate the framework with existing information, and incrementally add new data as they become available.
- The ability to accommodate pedestrian accessibility as an integral part of the framework.

Although we have an emphasis on modelling the complexities of the urban built-environment, we anticipate that rural areas (with potentially simpler geometries) could be described in the same framework.

3.2 CASE STUDIES OF EXISTING FRAMEWORKS

The products and experiences of Ordnance Survey®[1], the national mapping agency of Great Britain, will be discussed. Ordnance Survey was the first mapping agency of its size to digitize its nation-wide large-scale mapping[2], a process completed in 1995. It is these large-scale topographic maps which will be considered here (OS MasterMap® is captured at and designed to be viewed at 1:1250 for urban areas). These maps depict 'addressable objects' with a correspondence to conceptualizations of real-world features (e.g., buildings, trees, parkland, and post boxes).

3.2.1 NTD and Land-Line® – 'Digital Drawing'

The result of completing the wholesale digitization of Ordnance Survey's paper maps in 1995 was the National Topographic Database (NTD), on which the digital data product Land-Line® is based (Figure 3.1). NTD is a very map-centric framework, in that it is the linework that has been digitized. The linework corresponds to the boundaries of (implicit) area and linear features[3]. Points corresponding to particular positions of interest (e.g., spot heights, telephone boxes, wells and milestones) complement the linework. Each of these points and lines is termed a 'feature' and is allocated a 'feature class' which corresponds to a type of linework or symbology on a map (e.g., building outline, base of a slope, cliff edge or a point feature). Land-Line does not have areal features. This is a major limitation because many geographical features are conceptualized by their areal extents rather than their boundaries: examples being fields, building footprints and land parcels.

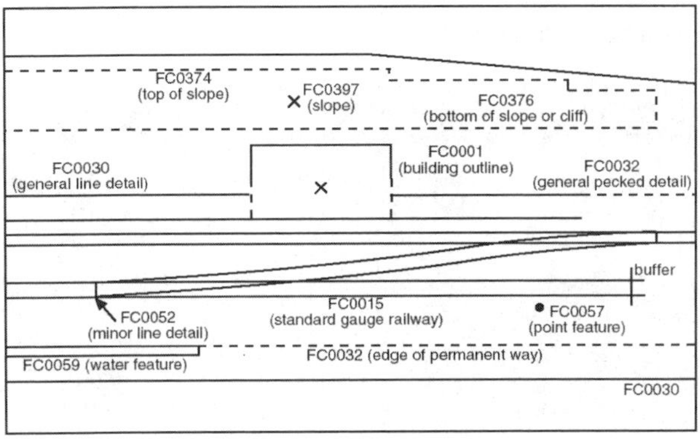

Figure 3.1 An extract from the Line-Line® users' guide showing part of a railway with points and lines classified by feature class (FC). Source: Ordnance Survey[2], Figure 4.8, p.4.18. Ordnance Survey © Crown copyright. All rights reserved.

Since Land-Line features refer to linework on a paper map rather than conceptualizations of 'real-world' features, the product is most appropriate for the production of customized maps for printing and display. Where areal features are required (such as individual building footprints for a GIS analysis task), polygons have to be built from the bounding 'building outline' lines, an achievable task in many GIS packages.

3.2.2 DNF® and OS MasterMap® – 'Feature-Based Mapping'

Some of the above limitations have been addressed in the Digital National Framework™ (DNF®), which replaces NTD. DNF was launched in 2001 and is the framework upon which the product OS MasterMap® is based (Figure 3.2). Two of OS MasterMap's important characteristics are:

- Area (polygon) features exist in addition to the line and point features present in Land-Line. Polygons are exhaustively tiled (they do not overlap and there are no gaps), so that road and pavement segments are represented as polygons too. Since many geographical features are often conceptualized as land parcels or footprints on maps, these new polygon features more closely represent these addressable objects, which are conceptualizations of real-world features.
- Since OS MasterMap features have a closer match to addressable objects in the real-world, they are each allocated a unique identifier called a

TOID® (topographic identifier). A substantial change in a real-world feature may result in it being allocated a new TOID. However, OS MasterMap features do not necessarily correspond to real-world entities, particularly those that are lines and points, so a TOID may not correspond to a real-world feature at all.

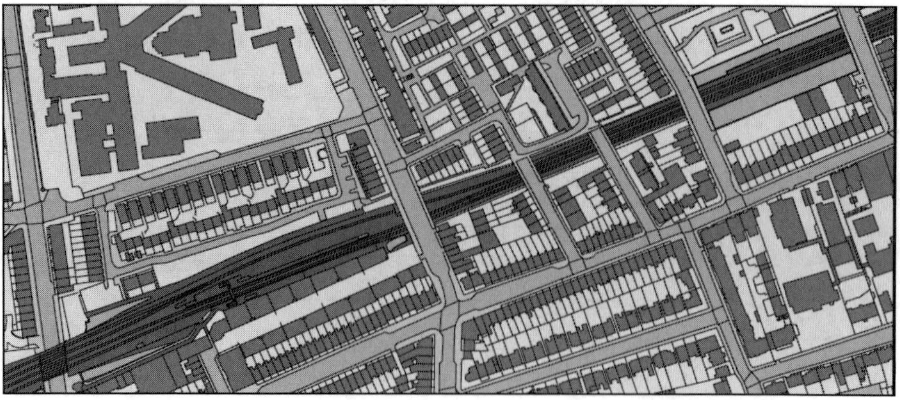

Figure 3.2 The polygon features of the Topographic Layer of OS MasterMap®. Each polygon has a unique identifier (TOID®) that usually corresponds to a 'real-world' areal feature such as a building or segment of road. In this extract, the polygons are shaded according to the category of real-world feature. Source: OS MasterMap® data displayed in ArcGIS 8.

OS MasterMap® is split into several themes called 'layers'. The point, line and polygon features are contained in the Topographic Layer. Road accessibility and connectivity information is provided as a fully connected 2D geometrical network of roads in another layer called the Integrated Transport Network™ (ITN). Nodes of the network represent the positions of junctions and link the topologically-connected lines. Lines represent the 2D geometry of road centerlines, which are allowed cross without being connected (this is the case when they are on different levels). Nodes and lines of the network have traffic access and restriction information associated. This network connectivity cannot be derived from the road polygons in the Topographic Layer because the 2D depiction of geometry results in over-passes cutting the connectivity of roads underneath.

3.2.3 Current Problems

The two main limitations of existing OS data which restrict the range of supported applications are:

- Addressable objects are defined by a particular and restricted set of conceptualizations of real-world features and geometry is defined in these terms.
- The 2D character of the geometry cannot adequately represent the multiple tiers common in the built environment.

Different organizations, individuals and research projects require geometrical data of the built environment and they would generally look to a national mapping agency for this. However, national mapping agencies may not provide the geometry adequate for their needs. For example, the UK Valuation Office requires the geometries of taxable units in the built-environment. As discussed below, these geometries tend to be subdivisions of those provided by national mapping agencies and they often need to be described on multiple stories of buildings.

The level of geometrical detail in OS MasterMap® reflects the remit of a national mapping agency, which primarily deals with information over large regions. While we do not propose that the remit of national mapping agencies should necessarily change, we propose that a common framework should exist, allowing many data providers to be able to add and retrieve data according to their requirements. This would also facilitate the sharing of geospatial data. A framework which supports the output of the feature sets of the Ordnance Survey, the Valuation Office and other data providers would become very powerful indeed. Case studies will now be used to illustrate the challenges and possibilities.

3.2.4 Non-Domestic Building Stock Project (NDBS)

The NDBS Project[4] is a research programme based at the Bartlett School of Graduate Studies at University College London. Its original aim was to achieve a statistical picture of the uses of energy in non-domestic buildings in order to help form national and international policies on energy use[5]. It has data on approximately two million non-domestic properties which were collected from surveys carried out in four English towns and details provided by the UK Valuation Office (VO). Relevant addressable objects for the VO are taxable units of non-domestic properties (offices, shops and other industrial buildings) which are usually subunits of the addressable objects defined by national mapping agencies.

The database which holds this information (SmallWorld™ GIS[6]) uses stacked polygons to model units on different stories (Figure 3.3) collected from street surveys[7]. Each polygon represents the floor area of each unit and, by extension, the unit itself. Each floor polygon has attributes corresponding to characteristics of the unit: attributes relevant to energy efficiency modelling (type of activity, materials, ceiling height, glazing-to-wall area ratio and for the uppermost stories, the roof type and roof pitch angle). The vertical separation between different stories was estimated from photographs. The 2D geometry was derived by subdividing Ordnance Survey building footprints, forming units with unique attribute sets (for example, a room with a lower floor-to ceiling height portion than the rest of the room would be divided into two units with different floor-to-ceiling heights). As a database holding three-dimensional geometry, this presents a rather simple, but adequate means of storing attributes for different parts of buildings and a basis for calculations such as extents of exposed walls and hence an input to energy efficiency modelling.

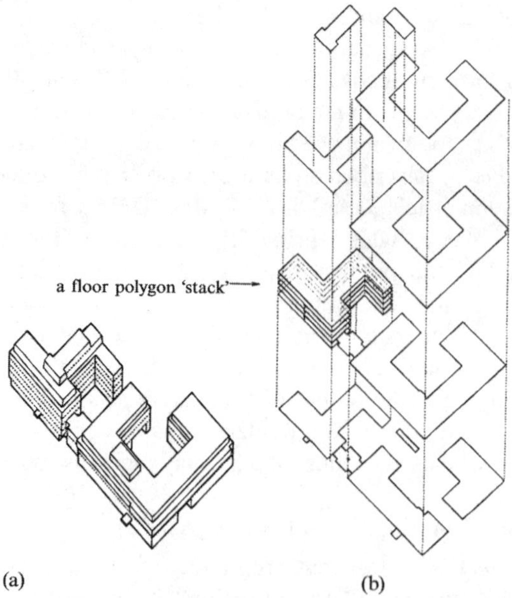

Figure 3.3 Office building represented in the NDBS database. Floor polygons can be different sizes on different stories. Source: Holtier et al.[5], Figure 1, p.53. With permission.

To create the NDBS, it was necessary to take digital national mapping data and enrich it both with geometrical data and attribute data gathered from surveys, photographs and VO data. The addition of attribute data is easily done in a GIS. What is not so straightforward is adding the further geometrical information, effectively producing 'property' features on multiple stories, rather than whole 'building' features. It is for this reason that a customized framework was designed to accommodate the requirements of the project. The resulting database is richer than that of the original national mapping data. If there were a way of incorporating this enriched dataset into the original national mapping dataset, other similar studies could be done without the time and expense of producing a specialized framework and dataset.

3.2.5 National Land and Property Gazetteer (NLPG)

The National Land and Property Gazetteer (NLPG) is an implementation of Part 2 of British Standard BS7666[8], the aim of which is to create a standard method of referencing property and land parcels. NLPG at the national level is a patchwork of standardized Local Land and Property Gazetteers (LLPGs) provided by local authorities. The specifications for the definition of properties are not strictly defined, so the details tend to vary slightly between LLPGs.

The central concept in NLPG is the addressable 'basic land and property unit' (BLPU). A BLPU corresponds to a property (real estate). It is defined as an *'area of land, property or structure of fixed location having uniform occupation, ownership or function'* [8, p1]. Every BLPU is identified by one or more textual postal addresses called 'land and property identifiers' (LPIs). Unlike in OS MasterMap® where addressable objects are tied to geometries, BLPUs are non-geometrical concepts of which geometry (2D extent) is a non-mandatory attribute. Relationships between groups of BLPUs can be modelled through the use of 'child' BLPUs which nest in 'parent' BLPUs. This provides support for a building hierarchy such as the relationship of individual flats to the block which contains them.

To give a brief example of the difference in emphasis between OS MasterMap and NLPG, consider a house with a front garden and back garden containing a shed. In OS MasterMap, four polygons with TOIDs would correspond to the house, the two gardens and the shed (if within the capture specifications). In NLPG, one BLPU representing the entire property would be uniquely identified with a postal address (sometimes also by alias addresses), since it is the entire property that is of interest. If present, the geometry of the BLPU would encompass the entire land parcel. If more than one property was contained within this BLPU (such as flats), each would be allocated a child BLPU.

Currently, NLPG does not hold geometrical extents[9] (an illustration of the fact that geometry is not of central importance to gazetteers). However many Local Authorities include such information for their own use. The Royal Borough of Kingston-upon-Thames' LLPG holds geometrical extents for parent BLPUs (but not for child BLPUs except where the extent is particularly required). The geometries were originally derived from maps, aerial photographs and the use of local knowledge[10]. Updates are submitted to NLPG every two weeks. Their website[11] provides maps of properties in the borough and their associated planning applications, demonstrating one of the uses of LLPG and NLPG. Figure 3.4 shows an example, illustrating concepts of nested BLPUs and alias (alternative) addresses.

OS MasterMap and NLPG are therefore two related products which have different approaches and emphases. The former is a digital map where geometry is the emphasis; the latter is a gazetteer where non-geometrical features are the focus.

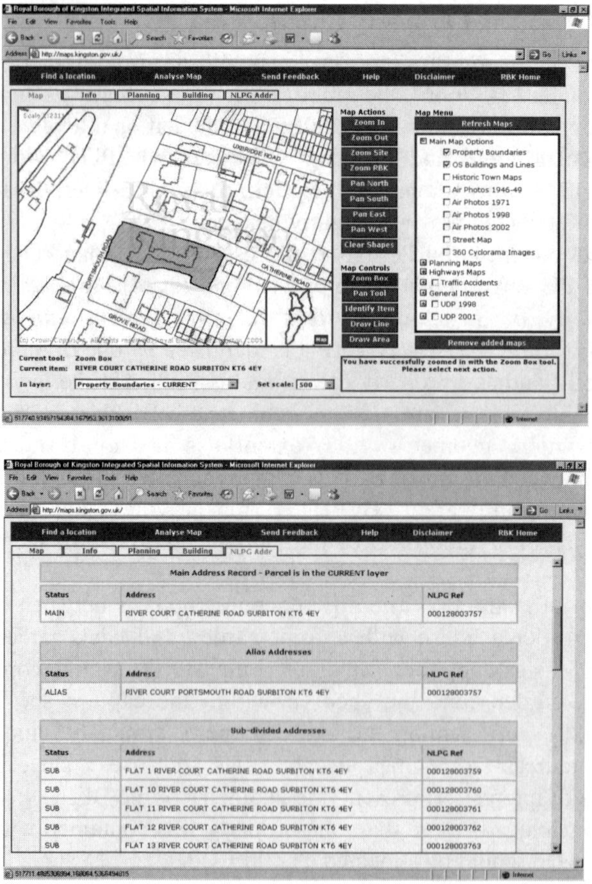

Figure 3.4 Screen captures from Kingston-upon-Thames' mapping website showing maps, addresses and planning applications for properties using data from their LLPG. The map in the upper image shows property boundaries (from the LLPG) and building outlines (from Ordnance Survey). A block of flats (Catherine Road) has been queried in this example and the extent of this BLPU is shown on the map. The LLPG address and alias addresses are shown in the lower image, as are the addresses of child BLPU units. Source: http://maps.kingston.gov.uk. Ordnance Survey © Crown Copyright and Royal Borough of Kingston. With permission.

3.2.6 Construction Industry

The construction industry has traditionally used computer aided design packages (CADs) as tools for the design of buildings. CADs provide general purpose drafting tools for engineers where libraries of objects can be added for different sectors of engineering.

DXF is a widely supported CAD file format, originally developed by AutoDesk® in the 1980s. Parallels can be made between DXF and Land-Line in that they are both based on linework with attributes, rather than 'real-world' features. As CADs have become more sophisticated, DXF has become a less useful exchange format, since it only deals with groups of attributed geometry.

Within the past decade, there have been moves within the construction industry to develop the concept of a unified model to support all stages of project management including initial specifications, geometrical models, analytical models (e.g., loading, lighting, heating and evacuation) and building construction[12]). Software packages have been developed with this in mind – an example being Autodesk® Revit®[13,14]. Instead of being based on a model where geometries are grouped and classified as 'real-world' features, products such as Revit are databases of objects which have a geometry plus built-in size and placement constraints that correspond to the real-world features.

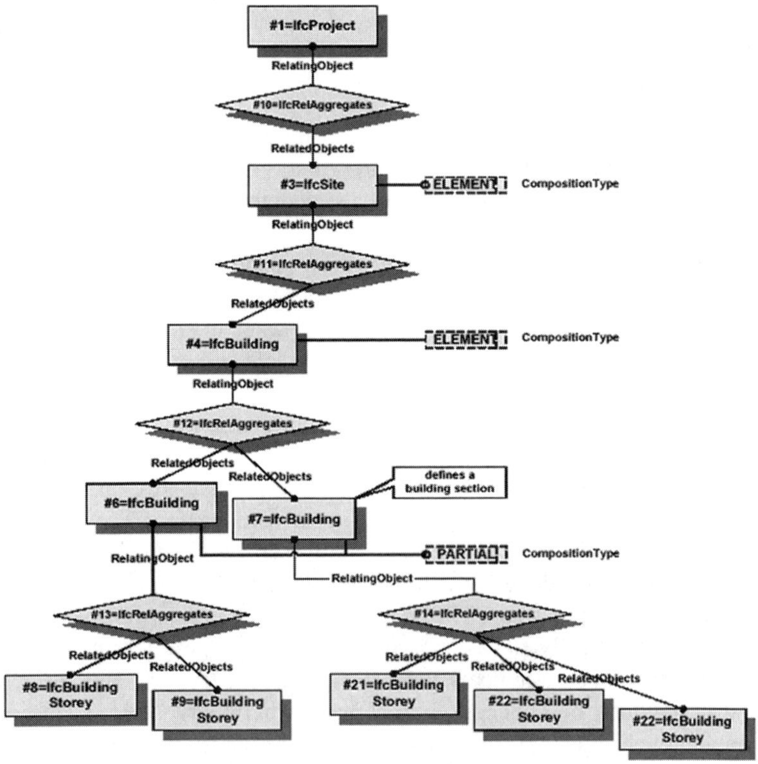

Figure 3.5 The hierarchical nature of the IFC standard. The highest level object is IFCProject, within which is IFCSite, then IFCBuilding. The IFCBuilding is composed of two wings, each of which has a number of stories. Many more levels also exist. Source: Liebich[8], Figure 104, p.102. With permission.

The Industry Foundation Classes (IFC) provide a standard for describing the 3D geometry, meaning (semantics) and topological properties of buildings and their parts[12,15]. IFC was developed by the International Alliance of Interoperability (IAI), a consortium of industrial partners including CAD vendors. Classifications of objects within IFC are arranged in a hierarchical fashion (Figure 3.5) allowing objects to be generalized to a parent class if required; for example, specific door types can be generalized to more generic 'door' types. Software that can interpret IFC data is able to abstract a building, extracting the relevant information for particular types of analyses (e.g., the surface geometry and reflectivity for lighting models).

A small project within IAI entitled Industry Foundation Classes for GIS (IFG) has been looking at ways in which models of buildings can be linked to their wider geographical context[16]. This would help streamline the process of submitting plans to planning authorities so that compliance with building regulations can be evaluated more easily. IFC may also represent a useful format for the data entry of buildings into a broader framework.

3.2.7 Virtual Cities

The term 'virtual city' is used in a wide variety of contexts. Implementations vary widely, but it is usually a digital representation of a city which acts as a basis for holding information about the city, its people and its interactions, real or imaginary.

It is 3D virtual cities that are of interest here, some of which are reviewed by Hudson-Smith and Evans[17]. 3D virtual cities are often designed to support applications of urban planning; for example, disaster management. Examples of uses of virtual city models for disaster management are: keeping a record of the state of a city before a disaster for the purpose of reconstruction, 3D visualizations and navigation for augmented reality systems, escape routes and flooding scenarios[18].

Different implementations of virtual cities are diverse in character and tailor-made for particular applications. There is no standard for the design or data exchange of virtual cities. This is being addressed by CityGML, an XML-based open data model under development by consortium of companies, municipalities and research institutes[19]. CityGML has characteristics in common with IFC: it is a detailed and semantically-rich standard for the exchange and interoperability of data for variety of uses; it deals with the 3D geometry, the semantics and the topological relations of concepts; and it has a dictionary of concepts. Among concepts addressed by CityGML are buildings, terrain, tunnels and support for the referencing of external models. The scope and set of concepts of CityGML are much more appropriate to national mapping than those of IFC.

3.3 DESIGN ISSUES

This section discusses the design principles listed in the introduction; namely the ability to describe different conceptualizations of features, the seamless treatment of space exterior and interior to buildings, the concept of a data repository to which information can be added to in an incremental fashion, the importance of pedestrian accessibility and the storage of 3D geometry.

Some of these design issues are present in the frameworks described in the previous section. The NDBS model holds enough information to construct a simple 3D geometry (prismatic building through the aggregation of constituent extruded floor polygons), but does not deal with access, nor does it deal with multiple conceptualizations of features. The latter is dealt with to a certain extent by NLGP through parent and child BLPUs, but geometry tends to be poorly described. Building models in the construction industry are the richest in terms of detail, but since they have been primarily developed for individual projects, there is a tendency for buildings to be treated in isolation from their surroundings. This is a problem for some energy modelling analyses; for example, a study by Howard et al.[20] found that only ten out of 33 building energy model software products took into account overshadowing effects of other buildings. The IFG hopes to address this problem; however, it stops short of dealing with detailed topographic maps, roads, bridges and transport[21]. CityGML is still under development, but it is a very interesting initiative from the point of view of our work. We are not attempting to build a catalog of semantic concepts, rather to provide support for their attachment with the possibility of exporting to CityGML in future.

3.3.1 Conceptualization of Real-World Features

All large-scale topographic maps depict addressable objects which correspond to real-world features in some manner. There are two issues relating to the conceptualization of a real-world feature; how to discretize it (geometrical extent) and how to classify it (describe what it is). The data provider (e.g., national mapping agency or the LLPG custodian) normally handles these issues. However, as discussed earlier, other organizations may require different conceptualizations of real-world features.

3.3.1.1 Classification of Addressable Objects

This section assumes that the addressable object's geometrical extent has been fixed and concerns how it is classified and identified.

Natural language words and expressions are in common use to express concepts of real-world features (e.g., 'house', 'garden'). The common understanding of these terms is not rigorously defined but is suitable for everyday conversation. Formally, the use of such words is problematic without an ontological model which explicitly defines concepts (incorporating a semantic model mapping concepts to words).

Bennett[22] explores aspects of classification for buildings and identifies the following breakdown of modes of classification:

- *Physical*: geometry, material properties
- *Historical*: origin (how the object came into being), initiated (some event that the object has undergone), periodic (types of event that the object repeatedly undergoes)
- *Functional*: actual, potential or intended functionality
- *Legal/Conventional*: ownership; associated rights and responsibilities, traditional convention

A 'church' can be defined physically (e.g., with a steeple), historically (e.g., an object was built as a church), functionally (e.g., as a Christian place of worship) or legally (e.g., as the workplace of a priest/vicar). Functional properties of buildings may be problematic because a range of activities may take place at different times and places within buildings. For example, a church building may have a basement in which meetings, workshops or social events take place; the priest or vicar may live in rooms physically encompassed by the church building; or a decommissioned church may become a residential or retail property even though its appearance is that of a church.

Steadman et al.[23] describe a physically-based classification scheme, not for the purposes of attaching semantic terms, but to assist in the statistical energy analysis of buildings. The scheme was developed using empirical data from the surveys of the four towns used in the NDBS project (Section 3.2.4) and is based on a classification of the geometrical forms of buildings, the arrangement of spaces (individual rooms, hall or open-plan) and whether space is naturally or artificially lit. Classifying buildings in these terms can partition them into groups useful for studying the energy efficiency of buildings.

As stated, the data provider usually uses a scheme to fix a classification and identity to each object. A more flexible approach would be to allow data users to apply their own semantic and ontological models, achievable by giving each object a set of attributes and allowing these to be used in conjunction with rules sets to classify objects into customized classification, or at least to attach semantic terms to them, the approach used in GIS ontological research. For buildings, the set of attributes might include color, architectural style, number of stories, height, date of construction, dates of refurbishment, function at different times of the day, ownership and tenants. This would support ontological models which define feature concepts in these terms; for example, a 'skyscraper' could be defined as a building being over 50 stories or over 150m high. Clearly, the range of supported ontological models is directly dependent on the choice of attributes available and the manner in which the attributes are expressed.

3.3.1.2 Feature Geometry

We will now turn our attention to the more fundamental problem of how to discretize addressable objects. This is especially a problem for natural features. For example, the boundary of a lake may be dependent on its water level, which may vary between seasons and with different amounts of rainfall. A related issue is to identify where a river ends and a lake begins. Some rivers have ponds and small lakes along their length; in some cases this might be treated as a single stretch of river; in others, as a series of river segments and ponds.

In the built environment, the uncertainty problem exemplified by a lake is less of a problem because man-made objects tend to have distinct boundaries. However, as discussed in Section 3.2, there are many different conceptualizations of features with different geometrical extents. The geometry of a detached 'house' is fairly unambiguous; it can be expressed as either a polygon or a polyhedron corresponding to its well-defined extent. However, in the case of a semi-detached house or a terrace of housing, the word 'house' may refer to the extent of the envelope of the combined semi-detached or terraced unit or an envelope of an individual residential unit within. Similarly, a block of flats is a stand-alone unit containing residential properties.

Geometry-based features are traditionally the focus of data provided by national mapping agencies. Using a combination of aerial photography and ground surveying, the identity and extent of features are recorded according to the capture specifications. The result is one feature with a geometry for every geometrical object (conforming to the capture specifications).

In gazetteers, geometry is not of primary importance. In address gazetteers such as NLPG, there may be one feature with geometry for each postal address delivery point. Since there may be more than one delivery point for one 'building', there will be a many-one relationship between 'postal delivery points' and 'buildings'.

In order for our framework to support these multiple feature types, we need to separate the geometrical description from features. We suggest that geometry be modelled as a set of primitives with no 'real-world' meaning and that these should be 'atomic' (indivisible) to the end user representing the smallest geometrical units possible. An addressable object's geometry will be composed of the union of one or more geometrical primitives. At the scale at which geographers work, there are no obvious natural discrete 'atomic' units; thus we propose to use the smallest units needed to support all the addressable objects currently required. Where the geometrical primitives are not atomic enough to describe the geometry of new addressable objects, the data custodian would simply subdivide the primitives in such a manner that they would. As more data are added, the primitives would become smaller and denser and would be capable of supporting the new and previous object geometries through aggregation.

The simplest way of describing the geometry of an addressable object is to assign a list of primitives to it, the union of which forms its geometry (when primitives are subdivided, this will have to be updated). We suggest that another way of doing this is to use a pedestrian access-based query. The urban environment is designed and built for human activity and groups of spaces (rooms) of closely related activity in the built-environment tend to be accessible by those groups of people involved in the activity. This is why we suggest access might be a useful property for the conceptualization of the built environment. The geometry of an addressable object could be defined as the union of all the geometries with access to a particular type of pedestrian from a particular point.

To this end, we have developed a conceptual model for describing access in the built environment[24]. Simple conceptualizations of 'wall', 'door' and 'space' features have access related attributes attached to them. For example, doors and spaces have pedestrian- and time-dependent access permissions attached. Using these attributes, an area of space can be delineated for a particular pedestrian and time. The flexibility of this framework is directly related to the attributes encapsulated in the access model.

To illustrate the conceptualization of units of a building, consider a block of flats containing three groups of spaces (in terms of access):

- Space communal to all residents of the block
- Space available only to the residents of a flat
- Space accessible only to the inhabitant of a room within a flat

One can identify a hierarchy of three levels here, corresponding to different groups of people: access to communal parts of the block from the main door, access to this plus the communal space available only to the residents of a flat and access to these spaces plus the room of a resident within their flat. (Such hierarchies of access have a highly variable depth and form.) The geometrical extent of space accessible by a particular pedestrian could be the basis for defining the geometrical extent of the building unit. The classification of the pedestrian could also provide the basis for that of the building unit.

3.3.1.3 Feature-Model Framework

Clearly the various definitions of features (and more specifically, building units) are strongly dependent on the type of attributes that are stored by each object and the granularity of the geometry that is stored. Our aim is not to implement a framework able to delineate every possible conceptual idea of a building – this would be an impossible task because of the vast data requirements. Rather, our aim is to design a framework in which it is possible to add the necessary granularity of data and attributes to support particular conceptualizations where the framework would continue to support all previously defined conceptualizations.

3.3.2 Seamless Treatment of Space Exterior and Interior to Buildings

A popular approach to modelling cities is to embed individual 3D models of buildings (ranging from simple prismatic to highly detailed) within 2D street maps. These 3D scenes can be rendered in a visually appealing manner, but little more functionality is offered. The individual buildings are normally modelled as hollow external shells.

The inclusion of hollow external shells of buildings does not add much new functionality to an urban model (except visual effects and viewshed-based analysis of the external envelope of the built environment). If, rather than being hollow, the buildings had their interiors modelled, this would add a great deal more functionality (although it is more difficult in terms of data acquisition), because it would be possible to identify spaces within buildings.

When inside and outside spaces are modelled separately from each other (as they usually are), a choice has to be made about whether a space is one or the other. As should be clear from the previous discussion, the concept of 'inside' or 'outside' is not compatible with our requirements because these are natural language terms which are defined with reference to a 'building' concept; thus dependent on a particular building conceptualization. The sometimes-blurred distinction between interior and exterior space is illustrated in Figure 3.6. This dependency can be removed by treating all space seamlessly, modelling the geometry (with no real-world meaning) in one framework.

Figure 3.6 Views of a small shopping center showing a blurred distinction between 'exterior' and 'interior' space. The galleries on the different levels are mostly uncovered. The escalator (left image) is in a covered but not walled area. There is pedestrian access through the shopping mall (center image) and all the shops form distinct units. Photographs: Aidan Slingsby.

There are also more practical reasons why we wish to treat space as seamless. In some circumstances, we would like to be able to treat all functional retail units with equivalence. This may be difficult if some of these are interior to a 'shopping center building' and others are accessible to the high street in a city center. In other circumstances, we may wish to treat pavements, uncovered pedestrian areas and corridors within an indoor shopping center as equivalent.

In summary, the concept of inside and outside is often ambiguous, is often not useful and, most importantly, is tied to 'building' concepts.

3.3.3 Data Repository and Incremental Updating

Our framework is designed to be a repository containing all available data. This is an important function because it allows data to be incorporated into the framework as soon as they become available rather than waiting until they fit a particular specification.

The incremental updating approach is also appropriate where there is no initial knowledge of exactly what needs to be captured. This is the case for two reasons:

- The framework aims to support different agencies' definitions and subdivisions of building units. Since there are no natural and universal discrete units, we use the smallest primitive units which are required to describe the geometry of current features. When new feature geometries are defined which the existing primitives cannot describe, the latter are subdivided to achieve such representation. Thus, we cannot predict a full and universal set of geometrical primitives without a predefined set of features.
- It is not appropriate to capture interiors of all buildings. Buildings within which there are public spaces should be part of the model, but private or domestic building interiors probably should not be. More generally, public spaces and those already depicted on maps should be modelled, whereas private spaces hidden from view do not necessarily need to be modelled in fine detail. However, it is unlikely that one can produce a definitive list of which spaces should be captured. One reason is that whether spaces are 'public' or not is open to some debate since this may depend on the group of people involved and management policies[25]. In practice, it is likely that spaces in certain buildings will be captured in fulfilment of specific requirements and then the representation may be available for others to use (perhaps with some restrictions). To some extent, this is what happens in Kingston-upon-Thames' LLPG. If the geometry of a child BLPU is required for a particular reason, its geometry is surveyed, retained in the database, and may be used again later.

The proposed framework is not intended to support direct 3D topological queries or rapid visualization. As a data repository, it is able to reconstruct 3D geometry through interpolation from spot heights, relative heights and access based on rule sets (see below). The output from the framework is then used to populate other data models capable of 3D spatial analysis (i.e., 3D GIS) and visualization (3D graphical modellers).

3.3.4 Representation of 3D Geometries and Pedestrian Accessibility

3.3.4.1 2½D Geometrical Models

2½D models are those where information about the third dimension is stored as non-spatial attributes and 3D geometries are reconstructed using special rules. The most simple and common example is where a polygon is extruded along a path perpendicular to the polygon's plane (along the z axis in 3D space) for a distance specified by an attribute. An attribute indicating the base height is also a common addition. Further attributes and rules can be defined to allow more complicated types of extrusion (e.g., sweeps along curved paths) or parameters for solid primitives (e.g., pyramids for roofs).

2½D models are widely used for urban modelling in GIS; many software products (e.g., ArcGIS®[26]) have these capabilities built in. The reason that these apparently crude methods of representation have been used so successfully in urban geography is that the geometries which exist in the built-environment are dominated by horizontal and vertical surfaces (for reasons of gravity and close-packing). 2½D models exploit these inherent symmetries for simplicity and lower storage requirements at the expense of the ability to represent more complex geometries. The NDBS project uses a simple 2½D modelling strategy, but it extrudes footprints for sections of individual stories rather than those of entire buildings (see Figure 3.3) so is able to model overhangs, bridges, variations in story extents within the same building and complex juxtapositions of building units – characteristics which simpler 2½D models are unable to represent.

3.3.4.2 Proposed 2½D Model with Height Constraints

Over the past few decades, the decreasing cost of data storage and increasing computer processing speeds have gradually enabled the routine use of full 3D models (where the z component is treated in the same way and with the same prominence as the x and y components). However, for our framework, since we are starting from existing 2D and incrementally adding detail, and we want to avoid mixing two data models, using a full 3D model is inappropriate. Instead, we propose a multilayered 2½D model in which layers are topologically linked to each other by access routes and three types of height constraints are used to allow the 3D interpolation of the model: absolute heights, relative heights and (pedestrian) topological consistency.

- *Absolute heights* are spot heights which are scattered within the 2D model. They effectively 'pin' the surface heights to a national mapping datum.

- *Relative heights* enforce a specific vertical separation between geometries. Examples include describing a bridge height over a road, curb and step heights. We consider that relative heights for bridges and curbs are more appropriate than the absolute heights of their upper and lower components. One reason for this is that curb heights (typically ten centimeters) are likely to be below the precision of the height capture specifications, yet in terms of access such steps may be significant obstacles.
- Access is an integral part of our framework because it is an essential aspect of the built environment. It has two key roles, one being feature conceptualization (Section 3.3.1.2) and the other is providing *constraints for topological consistency* in the geometrical model. Even if only sparse height data are available, ensuring the topological consistency of pedestrian access between geometries will mean that the resulting model is at least topologically consistent. Two examples are ensuring that a bottom of a door geometrically coincides with the pavement to which it allows access and that a ramp geometrically matches the two levels it connects. In addition, requiring topological consistency between geometries can be seen as enforcing a relative height of zero between layers.

Using these three types of constraints, a 3D model can be interpolated. Its quality is dependent on the number of height constraints incorporated. Following the 'incremental updating' principle (Section 3.3.3), it is anticipated that absolute and relative heights may be rather sparse at first, but as areas with poorly resolved heights are identified and more data are obtained, these can be incrementally added to improve the result of the interpolation.

Figure 3.7 shows complex multi-level streets in Edinburgh. The left and center images show a steeply sloping road and a horizontal elevated walkway above the lower few stories of the buildings. The top of the road and the walkway join on the same level at the George IV Bridge (not shown). In our framework, this topological connection would be used to ensure that the geometries of these elements were consistent without the need for a full 3D description. The image on the right shows a road bridge adjacent to buildings. The ground floors of the buildings have access to the lower road whereas access can be gained to the first floor from the upper road on the bridge itself. In our framework, we ensure that door access to the relevant sections of pavement and the relative height between the top and bottom of the bridge is maintained, rather than being concerned with absolute heights. (In this case, the bridge height would have also direct effect on the height of the lower story of the building). These photographs illustrate a complex multilayered urban environment where the pedestrian connectivity and relative heights between levels are given precedence over precise absolute heights and full 3D geometry; this is a very important principle of our framework.

Figure 3.7 Streets in Edinburgh showing multi-level complexity. The left and center images are of Victoria Street – a steeply sloping crescent-shaped street with a horizontal walkway at a higher level. The image on the right shows the George IV Bridge as seen from Cowgate. The buildings next to the bridge have access to both George IV Bridge and Cowgate. Photographs: Mike de Smith.

The design principle of being able to interpolate z values from poor or incomplete height data and of incremental updating is in marked contrast to most 3D models and surveying projects where every point has x, y and z values. For modelling cities over such large areas, we believe that full 3D models are inappropriate, as the data capture and storage requirements are usually prohibitive. Our concept of the incremental addition of detail also appears to be in contrast with the strict mapping specifications enforced by national mapping agencies. However, capture specifications are policy rather than a fundamental part of a framework and such a policy could be implemented with our framework. Data capture specifications are separate from the framework design.

3.4 PROPOSED MODEL

Taking into account the design principles discussed in the previous section, we will now present our model. The built-environment is abstracted to the spaces accessible to pedestrians; at this stage we are not concerned with the other geometries such as the spaces between ceilings and roofs. Figure 3.8 shows an output of a scenario using our model. Figure 3.9 illustrates the geometrical model and is further explained below.

3.4.1 Basic 2D Geometry and Topology

Geometry is modelled as a set of 2D geometrical primitives (we ignore height for the moment) with no real-world meaning. The geometrical primitives are represented in a classic topological 2D data structure, consisting of nodes, edges and polygons. Nodes have the geometrical information of x and y values. The rest of the geometrical information is built from the topological information held by the primitives. A node has a list of references to its connected edges. An edge (a single line segment) has references to its two end nodes and its two adjacent polygons. A polygon has a reference to one edge in each of its rings (inner and outer boundaries); the number of edge references a polygon has is equal to the number of islands it has plus one. From all this information, the full 2D geometry can be built (Figure 3.9a).

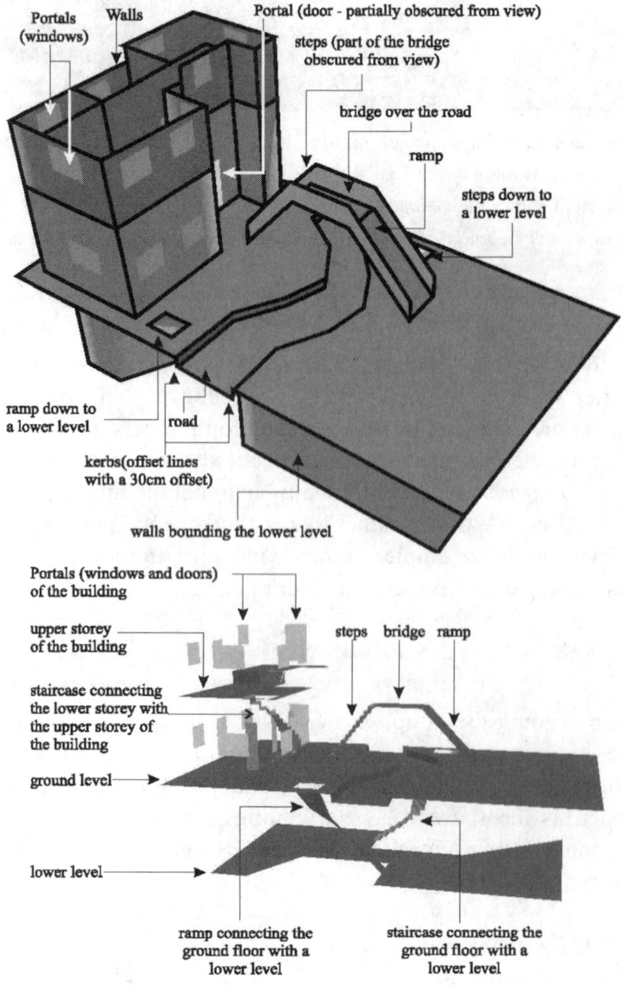

Figure 3.8 Annotated example of a small piece of the built environment modelled in the proposed framework. The upper image shows a section of road, over which is a bridge. The pavements have holes leading to a ramp and staircase down to a lower level. The building and its interiors are modelled seamlessly. The lower image shows the same scenario with the walls omitted to display the internal geometrical structure.

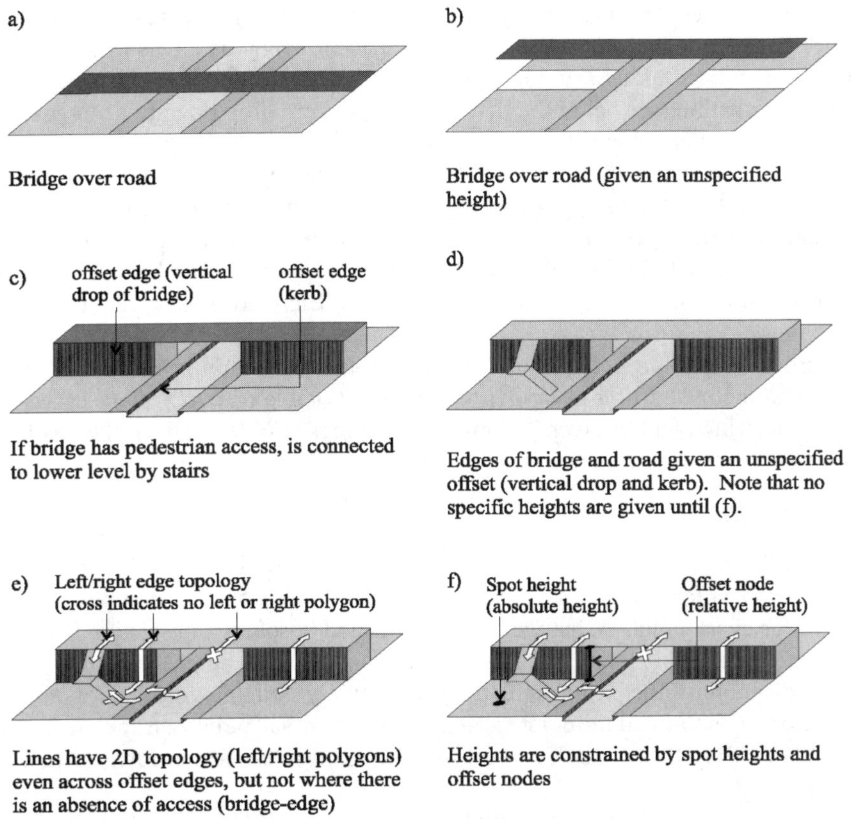

Figure 3.9 Illustration of the proposed model. See the labels and main text for further explanation.

3.4.2 Multi-layering

In order to support multi-layering, we make a number of modifications to the classic 2D model described above.

A layer is any group of geometries where the primitives do not self-intersect in 2D. Where this does occur (e.g., an over-pass), the geometries must be split into more than one layer (Figure 3.9b). This is done when the data are inputted (because data are added to the model as a series of 2D layers). Where layers topologically connect to each other, an edge will often have three bounding polygons (e.g., bottom of ramp in Figure 3.9e). Our solution in such a case is to have two coincident lines sharing the same nodes; this is how we connect separate layers. Polygons can be traversed as if there was no interface between two layers. The topology describes both how polygons are constructed and the pedestrian access between layers (Figure 3.9e).

Several edges are marked as 'offset edges'. These represent a (usually unspecified) vertical drop between the two adjacent polygons. This offset (vertical drop) may change along the edge; if unspecified, it depends on the surrounding height information (see below). Offset edges with zero drop operate as 'breaklines' indicating a break of slope.

Figure 3.9c shows the use of vertical offsets to represent a curb and bridge. Small offsets (e.g., curbs and steps) are common in the built environment and we envisage that all edges corresponding to them would be marked as such. Their heights are likely to be below many data capture specifications but are important features with respect to access and can be described using relative height constraints.

Vertical offsets do not apply to inclined drops (e.g., ramp in Figure 3.9d). As soon as an inclined drop has a polygon footprint (visible on a 2D map), it becomes a polygon (which will be steeply sloping). The threshold depends on the resolution (the smallest resolvable distance).

3.4.3 Heights

Thus far, we have described topologically connected layers with offset edges, but no height information. We have already reasoned that we do not need to store heights for every point. What we do instead is add absolute and relative heights in key areas and apply some rules. Special nodes with height information are scattered amongst the layers and these effectively 'pin down' the topological-connected layers at certain points. Heights at any other points on the surface are interpolated from these nodes, subject to 'horizontal rules' which force some areas to remain flat.

The two types of height node used are absolute and relative heights (see Figure 3.9f). Absolute heights are those above the national mapping datum. Relative heights enforce a height separation between two geometries. Where a relative height forms part of an offset edge, it quantifies the offset at that particular point. Otherwise, it represents a spot height relative to geometry on another layer (e.g., a three-meter separation between two stories).

There are two rules which force part of the interpolated surface to be horizontal. One of these ensures that roads are horizontal, perpendicular to their direction. The other rule makes areas with under-resolved height information horizontal by default. This is a useful constraint for building stories – if there is one spot height on a story, the story will be horizontal.

Figure 3.10a illustrates 2D layers with no height information. Figure 3.10b shows the geometry resulting from the addition of an absolute height, three relative heights and some horizontal constraints. Note that the height information is under-resolved and it is the application of a horizontal rule which allows a 3D geometry to be output. The addition of further height information would result in a more accurate 3D solution.

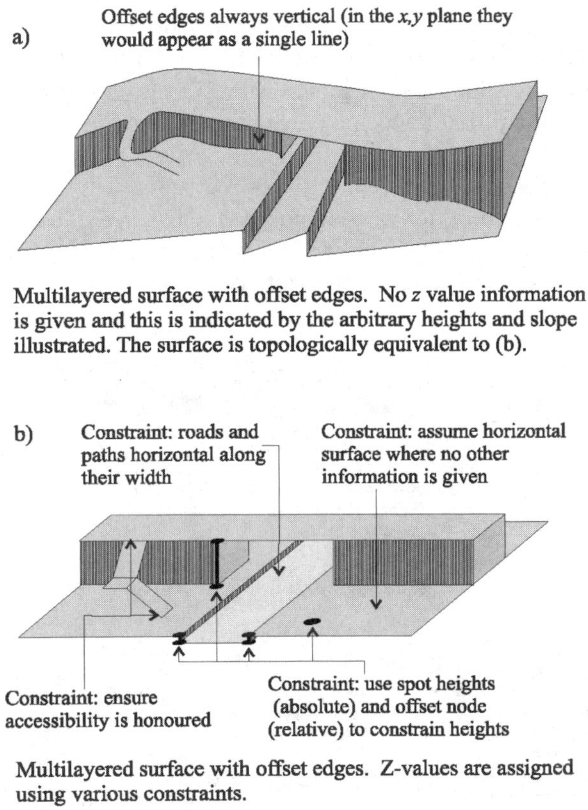

Figure 3.10 Example of how height and other constraints can be used to construct a 3D geometry.

The algorithms for constructing the 3D geometry are designed to produce a reasonable solution everywhere scant height data exist, but are able to accept additional information and use it to create a better 3D geometry. Although this does not have the deterministic rigor of more orthodox 3D models which require complete data, it fulfills the design criteria discussed earlier.

Figure 3.8 shows a scenario for a section of the built-environment modelled using our framework. Layers of topologically-structured polygons correspond to rooms or parts of rooms, some of whose edges are parts of walls, doors and windows. Amongst the polygons, points with either absolute or relative height information are used to provide heights for each layer. In this diagram all the layers are horizontal, but the presence of differing absolute or relative heights would result in the interpolation of a sloping or undulating surface.

3.4.4 Strategy for Managing Heights

In order to formalize the way in which the algorithm for constructing 3D geometries handles heights, we introduce the concept of 'patches'. A patch is a set of adjacent polygons which do not intersect in 2D, are in the same plane and are bounded by discontinuities of which there are three types (Figure 3.11).

- Offset edges
- Edges with a 'null' left or right polygon (e.g., edge of a bridge)
- Edges which close the former two types of edge (to make a patch) and which indicate a change in gradient.

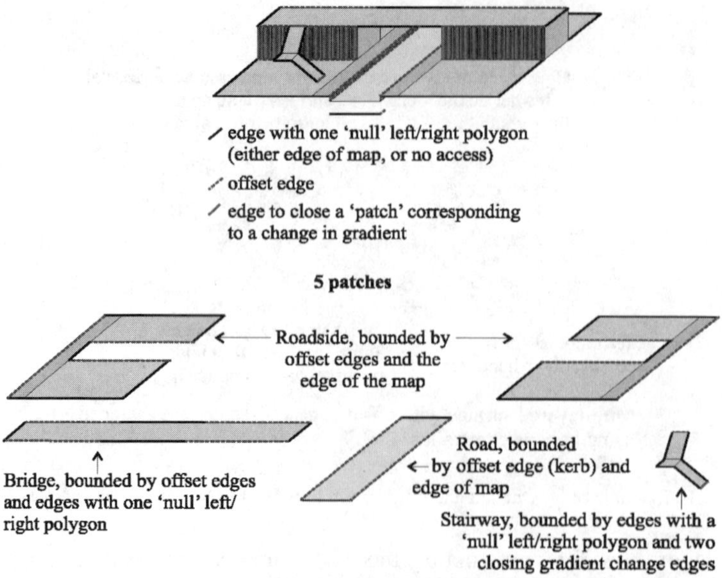

Figure 3.11 An example of how 'patches' are identified.

A patch has the property that it is a continuous surface and its geometries do not intersect in 2D. This enables its surface height (which may be undulating) to be modelled with a triangular irregular network (TIN). Each patch has its own TIN and this is usually independent from the TINs of the surrounding patches (with certain exceptions, see below). The concept that patches can be oriented independently of adjacent patches is important. Consider the railway in Figure 3.2 (dark gray polygons striped by lines, stretching diagonally from the bottom left of the map extract). It cuts a road in the bottom left of the figure (indicating that it passes above) and is cut by five roads towards the right. These roads are a separate patch to the railway and the railway patch weaves under five roads and over one.

Patches are not always independent of each other. At particular points, where there is a relative height between geometries on different patches or where there is pedestrian accessibility, the heights of the patches there are dependent on each other. If the relative height is zero, two patches appear to join, with no vertical offset between them. Patches representing ramps and stairs which link layers together do not need any additional height information as they take their height from the patches they connect.

One of the model's principles is to cope with under-resolved height information and one of the horizontal constraints applies here. If a patch only has one piece of height information, it will be assumed to be horizontal. For buildings, where this is generally the case, a story only requires one piece of height information (an absolute or relative height). If a patch has no height information, it will recursively request adjacent patches to build their TINs until it is able to obtain a height from an adjacent patch.

3.4.5 Real-World Meaning

We attach 'real-world' meaning to these geometrical primitives by describing various instances of feature concepts whose extent is defined by the union of a set of geometrical primitives. We define the feature concepts of 'wall' and 'portal' (i.e., door or window) which act as barriers in the built environment, though portals may have access permissions for certain types of pedestrian[24]. The extent of a wall or portal is described by a set of polygon or line primitives (Figure 3.12).

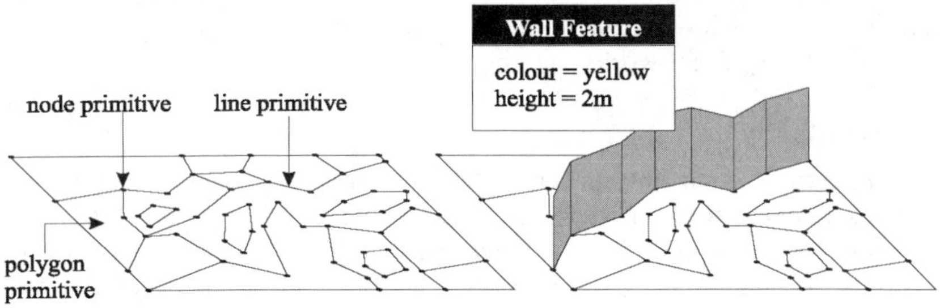

Figure 3.12 Example showing the geometry of an addressable object (a wall with attributes) being described by a list of seven line primitives.

3.5 IMPLEMENTATION

The prototype framework is being implemented as a Java application, using the object-oriented open-source database Ozone[27]. ESRI ArcGIS is used to help prepare the initial data. Using geometry from OS MasterMap (for the base map) and from user-drawn geometries (for building interiors), VBA scripts are employed to automatically build the required 2D topology for import into the model. The data are imported into the Java application and database using Shapefiles and DBF files. Topological linkage between layers is handled by the application using attributes set in ArcGIS. 3D output are then generated as 3D Shapefiles (using Multipatches) which can be viewed and inspected in ArcScene®[28] (Figure 3.8 is annotated ArcScene output).

3.6 INTERIM EVALUATION

The DNF framework of OS MasterMap has been designed for the distribution of geometrical 2D digital national mapping data, and it does this very well. However, if we take applications which deal with the building stock (e.g., land registration and facilities management), pedestrian modelling (e.g., retail modelling and evacuation), or those which require three-dimensional data (e.g., viewshed analysis, mast location and 3D visualizations), OS MasterMap has some important limitations.

3.6.1 Conceptualization of Features

Building unit extents in OS MasterMap are based on simple 2D geometries and include no concept of feature hierarchies (e.g., rooms, flats, storys, buildings and land parcels). These characteristics make OS MasterMap of restricted use to agencies such as the Valuation Office and the Land Registry. In our proposed framework, the geometrical extents of features are either prescribed or dynamically delineated according to the space accessible by a pedestrian in order to provide a flexible way in which a common geometrical database can be used for different conceptualizations of real-world features.

3.6.2 Accessibility

The Integrated Transport Layer in OS MasterMap provides a network with restrictions relevant to road transport (one-way streets, bridge heights). It only deals with road transport on prescribed routes. By contrast, pedestrian access connectivity is integral to our framework and customized pedestrian networks which seamlessly pass through and between buildings can be readily extracted. Attributes of access points specify different permission levels which control right of entry to the spaces between these points. This information can be used for pedestrian modelling, accessibility analyses, wayfinding and routing applications.

Making pedestrian accessibility integral to the framework is appropriate because the urban environment is one designed for people to work and move about in. Customized pedestrian routes can be extracted and spaces accessible by particular pedestrians can be delineated. Pedestrian topology is used to assist in the construction of a full 3D geometry and to help provide the geometrical extent of some addressable objects.

3.6.3 Geometry

In our framework we have abstracted the urban environment as a set of connected floor spaces, between which may be walls (barriers) and access points. Currently our emphasis is on the descriptions of features and pedestrian accessibility, rather than the accommodation of an accurate 3D geometrical representation – we do not currently deal with roofs or decorative appendages of buildings, both of which greatly affect the outward appearance and their impact on viewshed analysis. Support for these could be added in the future. Non-vertical walls could be described through the addition of more attributes and 2½D rules. Upper floor polygons of buildings could have parametrically-described roof geometries associated. Roofs and walls could have other appendages added. The presence of roof details would be important for visualization applications.

Rather than the requirement for a large amount of height data, we use rules related to pedestrian accessibility and surface characteristics (e.g., roads being horizontal along their width) to reconstruct 3D geometry. Additional height data can be added in the form of spot heights and then used to improve the 3D geometry. Since our framework is designed to reconstruct 3D geometry from sparse data, it may not be able to provide very accurate heights (though it would be able to evaluate the margin of error). Thus, the framework will not necessarily be suitable for applications (such as flood or viewshed modelling) which require a very detailed and accurate 3D geometrical description.

3.7 CONCLUSION

This chapter has discussed the rationale behind the development of a new framework for the storage of digital national mapping data in urban environments. Many applications need a richer set of data than national mapping agencies currently provide. We have reviewed some of the existing data products and identified applications for which we believe this framework is fit for purpose. The design issues which we consider important and which we have tried to address are:

- The need for a data repository which can store data in a compact and flexible form.
- The need to be able to provide 3D geometrical views of urban data.
- The need to be able to hold spaces both internal and external to buildings in the same framework.
- The need for a framework which can be incrementally updated.
- The need for information on pedestrian accessibility.
- The need to cope with alternative conceptualizations of real-world features.

We believe that our approach offers useful properties for the applications we identify, even though it stops short of accommodating a fully detailed 3D geometrical model. Further work is currently being carried out to implement our proposed model.

We hope that the innovation of incorporating the three aspects of geometry, flexible feature definitions and access for spaces both within and outside buildings in the same framework will widen the scope of research on the urban environment. This should also have positive implications for environmental decision-making, particularly with respect to issues of access which are extremely important in urban management. Information on pedestrian modelling or routing is central to a range of application from retail modelling to emergency management. At a more abstract level, accessibility patterns on a city-wide scale could be correlated to other aspects of human activity. This might highlight, for example, contrasts between areas of a city dominated by public parks compared to those dominated by gated communities, or between suburban car-oriented estates compared to inner-city mixed land-use areas, that would have wider implications for environmental management and planning.

3.8 ACKNOWLEDGMENTS

The authors gratefully acknowledge the support of the Economic and Social Research Council (ESRC) and the Ordnance Survey for their support of this project. They would also like to thank David Capstick for his useful comments on this chapter.

3.9 REFERENCES

1. Ordnance Survey, About us, http://www.ordnancesurvey.co.uk/oswebsite/aboutus/index.html, 2005.

2. Barr, R., OS MasterMap – the vision: How Ordnance Survey led the world in the creation of a new type of national geospatial dataset, *GI News*, March/April 2003, http://www.ginews.co.uk/0304_28.pdf, 2003.

3. Ordnance Survey, *Land-Line® User Guide*, Ordnance Survey, Southampton, 2002.

4. Non-Domestic Building Stock Project, http://www.bartlett.ucl.ac.uk/ndbs/home.htm, 2006.

5. Steadman, P., Bruhns, H., and Rickaby, P. A., An introduction to the national Non-Domestic Building Stock database, *Environment and Planning B: Planning and Design*, 27, 3-10, 2000.

6. GE Energy, Geospatial asset management, http://www.gepower.com/prod_serv/products/gis_software/en/index.htm, 2006.

7. Holtier, S., Steadman, J.P., and Smith, M.G., Three-dimensional representation of urban built form in a GIS, *Environment and Planning B: Planning and Design* 27, 51-72, 2000.

8. British Standards Institute, *Spatial Datasets for Geographical Referencing – Part 2: Specification for a Land and Property Gazetteer*, BS 7666-1:2000, British Standards Institute, London, 2000.

9. National Land and Property Gazetteer, About the NLPG, http://www.nlpg.org.uk, 2006.

10. Personal communication with the GIS Facilitator, ISIS Team of the Royal Borough of Kingston-upon-Thames.

11. Royal Borough of Kingston-Upon-Thames, Integrated Spatial Information System, http://maps.kingston.gov.uk, 2006.

12. Eastman, C.M., *Building Product Models: Computer Environments Supporting Design and Construction*, CRC Press, Boca Raton, Florida, 2000.

13. Autodesk, Autodesk Revit Building – product information, http://www.autodesk.com/revit, 2006.

14. Khemlani, L., Comparing Revit and Autodesk Architectural Desktop: Part 1, *CADENCE AEC Tech News*, 72, http://www.cadenceweb.com/newsletter/aec/0402_1.html, 2002.

15. Liebich, T., Ed., *IFC 2x Edition 2 Model Implementation Guide*, International Alliance for Interoperability Modeling Support Group, http://www.iai-international.org/Model/files/20030630_Ifc2x_ModelImplGuide_V1-6.pdf, 2003.

16. International Alliance for Interoperability, IFC for GIS introduction, http://www.iai.no/ifg/, 2006.

17. Hudson-Smith, A. and Evans, S., Virtual Cities: from CAD to 3D-GIS, in *Advanced Spatial Analysis – The CASA Book of GIS*, Longley, P. and Batty, M., Eds., ESRI Press, Redlands, California, 2003, 41-60.

18. Kolbe, T.H., Gröger, G., and Plümer, L., CityGML – Interoperable access to 3D city models, in *Proceedings of the International Symposium on Geo-information for Disaster Management*, van Oosterom, P., Zlatanova, S., and Fendel, E.M., Eds, Springer Verlag, Delft, The Netherlands, http://www.citygml.org/docs/Gi4Dm_2005_Kolbe_Groeger.pdf, 2005.

19. CityGML, Excahnge and storage of virtual 3D city models, http://www.citygml.org, 2006.

20. Howard, R., Wager, D., and Winterkorn E., Guidance on Selecting Energy Programs, Construction Industry Computing Association, Cambridge, 1994.

21. International Alliance for Interoperability, IAI New Project Proposal, http://www.iai.no/ifg/ifc_for_gis_proposal_form1_.doc, 2003.

22. Bennett, B., Foundations for an ontology of built environments, School of Computing, University of Leeds, Leeds, 2003.

23. Steadman, P., Bruhns, H., Holtier, S., Gakovic, B., Rickaby, P.A., and Brown, F.E., A classification of built forms, *Environment and Planning B: Planning and Design*, 27, 73-91, 2000.

24. Slingsby, A.D., Pedestrian accessibility in the built environment in the context of feature-based digital mapping, in Proceedings of Computers in Urban Planning and Urban Management (CUPUM), London, http://www.cupum.org, 2005.

25. Hwang, J. and Koile, K., Heuristic Nolli map: a preliminary study in representing the public domain in urban space, in Proceedings of Computers in Urban Planning and Urban Management (CUPUM), London, http://www.cupum.org, 2005.

26. ESRI, ArcGIS Desktop, http://www.esri.com/software/arcgis/about/desktop.html, 2006.

27. SourceForge, Ozone – Java OODBMS, http://sourceforge.net/projects/ozone/, 2006.

28. ESRI, ArcGIS 3D Analyst, http://www.esri.com/software/arcgis/extensions/3danalyst/index.html, 2006.

CHAPTER 4

From Electronic Logbooks to Sustainable Marine Environments: A GIS to Support the Common Fisheries Policy

J. Whalley and Z. Kemp

4.1 INTRODUCTION

Within the European Union (EU) there is a major focus on developing sustainable fisheries and addressing the environmental issues related to maintaining marine ecosystems. Many high value fish species have been over-fished or fully exploited. As a consequence fisheries are becoming increasingly subject to EU regulations[1]. In order to ensure the effectiveness of EU fisheries policy the catch and environmental data must be collected and analyzed accurately.

In this chapter we report on FishCAM (Fisheries Computer Aided Management), a fully-fledged, flexible, modular, component-based computer system built using an object orientated conceptual framework for modelling spatio-temporal data. The combination of georeferenced data with aspatial attributes enables extraction and visualization of complex spatial relationships to support decision-making. The system addresses the requirement for accurate reporting of fisheries and environmental data and the provision of software tools for scientific analysis and assessment. It relies on a rigorous data model to enable fusion of divergent data resources and a component-based analytical module to facilitate flexible querying across space, time, scale and theme. Intuitive interfaces provide a means of exploring and understanding the structures, patterns and processes reflected in the data.

4.1.1 Overview of the Common Fisheries Policy

Fishing is one of the most important economic activities in the EU. On average, the sector accounts for approximately 1% of the gross national product of Member States and is an important source of jobs in areas where there are few alternatives. Annually seven million tonnes of fish is caught making the EU the world's second largest fishing power after China. The Community fleet is substantial, consisting of approximately 90,000 vessels covering a wide range of sizes, catching capacity and power[2]. The fishing capacity and activity of the European fleet is too great for the available resources; this excess capacity and activity reduces the profitability of fisheries and increases the danger of exhausting stocks.

As a consequence, in recent years the EU has sought to facilitate an improved balance between vessels and fish, with an overall reduction in the capacity of its fleet. The 2002 Common Fisheries Policy (CFP) reform provided for an increase in this trend. The CFP included technical measures to protect fisheries resources. In this context, its reforms have adopted a long-term approach, fixing annual Total Allowable Catches (TACs) on the basis of fish stocks, and introducing accompanying conservation measures. Legislation that details specific control and inspection activities has been introduced in order to achieve adherence to these new measures[1]. In addition, the Community Fisheries Control Agency was established in 2005[3]. The success of a TAC-based approach is dependent upon the accuracy of the scientific estimates of sustainable yield as well as a commitment from states to accept this research and the fishermen to abide by the quota limits honestly.

4.1.2 The FishCAM Software

FishCAM is a system to provide electronic logbooks for fisheries that is linked to a Global Positioning System (GPS) and a Geographic Information System (GIS). The research and development of the system was funded by the European Union through a CRAFT project grant under the Sustainable Agriculture, Fisheries and Forestry initiative. FishCAM is a flexible, modular system consisting of three main components:

- *The onboard module*: a fisheries logbook data capture module used on board fishing vessels.
- *The mobile vessel tracking module*: designed to be used as a real-time monitoring system (see Section 4.3.1) and for use by fishery scientists to analyze the spatial distribution of fishing activities.
- *The analysis module*: designed to be used by fishermen, fisheries authorities and researchers for the purposes of catch analysis, stock analysis and decision support (see Section 4.3.2).

These customizable bespoke software modules comprise the current FishCAM suite and were developed using C++. They incorporate not only space-time analysis, but also next-generation databases and GIS. Each module was developed and tested individually before integration into the full system. Every module was designed to act as a stand-alone application but may be integrated as required. The onboard module was the first module developed as the core data substrate design is part of this module. Later modules were built on this data substrate, but may be extended by fusion with other datasets. Once the onboard module had been developed several sea trials were undertaken to test the system. The trials were carried out on fishing vessels using a variety of fishing gear, such as purse seine, gillnet, pots and dredge. Further discussion of the sea trials for the onboard module is presented in Sections 4.3.2 and 4.4.

4.1.3 Environmental Factors

Marine species exhibit complex patterns of distribution and interaction. The entire marine ecosystem depends on a sensitive equilibrium between food providers and consumers. Any disturbance to part of the food chain can result in serious stresses in non-target species. For example, in surveys of European mackerel, Fives et al.[4] found that egg numbers halved between 1977 and 1986. Such declines are often attributed to over-fishing. However, there are wide natural variations in the abundance of plankton, including fish eggs and larvae, from year to year. The reasons are not fully understood, though it has been postulated that variations in temperature, sunlight and ocean currents all have an impact as, of course, do over-fishing and human environmental impacts[5].

Fishing activities influence marine systems in a number of ways:

- *Removal of target species*, and resulting changes in the size structure of their population
- *Mortality of non-target populations* of fish, seabirds, marine mammals and other forms of benthic marine life
- *Alterations of the seabed habitat*
- *Changes in the food web*, with impacts on the predators and prey of the species
- *Discarded or lost fishing gear* can generate a process known as 'ghost fishing', entangling fish and marine mammals for many years

Other main sources of marine ecosystem damage include:

- *Nutrient run-off*: residues of fertilizer and sewage get washed via rivers into the sea and stimulate algal blooms that can become poisonous tides. The result is local eutrophication and death of marine life due to a lack of oxygen.
- *Siltation*: sediments from mining, construction, clearance of mangrove swamps, clearance of wetlands and dredging add to natural siltation from river flow, effectively burying coastal ecosystems, destroying coral reefs and making the water too turbid to sustain marine life.
- *Toxic waste*: water and seafood are contaminated by discharges of industrial waste, pesticides etc. Some toxins accumulate in the fat of predators at the top of the food chain. Estrogen-like compounds, including phyto-estrogens, may affect the hormonal balance causing infertility in marine animals.
- *Oil*: runoff and spills damage marine life most prominently in coastal areas, affecting fish, molluscs and seabirds.
- *Plastics*: a mass of plastic containers endanger marine life

- *Introduced species*: thousands of marine species are transported around the globe in ballast water, which is discharged by ships at sea without regard for the local ecosystem.

Conservationists argue that fish have no chance against modern fishing technology[6]. However, geographical information technology can be harnessed to monitor anthropogenic activity and its impact on marine environments in several ways:

- Accurate targeting of fish and thereby a reduction in by-catch. In its broadest definition, by-catch is simply regarded as unintended fisheries catch[7]. It has also been defined as the harvest of fish or shellfish other than the species for which the fishing gear was set[8]. The Magnuson-Stevens Fishery Conservation and Management Act[9, Section 3] defines by-catch as '*fish which are harvested in a fishery, but which are not sold or kept for personal use, and includes economic discards and regulatory discards*'. Large-scale commercial fishing can be extraordinarily wasteful due to the high proportion of by-catch in a typical catch. It is estimated that 20% of landings are thrown back, either because they are damaged, undersized, the wrong species or unsaleable. In some areas discards can be as high as 90%, for example in shrimp fishing where trash fish includes a wide variety of marine creatures, larvae and fish eggs. The fisheries software suite we have developed includes an onboard module. This module is used to record real time information about fishing trips. An important function of the onboard module is to record by-catch and discard data, without which the total catch cannot be confidently determined. This is a pre-requisite for improved determination and implementation of quotas, TACs and Catch per Unit Effort (CPUE). It is now widely recognized that by-catch has a severe impact on exploited or protected populations and should be taken into account during stock assessment.
- Monitoring of vessel position, making it more difficult for vessels to get away with illegal and unreported/unregulated fishing and, as a consequence, reducing the misreporting of catches (see Section 4.3.1).

Observing and sampling catches on board commercial vessels are widely used methods of learning what quantities of fish are retained for landing and discarding. According to Cotter et al.[10], a quarterly catch sampling period is appropriate and helps to isolate seasonal variations for target species, gear and movements of vessels into and out of fishing areas. Detailed surveys of fish populations have been carried out in English waters since 1970[11]; these surveys continue to date. Annual indices of fish abundance derived from these surveys are used to determine information such as the size of stocks, size and location of nursery grounds, presence of infrequent migratory species, diversity of the population, temperature and salinity.

Collection and analysis of relevant data are essential for the creation of fisheries management controls based on quotas or effort control as suggested by Shepherd[12]. The collection of environmental data such as temperature and salinity enables researchers to relate catch volume and species to specific environment conditions, thereby generating information about the captured species' habitats. In turn, this should allow predictions of the probable location of commercially important species, which should reduce time at sea and fishing effort costs. The maintenance of a repository for this sort of data will make it possible to establish any correlations between variations in capture volumes and changes in environmental parameters.

It is well established that patterns of biological activity parallel those in the physical environment[13]; accordingly any analysis of fisheries data should be considered hand in hand with environmental factors. In the long term, it is proposed that the onboard module be linked to a range of sensors and thus allow the system to collect and archive all the data relevant to stock assessment.

4.1.4 The Future: Geographic Information Management Systems for Fisheries

Although the exact timing is not known, it is clear that the EU intends to introduce electronic logbooks on all commercial fishing vessels in the next few years. This would represent an extension of the existing Vessel Monitoring System (VMS)[14]. Some of the benefits of using electronic data collection include:

- A single data entry function that is performed very soon after each fishing operation is completed.
- Data is validated and verified resulting in more complete data records.
- Minimal data entry errors owing to interface design and automated data collection.
- Reports produced in hard copy and/or transmitted in any of a number of different ways including ship to shore, to the fisheries agency, or other interested parties (such as the fishing company).
- Improvements in timeliness and accuracy, along with reductions in data processing costs are possible. This is especially relevant for certain types of data, namely catch volume, effort, location, environmental conditions and type of gear.
- Records of the geographical co-ordinates of each catch and a history of where the vessel was at each fishing event, thereby offering a basis for planning future trips.
- Accurate data for use in decision support systems, leading to improved fisheries management and control[15].

The onboard module along with the analysis and vessel tracking modules has the potential to play an essential role in providing information for the assessment of fish stocks[16,17], for compliance purposes and for the implementation of suitable management practices. Although similar GIS have been developed[18], they tend to have limited scope, deal with one particular type of fishery and do not integrate a data collection module. FishCAM on the other hand copes with many different types of gear and fishing activities, has an integrated data collection module and GI solutions for spatial decision support. Previous exploratory development work was carried out by Kemp and Meaden[19] and resulted in an early prototype of the FishCAM system from which the fully-fledged system described in this chapter emerged.

4.2 SYSTEM DESIGN

The challenge for scientists and decision-makers working with fisheries information is intelligent use of the disparate datasets, which are rich in the attribute sets they encompass, and extensive in their spatial and temporal coverage. We have met this challenge by constructing an integrated data substrate that provides the basis for the spatial analysis and visualization functions required for environmental management. Figure 4.1 illustrates the overall data model while Figure 4.2 summarizes the main information flows involved.

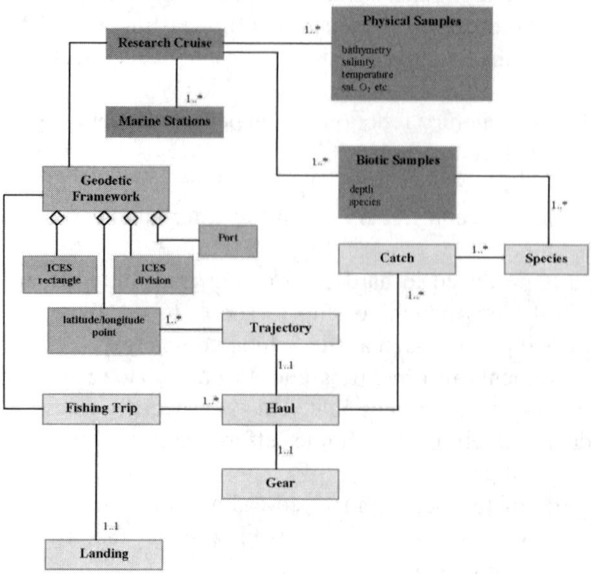

Figure 4.1 An overview of the full systems data model.

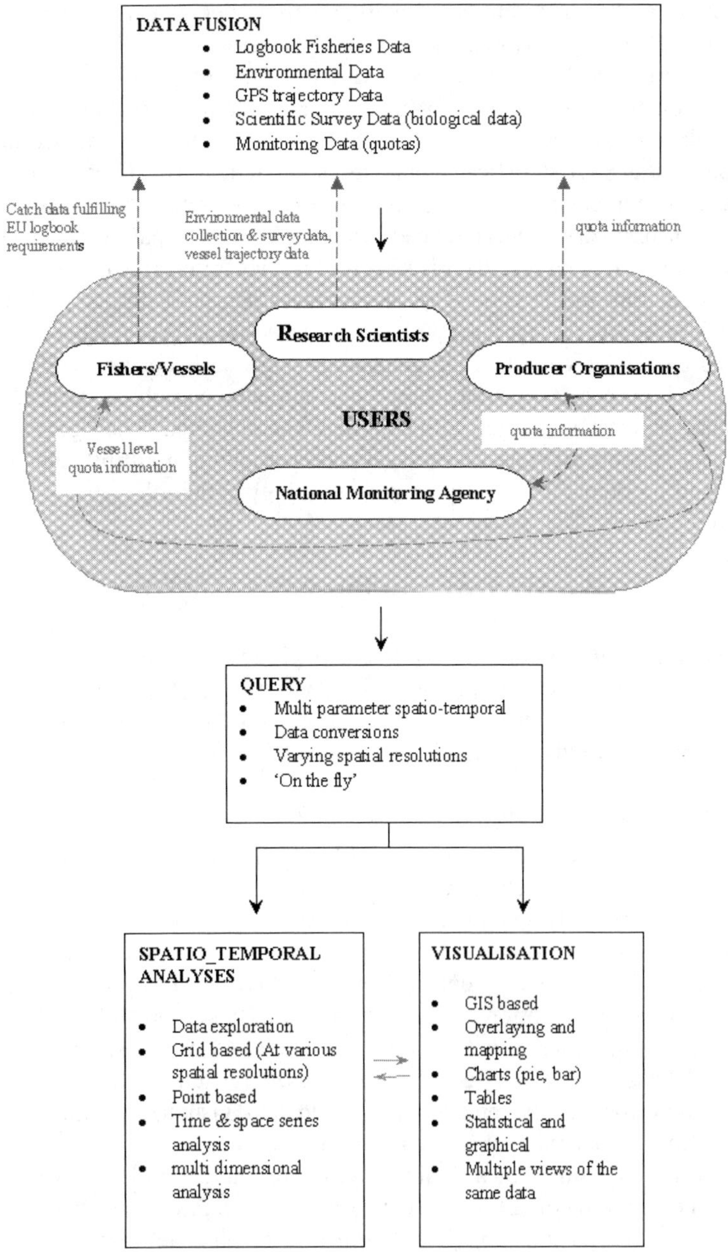

Figure 4.2 A schematic diagram of the system and information flows.

In the FishCAM system the commercial catch data set is generated and archived by the onboard module and consists of fishing activity details captured in real time. The catch data is dynamically geo and temporally referenced.

The georeferencing is carried out at a variety of customized spatial resolutions depending upon the requirements for analysis and monitoring. For example, the fishery catch data is georeferenced to the ICES (International Council for Exploration of the Sea) statistical rectangles specified by the CFP. On the other hand, the environmental data captured by marine biologists, using multiple parameter sensors, are georeferenced at much finer spatial and temporal resolutions. This provides the scientists with more accurate information for spatial predictive modelling in the context of stock assessment and time series analyses. The design of the diverse fisheries data sets is based on an extendable data model that enables data fusion for flexible analysis of various fishery related activities. For example, Figure 4.1 illustrates how environmental variables, sampled at set research stations, may be merged with commercial catch data. The rigorous dataset specifications support flexible data visualization allowing multiple views of the same data and on demand mapping as well as a highly flexible 'on-the-fly' query facility (Figure 4.2).

4.3 SPATIAL DECISION SUPPORT

The integrated data model can be used to generate visualizations of fishing activities and their effect on the marine environment. Here we present two illustrative examples of such visualization capabilities, vessel tracking and multivariate interactive visualization.

4.3.1 Mobile Vessel Tracking

Positional data are collected under the EU VMS[14] scheme, but the legislation still does not apply to vessels under 15 m or to inshore fleets. In addition, the current VMS is primarily intended to act as a tamper proof 'blue box', so that integrating anything else, such as details of catches, compromises the integrity of the VMS itself. The FishCAM onboard module therefore has an inbuilt position data logger retrieving a position and time from the GPS string (NMEA 0183) at an operator defined time interval between 2 seconds and 60 minutes. Output data are stored in a database table. These data can be used to help in the monitoring of fisheries as described below and are processed and visualized with one of the FishCAM GIS components dubbed the trajectory module.

The start and end times of a fishing event (i.e., the set and retrieval times of gear, e.g., a haul) are recorded by the onboard module and it is possible using this information plus the position logging information to plot the speed vs. time of a haul. In addition, the continuous recording of GPS position allows the entire fishing operation to be plotted, starting from port until arrival (Figure 4.3). One of the problems in creating complete tracks of a moving object is that if the data capture fails there are segments of the track where positions are unknown. These missing positions need to be estimated when the data capture method fails (e.g., in the case of FishCAM when the GPS signal is unavailable). The interpolation technique selected was a linear weighted distance or 'straight line method'[20] chosen for its simplicity. The method assumes that the locations of the missing intermediate track points are distance weighted averages of the data points occurring directly before and directly after the mapped point. Equation 4.1 was used to interpolate the location (x_i, y_i) of the intermediate point i at time t_i, where location point a is the point prior to the missing intermediary point i, and b is the location point directly after the intermediary point i.

$$x_i = x_a + (x_b - x_a)\frac{t_i - t_a}{t_b - t_a}$$
$$y_i = y_a + (y_b - y_a)\frac{t_i - t_a}{t_b - t_a} \qquad (4.1)$$

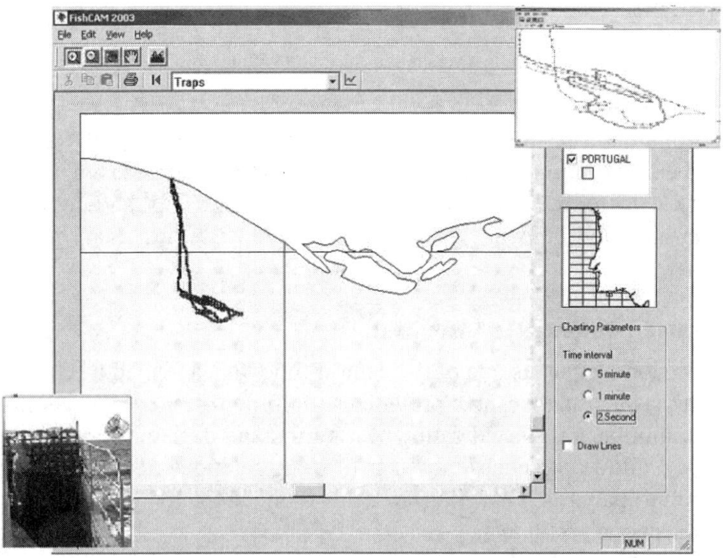

Figure 4.3 Plot of octopus fishing trip starting from port. The left inset picture is a photo of the fishing gear being set and that on the right is an enlarged view of the key activity portion of the trip.

The accuracy of the trajectory shape is largely dependent on the position sampling interval chosen by the user. While the sampling frequency has little bearing when a vessel is steaming to and from a fishing ground, when the actual fishing operations are in progress the sampling frequency is critical. If the sampling is too infrequent the shape of the track becomes oversimplified and therefore inaccurate. An ideal sampling frequency during a typical haul is anywhere between every two seconds to every five minutes.

Plotting the trajectory (Figure 4.3) can be, with some gears, used to estimate the real length of the gear, in the case of a net (such as a purse seine), as well as the fishing set duration. Plotting of time versus speed (derived from distance calculated between two consecutive GPS positions) allows the determining of travel time, seeking time, number and duration of fishing sets (Figure 4.4). As most vessels tow gear at speeds between 1.5 and 3.2 knots[21] it is possible to identify where the vessels were undertaking trawling activities as opposed to just cruising.

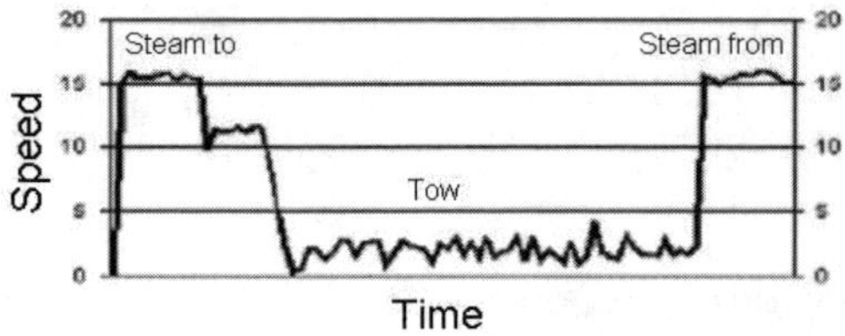

Figure 4.4 Plot of gillnet fishing trip time versus ship speed. Position sampled at five minute intervals.

Thus by collating a position-time log, it is possible to confirm the accuracy of the data supplied by a skipper in terms of fishing location and gear length, allowing more accurate data to be collected from the logbooks. This should lead to an improved marine environment via more effective fisheries management and controls.

4.3.2 Multivariate Interactive Visualization

Spatial decision support is one of the central functions ascribed to GIS[22]. In the FishCAM analysis module we have provided a multiple parameter decision-making tool allowing simultaneous visualization of criterion and decision spaces. This type of twinned view allows the decision-maker to study relationships between the data and their spatial patterns providing a firm foundation for understanding the structure of a decision problem[23].

Within and between the datasets every process and pattern has a spatial and a temporal effect. These effects can be observed in patterns of fishing and species abundance, in relationships between environmental factors and species abundance, and in the statistical analysis of diversity and other biological indices.

Patterns and processes that occur at different scales of time and space are linked and this relationship can be visualized. Visualization of these patterns is a valuable tool in the provision of decision support in fisheries information systems. The analysis module therefore provides a multidimensional capability to set query parameters and visualization modes to examine variations by:

- Space
- Time
- Species
- Gear
- Statistical methodology (e.g., the Shannon-Wiener diversity index[24])

The current standard resolution used by the International Council for the Exploration of the Sea (ICES) for recording fishing effort and landings for stock assessment purposes is 1° longitude x 0.5° latitude (an area of over 3,500 km^2 at 55°N, see Figure 4.5). However fishing sets and fishing locations (like bivalve sea beds) occur on a smaller scale. In the case of the Portuguese south coast, where the FishCAM sea trials took place, the whole coast is covered by two ICES statistical rectangles. This provides insufficient data for stock assessment or the use of technical management measures such as closed areas[21]. Another problem with this resolution is that fishing effort is often concentrated in areas of high catch resulting in fairly localized impacts. Thus, for the results of fisheries research to be applicable to the activities of commercial fleets there is a requirement for accurate high-resolution spatial fisheries data[25].

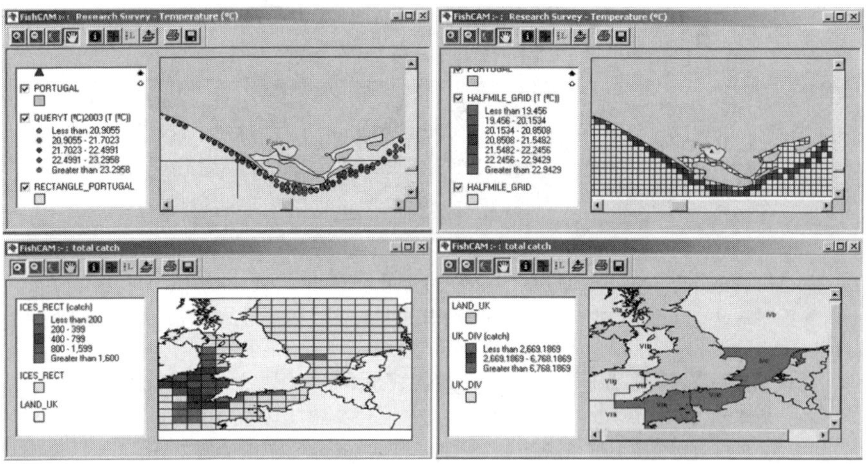

Figure 4.5 Examples of the different spatial resolutions required in fisheries analysis. Research scientists use (upper left) latitude-longitude and (upper right) a 30' x 30' grid, while fishers and authorities use (lower left) ICES statistical rectangles or (lower right) ICES divisions.

Owing to the diverse requirements of different user groups (Figure 4.2), we have integrated into the system a multi-perspective open-ended and flexible query facility for spatio-temporal analyses (Figure 4.5). National monitoring agencies typically analyze fisheries data at the ICES statistical rectangle spatial level and set quotas at the coarser ICES division level, while fishers and researchers require finer granularities. For example, Portuguese scientists work at three different levels on finer grids of a half-mile and a quarter mile as well as at the point data level provided by monitoring at fixed research stations. The fishers themselves also require the ability to analyze their activities at various levels; they need quota analysis at the ICES division level, landings and catch analysis at the ICES statistical rectangle level (for EU log-sheets) and finer resolutions depending on the type of fishing they are engaged in (e.g., dredging involves intensive fishing within a very small area whereas gill nets cover large areas in a single haul).

In addition, there are variations in analytical needs. National monitoring agencies hold huge archives of data and require analysis over large temporal ranges, whereas scientists may wish to compare environmental factors with abundance of a given species over a season or annually. Figures 4.6 and 4.7 provide examples of the types of analyses that can be conducted with FishCAM.

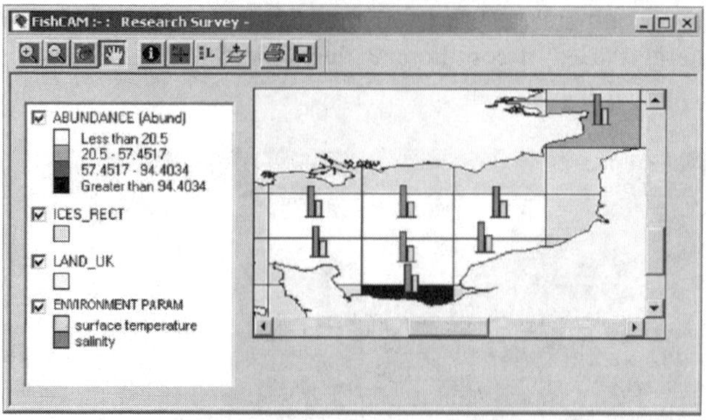

Figure 4.6 Lemon Sole research survey, abundance and environmental parameters at ICES statistical rectangle resolution and for the same spatio-temporal region.

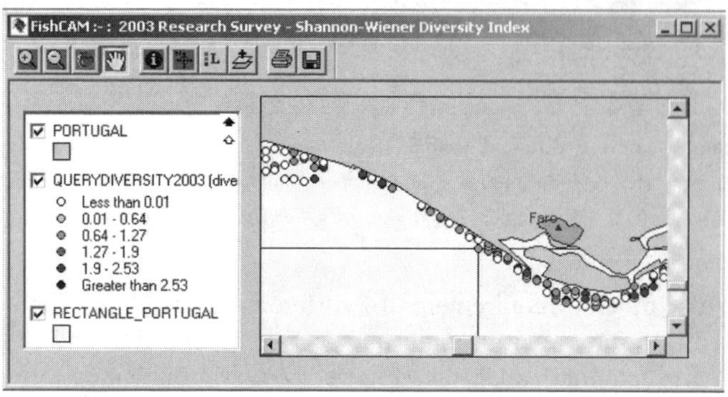

Figure 4.7 Shannon-Weiner diversity, samples collected at set research stations in the Algarve 2003.

One of the primary visualization challenges faced was how to represent spatio-temporal change. Patterns of change in environmental parameters, abundance or migration of a species are traditionally identified by viewing the data in a time-space series[26] (e.g., Figure 4.8).

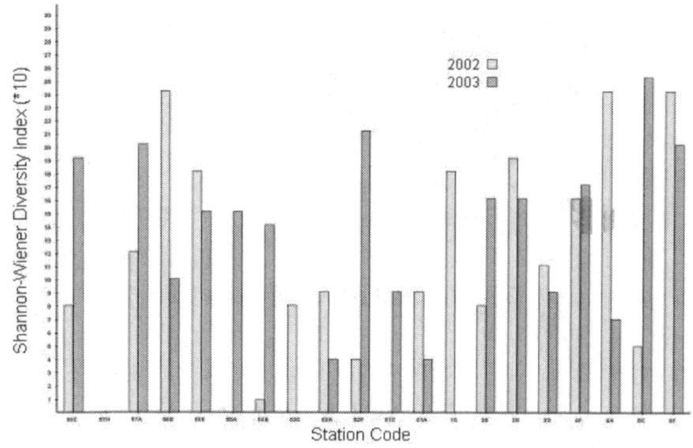

Figure 4.8 Spatio-temporal change: Shannon-Wiener diversity index for *Chamelea gallina* in 2002 and 2003 graphed by research station.

The analysis model incorporates the ability to view multiple views of the same data and hence assists the user in the recognition of spatial and temporal data trends and anomalies[18]. Figure 4.9 illustrates the change in the distribution of lemon sole between the first quarter and second quarter of a year. The distribution of lemon sole moves west in the second quarter, but they stay abundant in the western channel throughout the first half of the year.

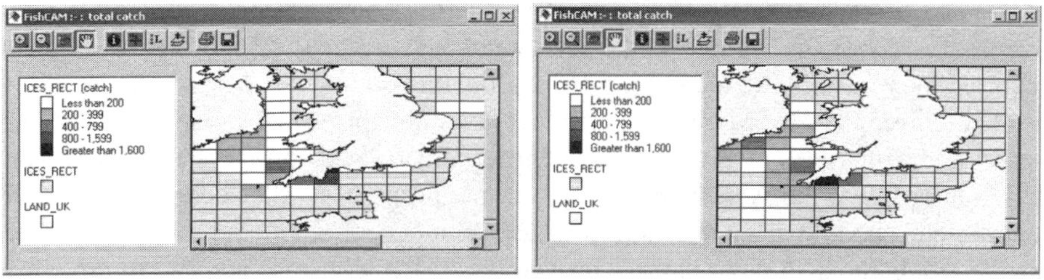

Figure 4.9 Distribution of Lemon Sole catch by ICES rectangle and two time periods.

Additional levels of information are available to the user by way of interactive maps[27]. Figure 4.10 shows the result of a mouse click on an ICES statistical rectangle within the map using the data exploration tool. In this manner the user can explore the data for areas of interest in finer detail.

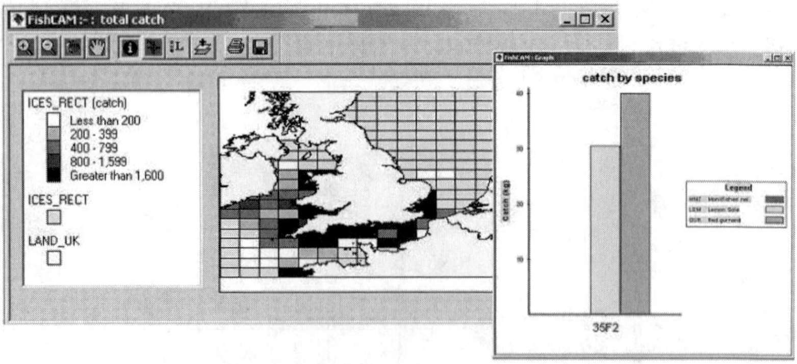

Figure 4.10 Selecting the rectangle 35F2: breakdown by species (right) of the catch for the selected ICES statistical rectangle.

4.4 CONCLUSIONS

In summary, a commercially viable integrated GI solution for Common Fisheries Policy and decision-making has been developed. We conclude with a few comments on the hardware and software aspects of FishCAM and its potential for monitoring and informing fisheries policies.

The experience gained from trialing the onboard module on smaller sized fishing vessels (under 20 m) leads us to identify some key issues as regards the hardware required for such a system and the usability of the existing module. While observing the fishermen it quickly became clear that fishing is a hard and labor intensive job with little time for data entry. We need to be able to minimize the data input tasks required by the fishers during the course of a haul. Integrating automatic sensors to nets would help trigger start and end of haul data records and weight sensors to record the weight of the total catch. However the financial investment required may be an issue for smaller fishing operations.

The current hardware comprises a PC or laptop that can be hooked up to the vessel's existing GPS receiver. While PCs are already used on large fishing vessels and integration of the system would be a simple process, smaller vessels often have insufficient cabin space for a laptop or PC. Furthermore, in rough conditions on small vessels the data input devices proved less suitable. Consequently, future work must include investigation of appropriate input hardware.

Certain generic conclusions can be drawn about the design and implementation of systems that have a strong basis in spatial analysis:

- 'Geography matters': the spatial (or spatio-temporal component) is a crucial dimension in many systems.

- The data models that support a GIS should, preferably, address the spatial, temporal and thematic dimensions.
- The system should enable multiple user requirements to be addressed, without undue effort on data fusion and interoperability for ad hoc analyses.

Thus the modular design of FishCAM includes capabilities to collect and verify data and to enable flexible extraction and display of different sets of fishery and environmental data. This includes the means to examine different time periods and temporal resolutions, areas and spatial resolutions, or species and gear combinations, as well as the ability to alter symbolization and display formats. Such an approach enables the industry as well as marine scientists to achieve a better understanding of the trends in fishery harvests, fluctuations in stocks and, in the longer term, supports an ecosystem-based approach to the exploitation of marine resources.

4.5 ACKNOWLEDGMENTS

This project was funded by the EU: CRAFT project, Key Action No. 1.1.1 - 5: Sustainable Agriculture, Fisheries and Forestry, *Ref: Q5CR-2001-70746*.

We would also like to thank: IPIMAR (Instituto de Investigação das Pescas e do Mar), Centro Regional de Investigação Pesqueira do Sul, Avenida 5 de Outubro s/n, P-8700-305, Olhão, Portugal, for the data from their 2002 and 2003 bivalve research surveys in the Algarve; Miguel Gaspar and Pedro Lino for their helpful suggestions during the development of the FishCAM software; and Geoff Meaden in the Fisheries GIS Unit at Canterbury Christ Church University College, Canterbury, UK, for his valuable input into the project with regards to the CFP and EU logbook requirements.

4.6 REFERENCES

1. European Commission, About the Common Fisheries Policy, http://ec.europa.eu/fisheries/cfp_en.htm, 2006.

2. European Commission, Statistics on the CFP, http://ec.europa.eu/fisheries/cfp/statistics_cfp_en.htm, 2006.

3. European Commission, Community Fisheries Control Agency, http://ec.europa.eu/fisheries/cfp/control_enforcement/control_agency_en.htm, 2006.

4. Fives, J.M., Acevedo, S., Lloves, M., Whitaker, A., Robinson, M., and King, P.A., The distribution and abundance of larval mackerel *Scomber scrombus L.*, horse mackerel, *Tachurus trachurus (L.)*, hake, *Merluccius merluccius (L.)*, and blue whiting *Micromesistius poutassou* (Risso, 1826) in the Celtic Sea and west of Ireland during the years 1986, 1989 and 1992, *Fisheries Research*, 50, 17-26, 2001.

5. Buckley, R., Ed., *World Fishing: Beyond Sustainability*, UGI Ltd, Cheltenham, 2002.

6. Zeller, D. and Russ, G.R., Are fisheries 'sustainable'? A counterpoint to Steele and Hoagland, *Fisheries Research*, 67, 241-245, 2004.

7. National Oceanic and Atmospheric Administration (NOAA), *Bycatch: A National Concern*, National Marine Fisheries Service, National Oceanic and Atmospheric Administration, Silver Springs, Maryland, 1997.

8. Wallace, R.K. and Fletcher, K.M., *Understanding Fisheries Management: A Manual for Understanding the Federal Fisheries Management Process, Including Analysis of the 1996 Sustainable Fisheries Act*, 2nd ed., Mississippi-Alabama Sea Grant Consortium, Ocean Springs, Mississippi, 2001.

9. Magnuson-Stevens Fishery Conservation and Management Act (MSA), Public Law 94-265, U.S.A., http://www.nmfs.noaa.gov/sfa/magact/, 1996.

10. Cotter, A. J. R., Course, G., Buckland, S.T., and Garrod, C., A PPS sample survey of English fishing vessels to estimate discarding and retention of North Sea cod, haddock, and whiting, *Fisheries Research*, 55, 25-35, 2002.

11. Rogers, S.I., Millner, R.S., and Mead, T.A., *The Distribution and Abundance of Young Fish on the East and South Coast of England (1981 to 1997)*, Science Series Technical Report Number 108, The Centre for Environment Fisheries and Aquaculture Science (CEFAS), Directorate of Fisheries Research, Lowestoft, 1998.

12. Shepherd, J.G., Fishing effort control: could it work under the common fisheries policy?, *Fisheries Research*, 63, 149-153, 2003.

13. Ricklefs, R.E., *Large Marine Ecosystems: Pattern, Processes and Yields*, 2nd ed., American Association for the Advancement of Science, Washington D.C., 1992, 170.

14. European Commission, Vessel Monitoring System (VMS), http://ec.europa.eu/fisheries/cfp/control_enforcement/vms_en.htm, 2006.

15. Food and Agriculture Organization (FAO), *FAO Technical Guidelines for Responsible Fisheries - Fishing Operations - 1 Suppl. 1 - 1. Vessel Monitoring Systems*, http://www.fao.org/documents/, FAO, Rome, 1998.

16. Kemp, Z. and Meaden, G.J., The management of commercial fisheries with the aid of a computerised catch logging system, in *EEZ Technology: The Review of Advanced Technologies for the Management of EEZs Worldwide*, ICG Publishing Ltd., London, 1997, 181-186.

17. Kemp, Z. and Meaden, G.J., Towards a comprehensive fisheries management system, in *Proceedings of the International Institute of Fisheries Economics and Trade, IIFET '98 Conference*, Tromsoe, Norway, 1998.

18. Pierce, G.J., Wang, J., Zheng, X., Bellido, J.M., Boyle, P.R., Denis, V., and Robin, J-P., A cephalopod fishery GIS for the Northeast Atlantic: development and application, *International Journal of Geographical Information Science*, 5, 763-784, 2001.

19. Kemp, Z. and Meaden, G.J., Visualisation for fisheries management from a spatiotemporal perspective, *ICES Journal of Marine Science*, 190-202, 2001.

20. Wentz, E.A., Campbell, A.F., and Houston, R., A comparison of two methods to create tracks of moving objects: linear weighted distance and constrained random walk, *International Journal of Geographical Information Science*, 17, 623-645, 2003.

21. Marrs., S.J., Tuck, I.D., Atkinson, R.J.A., Stevenson, T.D.I., and Hall, C., Positional data loggers and logbooks as tools in fisheries research: results of a pilot study and some recommendations, *Fisheries Research*, 58, 109-117, 2002.

22. Jankowski, P., Andrienko, N., and Andrienko, G., Map-centered exploratory approach to multiple criteria spatial decision making, *International Journal of Geographical Information Science*, 15, 101-127, 2001.

23. Malczewski, J., Visualisation in multicriteria spatial decision support systems, *Geomatica*, 53, 139-147, 1999.

24. Spellerberg, I. F. and Fedor P.J., A tribute to Claude Shannon (1916-2001) and a plea for more rigorous use of species richness, species diversity and the 'Shannon-Wiener' Index, *Global Ecology and Biogeography*, 12, 177-179, 2003.

25. Kaiser, M.J., Significance of bottom fishing disturbance, *Conservation Biology*, 12, 1230, 1998.

26. Pawson, M.G., *Biogeographical Identification of English Channel Fish and Shellfish Stocks*, MAFF Fisheries Research Technical Report No.99, Directorate of Fisheries Research, Lowestoft, 1995.

27. Andrienko, G.L. and Andrienko, N.V., Interactive maps for visual data exploration, *International Journal of Geographical Information Science*, 13, 355-374, 1999.

Part II

Tools to Support Decision-Making

CHAPTER 5

GIS and Environmental Decision-Making: From Sites to Strategies and Back Again

R. MacFarlane and H. Dunsford

5.1 INTRODUCTION

Spatial policy and planning are fundamentally about locating facilities, services, industry, housing, utilities and other land uses that are required by society and the state in such a way that benefits are justified against the foreseen costs. Many policies have a spatial dimension in that economic activities tend to cluster, and similar social groups exist in proximity to each other, so benefits and costs, for instance through the collapse of heavy industry, are unevenly distributed in space. However, spatial policies and land-use plans are much more explicitly about where things should be. There are 'winners' and 'losers', with NIMBYism representing one response when the 'losers' are proposed to be 'here' rather than 'there'. If policy is premised on achieving the greatest good (taking a wholly apolitical perspective!) and minimizing harm, we are left with the problem that Locally Unwanted Land Uses (LULUs) have to go somewhere and there can be opposition in almost every direction. Alternatively, environmental decision-making may be concerned with the allocation of scarce resources that are desired in a number of places. Public funding for regeneration, amenity or conservation associated investments such as woodland planting, access enhancement, wetland creation or 're-wilding' schemes[1] may be subject to what is in effect a bidding process, attempting to 'win' the proposed development or other measures. This might be described as NIMBYism without the N. In neither case of course are positions going to be universally held within any given community or locality, for instance in the case of wind farms where the landowner and development company may be united in their support, against the wishes of other local stakeholders. Determining which options are 'best' lies at the core of environmental decision-making.

This chapter draws on research and consultancy experience for a range of organizations in the UK. These include projects on wind energy, community forests and the mapping of tranquillity[2-7]. The experience spans what might be simplistically identified as a divide between 'development' and 'conservation' interests, enabling us to offer a commentary on some of the values that are often implicit, and sometimes explicit, in what may generally be termed environmental decision-making.

Conducting these studies has provided the authors with a viewpoint on a range of issues and debates which are pertinent to the application of GI and GIS to environmental decision-making at the policy and planning levels. Planning is used here in a relatively broad rather than a technically specific sense, focusing primarily on regional and sub-regional scale activities such as Regional Spatial Strategies[8] and sectoral policies such as the Regional Forestry Strategies[9]. Land-use planning within the statutory framework of Town and Country Planning is not the primary focus, although many of the issues are equally relevant and applicable at that level given the hierarchically consistent nature of the system, further tightened up with recent revisions[10].

A number of key themes are elaborated in the next section. This is followed by a review of several projects and the chapter concludes with a discussion that draws out some lessons learnt from them and their implications for research and practice.

5.2 BACKGROUND: KEY THEMES

Broadly speaking, there are different kinds of information (including GI) used in decision-making and different stages at which they may be sought or created:

- *Background*: various sources, experiences, networks and the media
- *Exploratory*: sought out, more structured and categorized
- *Analytical*: selected, a focus on defensibility
- *Confirmatory*: related to formal evaluation to confirm decision or to retrospectively support a position adopted.

Information is thus central to the social processes of arriving at and defending a position. However, if the information was unequivocal, if the evidence base was not open to challenge, if variable interpretations of the available data and information did not have the power to carry people to different conclusions, there would be no problem and no literature on the subject. More to the point, there would be uniformity and consensus instead of dispute, challenge and opposition. The convention is that both planning, taken in the broadest sense, and organized opposition needs to be 'evidence-based' to be effective. Evidence-based (good) practice is a key element in the lexicon of the current government, but evidence can be constructed and construed in many different ways. What then is the role of information, including GI, in environmental policy and planning? To answer this it is important to review, although in summary, some different views on the nature of planning and the role of technical information in support of it.

Early models of planning were procedural, systematic, rational and apolitical – things happened in a process that had a beginning and an end point, and where all options were considered on the basis of the available information, before selecting or making an optimal decision. Over time these have been challenged and replaced with models that emphasize the nature of planning as a more interactive, communicative activity.

Figure 5.1 illustrates the sequential, procedural model of planning and in each of its stages information plays a distinctive role. The name of the stage appears at the top, what happens during this phase is in [square brackets] and the role of information is in (brackets).

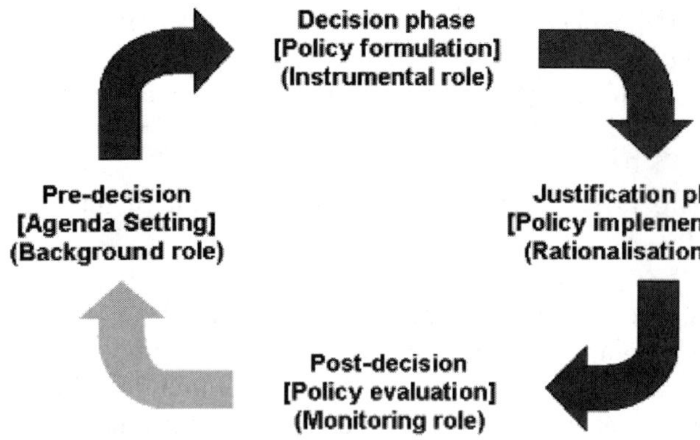

Figure 5.1 The role of information in the sequential model of the planning/policy cycle (after Danziger et al.[11], Barkenbus[12], and Stephenson [13]).

Stephenson[13, p241], commenting on the sequential, procedural model, noted that *'the notion of a process where information is collected, a choice of policy is made, the policy is adopted and then implemented was too simple when compared with reality, where the process was far more messy'*, yet it is an approach that is publicly adhered to as good practice.

Campbell and Masser[14, p153] critically observe that *'much of the folklore surrounding computers endows them with the capacity to compensate for the inadequacies of human intellect ... and [that] this sentiment is reflected in much of the current writing about geographic information technologies'*. Klosterman[15, p47] builds on this, arguing that *'planning practice ... takes current technology largely as given and moulds planning to fit the technology ... [something] ... that is particularly relevant to the current fascination with GIS among planning academics and practitioners'*. GIS, it could be argued, has the potential to reinvent planning as an applied science with its emphasis on locations that are optimized on the basis of data describing defined parameters and criteria. However, this would be naïve as the literature has developed an extensive critique of the neutrality of evidence and indeed the neutrality of planning. The planning as communication and planning as reasoning models[15,16] are essentially premised on the contested nature of planning processes and their outcomes, and they embed a more inclusive approach, at least to some degree, in the planning process. In turn GIS applications are developing to accommodate these new models, and the planning as reasoning together approach has underpinned the emergence of the grandly titled 'participatory geographic information science'.

> *'This surge of interest in collaborative spatial decision-making in particular, and participatory decision-making more generally has been spurred on ... by the realisation that effective solutions to spatial decision problems require collaboration and consensus building... [which] require the participation and collaboration of people representing diverse areas of competence, political agendas, and social interests'*[17, p2-3].

So, attempting to briefly summarize the literatures on planning, opposition, information and GIS:

- There is a geography to the impact of developments;
- There is a geography to the opposition to developments;
- Models which seek to rationally identify optimal locations have been discredited by many authors (e.g., Pickles[18]) and GIS have been re-cast as decision support tools;
- In spite of the above, GIS retains substantial credibility as a planning tool, and the emergent literature on participatory applications of ICTs in general and GIS in particular has established a tool for communicative planning (e.g., web-based mapping) and reasoning together models.
- It is questionable whether reasoning together moves us forward, however much the process is facilitated by evidence, new media and participatory digital frameworks for debate.

This chapter relates our experience to these issues and problems, based upon a series of case studies that all involved the Centre for Environmental and Spatial Analysis (CESA) at Northumbria University. Cutting across these we can identify a change in focus away from individual site proposals to regional or national frameworks which explicitly articulate objectives and establish priorities in a spatial form. Admirable as the latter can be, the problem remains that developments have to actually go somewhere specific. This engenders a need for strategic policies to mesh with more local frameworks to guide individual proposals to actual locations that satisfy (or perhaps more accurately, address) the various demands that the state, the developer, local communities and other interest groups have in relation to the planned activity and the site.

5.3 CASE STUDIES

5.3.1 Scottish Natural Heritage: Landscape Capacity Assessment for Wind Energy

The remit of Scottish Natural Heritage (SNH) is primarily to secure the conservation and enhancement of Scotland's unique and precious natural heritage. In keeping with their role SNH commissioned pilot studies to provide a strategic view of the scope for onshore wind energy developments in four selected districts. It was explicitly stated by SNH that these studies were not to identify suitable locations, but were to provide frameworks that could be used to minimize environmental impacts, as well as guiding developments to the most appropriate landscapes. It was also intended that the four studies would independently develop and apply a methodology to achieve this objective.

We were involved in the study for the Western Isles[2] which utilized three key methods:

1. An existing landscape character assessment was refined and adjusted to accommodate detailed physical and perceptual criteria associated with small (domestic) and large-scale (commercial) wind turbine developments.
2. Visibility analysis was carried out to derive 'scores' for relative visibility across each landscape unit and for turbines of varying heights.
3. A stakeholder study was used to assess the values associated with the types of landscape identified in the landscape character assessment.

Guidance on the conduct of landscape character assessment[19] and for evaluating the impacts of development on the landscape and visual resource of a particular location[20] was used alongside renewable energy policy documents[21-23] to inform an approach to landscape capacity assessment. In total 15 Landscape Character Types (LCTs) were identified in the study area. After considering the sensitivities and locations of individual LCTs, a basic distinction was made as to whether each LCT could accommodate domestic or commercial windfarms, based on a subjective but informed understanding of levels of habitation. A landscape capacity was then attributed to each LCT, ranging in five categories from low to high ability to accommodate development. The resulting product took the form of a table and a series of maps with information provided for different types of windfarm in each LCT. The output was not spatially disaggregated to specific LCT parcels, even though the intervisibility analysis techniques used were well suited to this, and consequently the information presented was strategic in nature

5.3.2 Welsh Assembly: Strategic Assessment for Wind Energy Developments

In March 2002, the Welsh Assembly Government (WAG) commissioned Arup Consultants to carry out research that would assist with updating existing planning guidance on renewable energy in the form of a Technical Advice Note (TAN 8)[24]. CESA were sub-contracted to develop constraint criteria within a GIS in order to identify 'strategic search areas' capable of delivering WAG renewable energy targets by 2010.

The criteria for the identification of strategic areas were initially developed by a steering group consisting of Arup staff and WAG officials. These criteria were categorized as being either technical (practical limitations to windfarm siting; wind speed, slope, Ministry of Defence requirements etc.) or environmental (landscape designations, nature reserves etc.). The criteria were compiled into separate layers in a GIS and assigned into three distinct categories:

- *Absolute Constraints*. Those which, for all intent and purposes (at the all-Wales level), would be likely to prevent large-scale wind energy developments.
- *Second Degree Constraints*. Factors likely to inhibit the development of large wind energy developments but for which there was either a) some variability/uncertainty in their spatial extent or b) the possibility to develop inside the area concerned with appropriate mitigation.
- *Positive Development Siting Factors*. Characteristics which, in general, were likely to be viewed positively for large scale wind developments by the wind industry (e.g., single landownership, reasonable accessibility and relatively simple land cover)[3].

Identification of strategic areas then involved the following three steps:

1. *An initial screening exercise*. Absolute and second degree constraint data (both environmental and technical) were overlaid with grid capacity information in a GIS to determine suitable areas.
2. *A refinement exercise*. The areas identified were subject to a more detailed review at 1:50,000 scale. Specific features such as access, isolated residential properties, land availability and ownership were taken into account.
3. *Testing and validation*. Analyses of visibility from National Parks, Areas of Outstanding Natural Beauty, National Trails, existing (or approved) wind-farms, and landscapes 'wild' in character were combined with details of wind speeds and the scale of development possible to further refine boundaries which extended 5 km either side of the perimeters for each strategic area.

Seven strategic areas were identified at the end of this process and TAN 8 was publicly launched in March 2004, with a subsequent review published in July 2005[24]. Inevitably, TAN 8 also resulted in studies commissioned by local authorities within the strategic areas to assess on a site by site basis the most suitable 'zones' within their territories. In one case this suggested a higher capacity in a strategic area, which in turn fed back into a review of national targets for renewable energy. A key point therefore is that because the initial strategic study took no account of administrative boundaries, the onus was placed on local authorities to fund studies in order to assess suitable sites for windfarm development.

5.3.3 North East Renewable Energy Strategy

In November 2002 the Government Office for the North East commissioned The Northern Energy Initiative, CESA and the Landscape Research Group at the University of Newcastle to prepare a Regional Renewable Energy Strategy (RRES). The need for this activity stemmed from a review prepared by the Cabinet Office Policy and Innovation Unit in February 2002 which recommended that regional planning bodies should give greater prominence to energy issues in regional planning guidance and become pro-active in planning for energy developments at a sub-regional level[25]. The pre-existing North East Regional Energy Group (NEREG), whose members spanned regional bodies, local authorities, developers and conservation interests, became the project steering group.

To support the development of the RRES, a wide range of digital maps were combined through a 'sieve mapping' approach[26] in a GIS to identify where windfarms could not be located. Many of the datasets were unanimously accepted by all parties on the steering group, including factors such as topple distance from A-roads and exclusion from designated nature conservation sites or on the basis of low wind speed. Amongst these accepted factors were GIS-based calculations of radar line of sight. In essence, both civilian and military radar can be compromised if any part of a rotating blade can be 'seen' by the radar array. The Ministry of Defence (MoD) is particularly assertive about this and adhered to a policy that no turbines were permissible within 74 km of a radar array where there is line of sight. There is no weighting or gradation of this opposition by distance; it is a black/white issue. As the military are not obliged to release technical information in support of their position (in itself a core requirement of transparent public sector working) and they have a critical role in policy, planning and siting decisions (which may be loosely summarized as having the power of veto) this is an absolutely critical input, yet it was not subject to close questioning in private meetings or at the closed committee stages of the process. In addition, although the line of sight calculations were carefully executed, and the results closely checked by re-running a sample of them, they contained two potential sources of error. Neither of these was hidden in any way during the process.

Firstly, no account was taken of known vertical error levels in the Ordnance Survey Panorama[27] terrain model used in the calculations, since this was not regarded as having a serious effect on the results. Secondly, although the MoD have a stated position of not having wind turbines located in line of sight from their radar installations, they would not inform the study group of the precise geographical coordinates and height above ground level of the radar arrays at Brizzlee Wood and Fylingdales. Through a careful process of estimation these locations were identified and the calculations then run on these parameters. As the study was strategic rather than focused on tactical siting of individual turbines, the implications are unlikely to have been in any way significant, but the absence of any critique is the key point.

In contrast to the perceived 'scientific' evidence represented by the radar calculations (see extract in Figure 5.2a), another input layer consisted of a map of landscape sensitivity, the vulnerability of landscape character to change (Figure 5.2b). This sensitivity was assessed by considering the physical and perceptual characteristics of landscapes with respect to different forms of wind energy development[28].

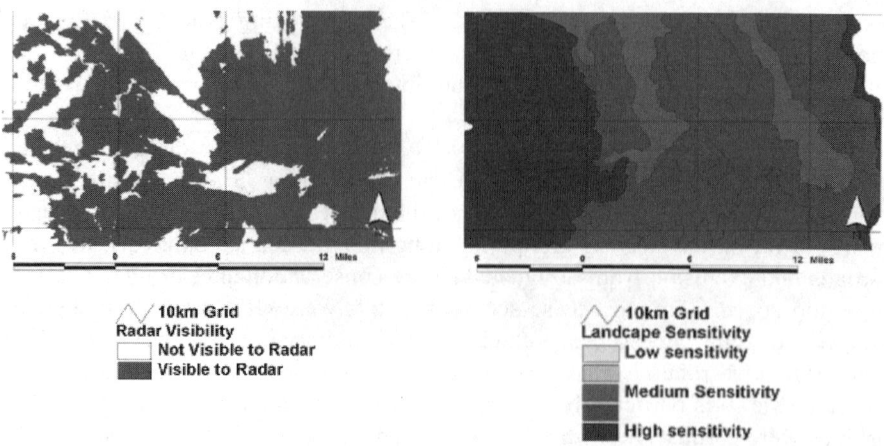

(a) Radar visibility calculation (b) Landscape sensitivity assessment

Figure 5.2 'Hard' and 'soft' GI for an area in North Northumberland.

Landscape sensitivity is an intrinsically more complex and contested concept than radar visibility. In the latter the visible (unacceptable) and non-visible (acceptable) zones were sharply delineated and there was no argument with them, even though there are grounds to challenge the results as outlined above. It was clear from the earliest presentation of the landscape sensitivity assessment, however, that the results were not acceptable to some of the stakeholders. Both the radar and the landscape analyses drew on 'expert' input, but the much 'softer' nature of the sensitivity assessment process caused a series of conflicts in the committee discussions. In essence there were three stages to this:

- Competing 'expert' views on the approach;
- Competing 'expert' views on the results;
- As a consequence of the above, disquiet from 'non-experts' on the committee on the nature of the information being presented.

The latter was perhaps most illuminating in the private committee meetings during which one wind energy developer, who had little to contribute to the discussion around the results themselves, was aghast at the realization that this was a map, using a numerical scale to display relative sensitivity and therefore suitability, which was (a) based on human judgement, and (b) wholly contingent on judgements that, when revised, could result in a different map. This second characteristic was something that was no more or less true of all the other map layers included in the exercise, for instance the decision to exclude National Parks but not Green Belts, an entirely social judgement. In such adversarial debates the authority of the sources/experts and the decision criteria become critical, illustrating how conflicts over the principles and details of spatial policy formulation are not simply confined to developers vs. planners or planners vs. local pressure groups.

5.3.4 North East Community Forests: A Regional GIS for Decision Support

North East Community Forests (NECF) commissioned a study to provide a framework for the future planting and sustainable use of community woodlands[5]. There were two key elements in the context for this study:

- The reorientation of community forests towards a more broadly based, less tightly spatially defined, approach to countryside management in the urban fringe;
- The explicit direction from government that new woodlands, and the management of existing woodlands, should take as their primary focus the potential for social benefit, through landscape enhancement, recreational opportunities, strengthened regional images and improved quality of life.

The first of these requirements dictated that the study should be regional (North East Government Office Region) rather than site-specific, even through realization of a strategic plan for the region would be ultimately achieved on a site by site basis. The second meant that the frame of reference (and consequently the data involved) was broad ranging – this was no simple land capability study.

Utilizing a 1 km x 1 km grid over the region a series of datasets were integrated in line with the criteria defined by the steering group. The main components of the GIS cartographic model were:

- *Where people live and work.* This reflected the need to target woodland establishment and development where there are concentrations of homes and workplaces.
- *Environmental gains.* Although ecological theory at the landscape scale tends to be at the level of general principles rather than specific 'rules', reducing woodland fragmentation and enhancing connectivity were generally accepted positive objectives.
- *People on the move.* Targeting woodland where it had the potential to benefit the lives of people who are travelling, for either work or leisure, locally or through the North East. A particular factor here was the regional image.
- *Nature conservation and landscape considerations.* These were not a core component of the model, but it was accepted that there were areas where woodlands could have a potentially negative impact on nature conservation interests or landscape character and should therefore be discouraged.

The principles underlying each of these factors were defined by the steering group and with reference to a wider set of stakeholders. Articulating these principles in a complex and differentially weighted cartographic model required that a large number of technical decisions were made by the researchers and there was minimal involvement in this by the steering group or stakeholders. The researchers consistently made it clear to the steering group that decisions made, for instance, with respect to the visibility analysis, or the distance-weighting techniques used for population concentration, would have an impact on the model output. Figure 5.3 shows the final map.

Although the criteria were agreed, the map underwent a process of revision once an earlier version, which of course reflected the initially agreed parameters, was shown to the steering group and key stakeholders. The 'look' of the map did not entirely accord with their ideas of what it should look like to fulfill its policy function. The relationship between the GI product (a map, illustrating a strategy) and the GI process (the weighted combination of datasets to produce the map) was therefore two-way. In simple terms it was a case of 'do the analysis, and let's see what it looks like, then we can take another look at the process', clearly spanning the analytical and confirmatory uses of information.

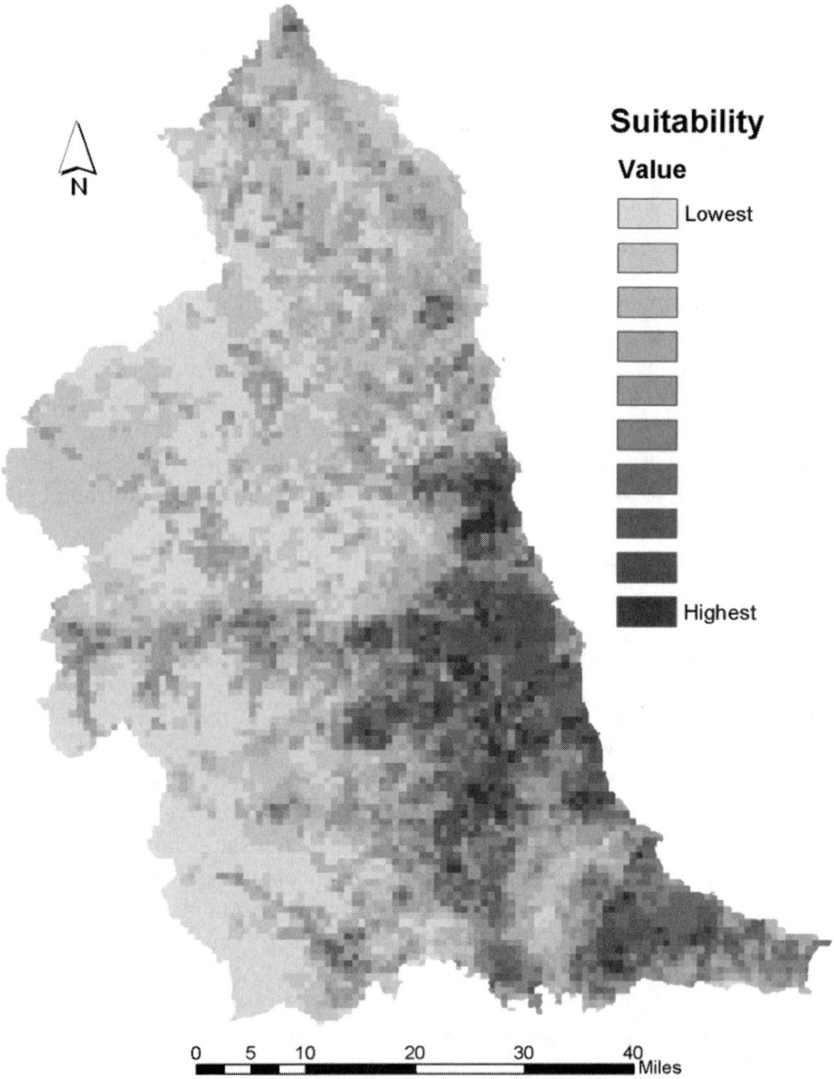

Figure 5.3 Final suitability map for the community forests study.

5.3.5 Tranquillity Mapping

The importance, value and need to protect tranquillity is clear from a range of policy documents, and from research on the health and social benefits to people of being in a tranquil place[29-31], but a systematic and robust method to document where these areas are has been lacking. MacFarlane et al.[7] were commissioned by Campaign to Protect Rural England (CPRE) and the Countryside Agency to develop a methodology to identify and map areas where people are more likely to be able to have a tranquil experience. Tranquillity is an experiential aspect of what may broadly be defined as landscape quality and assessing and mapping it poses a series of conceptual and methodological challenges.

The concept of tranquillity mapping is not new. Several studies were undertaken in the 1990s[32,33], based primarily on 'expert' judgments of factors that detract from tranquillity. In 2000, however, Levett[34] argued that what was needed was a measure of tranquillity that included all, and only, those sources of disturbance which people feel actually damage tranquillity; and which weighted them in proportion to peoples' perceptions of their relative impacts on tranquillity. This is what was done by MacFarlane et al.[7] through a framework based around the use of Participatory Appraisal (PA), an approach to consultation that focuses on exploring peoples' perceptions, values and beliefs, and is designed to allow participants to discuss what is important to them, and express this in their own words[35]. PA sessions were carried out across two study areas, the Northumberland National Park and the West Durham Coalfield area.

After the initial PA data collection, two 'verification' events were held. Prior to these meetings the responses made during the PA sessions were collated into themes (whether they were something 'you see', 'you hear' or 'you do' in a tranquil area, whether they were something 'of the mind', or whether they were something 'you do not see' or 'you do not hear'). People were then asked to select their top three responses within each theme, according to their perceived level of importance to tranquillity. These events quantitatively established the relative importance of different factors. In consultation with the project steering group, and with reference to published best practice, the distilled PA data were then associated with specific spatial datasets that could be used to translate the preferences and judgements expressed by respondents into mapped outcomes. The GIS analysis followed a raster modelling approach in ArcGIS with a 250 m grid cell resolution.

Three main categories of data formed the basis of the GIS model. These stemmed directly from information obtained during the PA and were as follows:

- *People.* In the consultation 'people' were associated with many kinds of behaviors (e.g., loud noise, litter, barking dogs and noisy children) and in some cases the very presence of people detracted from tranquillity. A modelling approach that identified the relative likelihood of people being in a given square was used to calculate a measure representing remoteness from other people.
- *Landscape.* Variables indicating the relative amount of exposure to visual elements (both positive — rivers, wide open views, the sea or broadleaved woodland, and negative — coniferous woodland, light pollution) in the

landscape were combined with the perceived naturalness of each individual square and the presence of rivers to generate a score for each individual grid square.

- *Noise.* GIS techniques were used to model the diffusion of noise away from sources such as roads, urban areas, railways and military training areas. Both the maximum noise at any time and the time-averaged noise exposure were estimated, to take account of the effect of intermittent but very loud noises, and low but constant background noise on the experience of tranquillity.

All of the data layers were weighted (using the PA responses) to reflect their relative significance (e.g., remoteness from people had a much higher rating than overhead light pollution). The positively and negatively weighted component data sets were then separately added together, and these were then combined to produce a total score in accordance with the relative significance of positive (0.44 weighting) and negative (0.56) factors from the PA data.

Figure 5.4 shows the overall score map and provides information that can be used for a number of land-use and landscape planning purposes. More specifically, the Regional Planning Guidance for the North East of England[36] states that development plans and strategies should identify, protect and work to increase tranquil areas. Using the methodology and maps produced by the study will assist planners in achieving this objective.

The research also provides another illustration of the judgements that are often involved in the generation and analysis of geographical information. GIS are tools that are rooted in quantified and/or categorized data as representations of reality. For over 20 years, a growing number of researchers and practitioners have begun to question the validity of many of the assumptions and constructs used in this representation process. This has begun to inform a growing, but still relatively limited, body of work that seeks to bridge the ability of qualitatively-based research to develop in-depth understandings of spatial phenomena and the ability of GIS as a set of techniques to represent these[37]. The tranquillity mapping study contributes to such work but, like several of the other case studies reviewed above, highlights the need for further research on this issue.

5.4 DISCUSSION

Table 5.1 compares the five case studies on a number of characteristics. Some broader themes arising from the studies are discussed below.

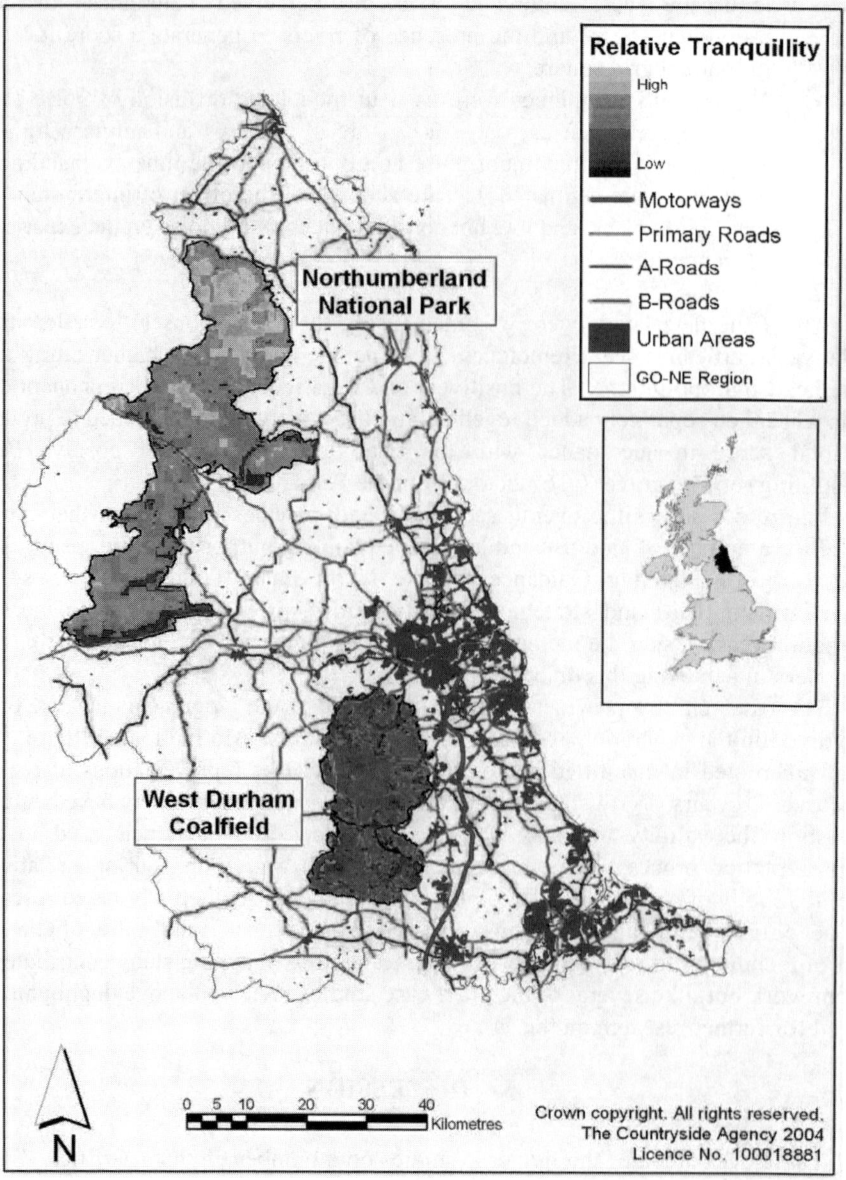

Figure 5.4 Overall tranquillity score map for two study areas in the North East region. For color versions see MacFarlane et al.[7] and http://www.cpre.org.uk/publications/landscape/tranquillity.htm. With permission.

Table 5.1 Comparison of case study characteristics

Theme	SNH	WAG	NEREG	NECF	Tranquillity
Focus	Landscape capacity	Regional strategy	Regional strategy	Forestry strategy	Policy
Ultimate objective	Resource assessment	LULU allocation	LULU allocation	Resource allocation	Resource assessment
GI product	Method / Map / Data	Map / Method	Map	Map / Method	Method / Map
Strategic ⇔ Site	Strategic	Initially strategic, then site	Both	Strategic	Primarily strategic
Selection of parameters	Steering group	Expert, informed by steering group	Steering group	Steering group and stakeholders	Public consultation
Review of product by stakeholders	Detailed and sustained	None, but resulted in a consultation document	Detailed and iterative	Detailed and through open stakeholder discussion	None
Role of GI/S experts	Technical advice and method development	Technical advice	Established framework for rational decision-making	Managed entire project	Managed entire project
Latitude for experts to determine details	Shared	Shared	Near complete control	Shared	Complete control
Overtly political nature of process	None	Medium	High	Low	None
Role of GI	Analytical	Analytical	Analytical / Confirmatory	Analytical / Confirmatory	Analytical
Data problems	None	None	Significant	Significant	Severe
Agenda of project	Advisory	Promotional	Promotional	Lobbying	Lobbying / Advisory

5.4.1 The Role of Information in Decision-Making

The use of GI and GIS as analytical and communicative tools is problematic. Checkland and Holwell, writing about information systems at large, observe that *'what such systems cannot do, in a strict sense, is provide unequivocal information; what they can do is process capta (selected data) into useful forms which can imply certain categories of information. They cannot, however, guarantee that the capta will be interpreted in this way by people making use of the system's outputs'*[38,p92]. The simplicity of the GIGO acronym (Garbage in, Garbage Out) has been significant in raising the profile of data quality issues, but more complex matters of context-dependency, the contested nature of information and the often political character of decisions are harder to distill and communicate.

The wind energy developer referred to in Section 5.3.3 appreciated the equivocal, contingent nature of one particular input into the model (landscape capacity), but failed to extend this appreciation to other factors where there were questionable social judgements, such as why B-roads were excluded from having to have a safety distance from turbines. They are less busy than A-roads but if there is a hazard from falling turbines, people, albeit fewer of them, are still exposed to it. This does illustrate that people have contrasting expectations of different kinds of information, something on which the related literature on risk and science has much to contribute.

There is no simple dichotomy between 'hard' information, with all its connotations of reliability and impartiality, and 'soft' information, with associated images of uncorroborated sources and corridor conversations, any more than there are sharp boundaries between 'safe' and 'hazardous'. Information is best seen as a rather more 'squashy' commodity, with inescapable implications of malleability.

5.4.2 Data Availability and Quality Issues

Most GIS applications are data hungry exercises and the criteria in models such as those outlined above have a tendency to outstrip the availability of suitable data. Although nationally available datasets are of a generally extremely high quality in the UK, most notably the OS range of spatial data products, the availability of locally created datasets that can be combined for a regionally consistent picture is often problematic. In the longer term initiatives such as MAGIC[39] promise greater availability, consistency and quality of spatial data for environmental decision-making, but at present such datasets are typically accorded lower priority than those relating to regeneration in regional observatory projects.

5.4.3 The Significance of Spatial Boundaries

Planners like sharp spatial boundaries. This is quite understandable as decisions based on imprecise spatial boundaries are as hard to make, and then defend, as those based on imprecise criteria. For this reason areas that have associated policy status and planning measures are almost universally precisely bounded, even when underlying phenomena such as level of deprivation or landscape quality are not satisfactorily captured in this way. Lines on maps are necessary, however sophisticated the critique or advanced the fuzzy modelling techniques become. For the purposes of opposing proposed developments, equally sharp boundaries are often not required. Indeed, one current aspect of research into NIMBYism is where the 'backyard' actually starts and ends. However, as a twist to this it is interesting to note that future developments of the tranquillity mapping approach may work towards defined zones (delineated by sharp spatial boundaries) rather than a continuum of values. Despite a recognition that landscape quality rarely varies in a way that can be accurately reflected in sharp boundaries, the requirements of a tool to protect high quality landscapes demands that they are delineated in such a manner; both development and conservation interests have a requirement for spatial boundaries, although their location is of course a subject of debate and conflict.

5.4.4 The Often Contentious Nature of GI Processes and Products

GI products emerge through processes, with some of the latter being subject to a sustained critique and others more widely accepted. In the case studies described, project steering groups all had a role to play in defining, scoping, overseeing and evaluating the GIS applications. Different steering groups played these roles in different ways, and this may be related to the nature of the applications involved. In the case of NEREG (Section 5.3.3) a government target had been regionalized and the immediate context was that the North East had to accommodate 'its share'. The approach was therefore one of finding sites that could fulfill this target. Within the steering group a range of interest groups was represented, from countryside conservation groups that sought to minimize perceived damage by steering development towards urban fringe areas on the general basis that they were already degraded in landscape terms, through local authority staff who sought to deflect developments away from areas that were, in their sub-regional context, the most valuable to them, to development interests which were tightly focused on economic imperatives. It is through such, often unhappy, marriages, that strategic plans are created and the way in which information is used in background, instrumental, rationalization and monitoring roles needs to be as transparent as possible.

5.4.5 The Crucial Nature of Site-Specific Analyses in the Pursuit of Strategic Objectives

Planning, in its broadest form, is a precursor to action, defining what and where things will take place. Spatial planning may be strategic, in which case precision about the where is less critical, or it may be locally-specific at which level the task is one of translating higher level aspirations and general spatial guidance, into site-specific outcomes. These outcomes may be an absence of development (e.g., within a nature reserve or a green belt) or approval of proposals (e.g., for new housing or woodland), but the point is that translation from the general to the specific is required and frameworks, tools and techniques are needed to support this and ensure consistency with higher principles.

5.4.6 Technical Latitude in the Application of 'Best Practice'

There is now a large body of work which seeks to establish and support 'best practice' for a number of methods relevant to environmental planning and policy[19-22,40]. These documents use a combination of site specific evidence-based work and a general checklist of applicable methodologies to provide guidance. However, the interpretation of these guidelines, i.e., how methodologies are implemented, is very much up to the practitioner. Using visibility analysis as an example (it was employed in all five case studies), the way in which parameters were agreed, how it was used, the scale at which it was applied and how the results were incorporated into the final product varied greatly, even when the aims were very similar (e.g., the NEREG and WAG wind energy projects).

5.4.7 The Innately Political Nature of Environmental Policy and Planning

At a strategic level environmental policy and planning is a Political (large P) activity in that it is conducted in line with the objectives of the elected government, and this holds to true to a degree at the regional and local levels as well where elected members determine the framework within which council officials operate. Opposition is also an inherently political (small p) activity in that it is concerned with the (spatial) allocation of resources and LULUs. Opposition groups have been crudely characterized as NIMBYs, NIMPOOs (Not In My Period Of Office) or CAVEs (Citizens Against Virtually Everything). Planning decisions, from the refusal of a development control committee to allow a new extension to a single building, through to RSSs are inherently political, and the way in which GI is utilized to develop, build, question or support positions and decisions can be equally politically charged or motivated.

5.5 CONCLUSION: FROM LOCAL TO STRATEGIC AND BACK AGAIN

Although strategic land use and regional planning was extremely significant in post-Second World War Britain, the laissez faire politics of the 1980s saw it rapidly fade. This context saw the emergence of a more 'predatory' role for development interests and the relatively de-regulated landscape became increasingly subject to urban sprawl and the diffusion of urban developments. Developers sought to secure sites and then develop them and the ability of local government to manage this growth within a strategic framework was limited.

Under the post-1997 Labour government, regionally-organized delivery of public policy agendas, strategy formulation and planning has become much more significant. This period has also seen a proliferation of targets for economic, social and environmental progress, from regional to national and international in scale. Such targets (e.g., for employment or renewable energy generation) have required developments of different kinds; previously developed land has been re-used for new business parks and the increasing number of site-by-site windfarm developments have began to have cumulative impacts on certain landscapes. With increasing emphasis on efficiency (targeting supply on areas of demand), effectiveness (reaching targets), equity (ensuring that opportunities are available in areas of relative deprivation) and sustainable development (balancing economic, social and environmental factors) GIS has emerged strongly as a tool to try and juggle the differing criteria. Many of the case studies presented in this chapter are clear examples of such endeavors.

However, at the end of the day government, at whatever level, is not usually directly involved in building business parks, planting trees or establishing windfarms, and its role is largely restricted to policy formulation, strategic planning, the design of incentive schemes and regulation. By providing strategic direction, developers can avoid costly, lengthy and probably unsuccessful applications in 'the wrong area', government is seen to be transparent and the outcomes are more consistent with frameworks that will have been subject to public consultation. Nevertheless, we are still left with the problem that developments actually take shape in specific places, on actual parcels of land. The increasing body of good practice at the strategic level must now be supplemented with GIS-based site assessment methodologies for Local Development Frameworks and development control planners that retain consistency with the strategic principles and are suitably transparent and efficient in their operation. At present, it is the private sector that is accelerating most rapidly in this respect as developers seek to utilize GIS to retain their competitive advantage in the search for sites that are 'right'.

5.6 REFERENCES

1. The Wildland Network, What is wildland?, http://www.wildland-network.org.uk, 2005.

2. Benson, J.F., Scott, K.E., Anderson, C., MacFarlane, R., Dunsford, H., and Turner K., *Landscape Capacity Study for Onshore Wind Energy Development in the Western Isles*, Scottish Natural Heritage Commissioned Report No. 042, ROAME No. F02LC04, Scottish Natural Heritage, Edinburgh, 2004.

3. Arup Consultants, Facilitating Planning for Renewable Energy in Wales: Meeting the Target, Final Report Research Contracts 105/2002 and 269/2003, Arup, London, 2004.

4. Dunsford, H., MacFarlane, R., and Turner K., The Development of a Regional Geographical Information System for the North East Renewable Energy Strategy 2003, North East Assembly, Newcastle, 2003.

5. MacFarlane, R. and Roe, M., Strategic Planning for the Development of North East Community Forests: A GIS Study, North East Community Forests, Stanley, Durham, 2004.

6. Davies, C., MacFarlane, R., McGloin, C. and Roe, M. Green Infrastructure Planning Guide, http://www.greeninfrastructure.eu, 2006.

7. MacFarlane, R., Haggett, C., Fuller, D., Dunsford, H., and Carlisle, B., Tranquillity Mapping: Developing a Robust Methodology for Planning Support, Report to the Campaign to Protect Rural England, Countryside Agency, North East Assembly, Northumberland Strategic Partnership, Northumberland National Park Authority and Durham County Council, Report available at http://www.northumbria.ac.uk/tranquillity, Centre for Environmental & Spatial Analysis, Northumbria University, Newcastle upon Tyne, 2004.

8. Planning Portal, Regional Spatial Strategies and Regional Planning Guidance – England, http://www.planningportal.gov.uk/england/professionals/en/1020432878443.html, 2006.

9. Forestry Commission, Regional Forestry Frameworks, http://www.forestry.gov.uk/forestry/infd-5llet7, 2006.

10. Cullingworth, J.B. and Nadin, V., *Town and Country Planning in the UK*, 14th Edition, Routledge, London, 2006.

11. Danziger, J.N., Dutton, W.H., Kling, R., and Kraemer, K.L., *Computers and Politics: High Technology in American Local Governments,* Columbia University Press, New York, 1982.

12. Barkenbus, J., Expertise and the policy cycle, Unpublished paper, Energy, Environment and Resources Centre, University of Tennessee, 1998.

13. Stephenson, R., In what way, and to what effect is technical information used in policy making? Findings from a study of two development plans, *Planning Practice & Research*, 13, 237-245, 1998.

14. Campbell, H. and Masser, I., *GIS and Organisations*, Taylor & Francis, London, 1995.

15. Klosterman, R., Planning Support Systems: a new perspective on computer-aided planning, *Journal of Planning Education and Research*, 17, 45-54, 1997.

16. Orland, B., Budthimedhee, K., and Uusitalo, J., Considering virtual worlds as representations of landscape realities and as tools for landscape planning, *Landscape and Urban Planning*, 54, 139-148, 2001.

17. Jankowski, P. and Nyerges, T., *Geographic Information Systems for Group Decision Making*, Taylor and Francis, London, 2001.

18. Pickles, J., Ed., *Ground Truth: The Social Implications of Geographic Information Systems*, Guilford Press, New York, 1995.

19. Countryside Agency and Scottish Natural Heritage, *Landscape Character Assessment: Guidance for England and Scotland*, CAX 84, Countryside Agency, Cheltenham, 2002.

20. Landscape Institute and Institute of Environmental Management & Assessment, *Guidelines for Landscape and Visual Impact Assessment*, 2nd Edition, Spon Press, London, 2002.

21. Scottish Executive, *Planning Advice Note 45: Renewable Energy Technologies. Revised.* Scottish Executive, Edinburgh 2002.

22. Scottish Natural Heritage, *Strategic Locational Guidance for Onshore Wind Farms in Respect of the Natural Heritage*, Policy Statement 02/02, Scottish Natural Heritage, Edinburgh, 2002.

23. Department of Trade and Industry, *Future Offshore – a Strategic Framework for the Offshore Wind Industry*, Department of Trade and Industry, London, 2002

24. Welsh Assembly Government, *Technical Advice Note (TAN) 8 Renewable Energy*, Welsh Assembly Government, Cardiff, 2005.

25. The Northern Energy Initiative, North East of England Regional Renewable Energy Strategy. Summary Report and Input to Regional Planning Guidance, Report to the North East Assembly RSS Forum 28th March 2003, North East Assembly, Newcastle, 2003.

26. Heywood, I., Cornelius, S., and Carver, S., *An Introduction to Geographical Information Systems*, 3rd Edition, Pearson, Harlow, 2006, 282.

27. Ordnance Survey, Land-Form Panorama®, http://www.ordnancesurvey.co.uk/oswebsite/products/landformpanorama, 2006.

28. Benson, J., Scott, K., and Anderson, C., Landscape Appraisal for On-Shore Wind Development, Consultation Draft Report (Part), 11th March 2003. Report to North East Renewable Energy Group, Landscape Research Group, Newcastle University, 2003.

29. Hartig, T., Mang, M., and Evans, G. W., Restorative effects of natural environment experiences, *Environment and Behaviour*, 23, 3-26, 1991.

30. Mace, B. L., Bell, P. A., and Loomis, R. J., Aesthetic, affective, and cognitive effects of noise on natural landscape assessment, *Society and Natural Resources*, 12, 225-242, 1999.

31. Morris, N., Health, Well-Being and Open Space Literature Review, OPENspace Research Centre for Inclusive Access to Outdoor Environments, Edinburgh, 2003.

32. Council for the Protection of Rural England and the Countryside Commission, Tranquil Areas – England Map, Council for the Protection of Rural England, London, 1995.

33. Association for the Protection of Rural Scotland, Mapping Tranquillity in Scotland (Inverness to Aberdeen): Concept, Methodology and Potential, Association for the Protection of Rural Scotland, Edinburgh, 2000.

34. Levett, R., A Headline Indicator of Tranquillity: Definition and Measurement Issues, Interim Report to CPRE by CAG Consultants 7 July 2000, CAG Consultants, London, 2000.

35. Chambers, R., The origins and practice of Participatory Rural Appraisal, *World Development*, 22, 953-969, 1994.

36. Government Office for the North East and the Office of the Deputy Prime Minister, *Regional Planning Guidance for the North East (RPG1)*, The Stationary Office, London, 2002.

37. Bell, S. and Reed, M., Adapting to the machine: Integrating GIS into qualitative research, *Cartographica* 39, 55-66, 2004.

38. Checkland, P. and Holwell, S., *Information, Systems and Information Systems: Making Sense of the Field*, John Wiley, Chichester, 1998.

39. Multi-Agency Geographic Information for the Countryside (MAGIC), http://www.magic.gov.uk/, 2006.

40. English Nature, Nature Conservation Guidance on Offshore Windfarm Development, http://www.defra.gov.uk/wildlife-countryside/ewd/windfarms/windfarmguidance.pdf, 2005.

CHAPTER 6

Creating a Digital Representation of the Water Table in a Sandstone Aquifer

P. Posen, A. Lovett, K. Hiscock, B. Reid, S. Evers and R. Ward

6.1 INTRODUCTION

Groundwater is an important resource in England and Wales and provides on average 33% of the total public drinking water supply[1]. This figure rises to around 80% in the southeast of England, where large areas of chalk and limestone aquifer outcrop in regions under intensive cultivation. Consequently, protection of the resource from diffuse agricultural pollution is a primary concern in groundwater management[2-4].

Groundwater vulnerability assessment began in the 1960s and, over the last 15 years, the use of methodologies based on GIS techniques has become quite widespread[5-10]. However, ongoing implementation of the EU Water Framework Directive has made the need for further refinements to groundwater vulnerability assessment systems more pressing[11-13].

Current groundwater vulnerability assessment methods, many of which utilize a GIS[14,15], combine parameters related to the nature and source of contaminant, physico-chemical properties of the topsoil and unsaturated zones, hydrogeology and climatic conditions to produce a variety of contaminant fate models.

One important influence on groundwater vulnerability is unsaturated zone thickness[16], which governs the time taken for contaminants to travel through this layer, and therefore the degree of potential degradation that may occur before introduced compounds reach the water table. The importance of water table depth is emphasized in the widely-used DRASTIC model[17], which assigns the highest weight to this parameter. To date, depth to water table has been estimated in the UK at a local scale, due to the extent of seasonal and spatial variability. However, an initiative by the Environment Agency (the main public body for environmental protection in England and Wales) to produce a new national assessment framework for groundwater vulnerability[13] gave impetus to the current study as a pilot to develop an automated method for generating a nationally consistent database of water table depths.

One approach to improving the estimation of unsaturated zone thickness is to create a digital representation of the water table, which can then be subtracted from surface topography within a GIS to create a 'depth to water table' map. As a demonstration, it was decided to create such a map for a sandstone aquifer unit in the Midlands region of England by using digital maps of surface topography in conjunction with groundwater level monitoring data. With the help of a hand-contoured reference map of groundwater levels in the study area, three different methods of interpolating the water level data were appraised, and the most representative model then applied to calculate the depth to the water table.

6.2 BACKGROUND

The groundwater unit used for the purposes of this study (the Triassic Sherwood Sandstone) is located in the River Trent catchment of the Midlands region of England, and comprises an easterly dipping sandstone aquifer bounded by a Permian Magnesian Limestone aquifer to the west and confined by Triassic mudstones to the east (Figure 6.1).

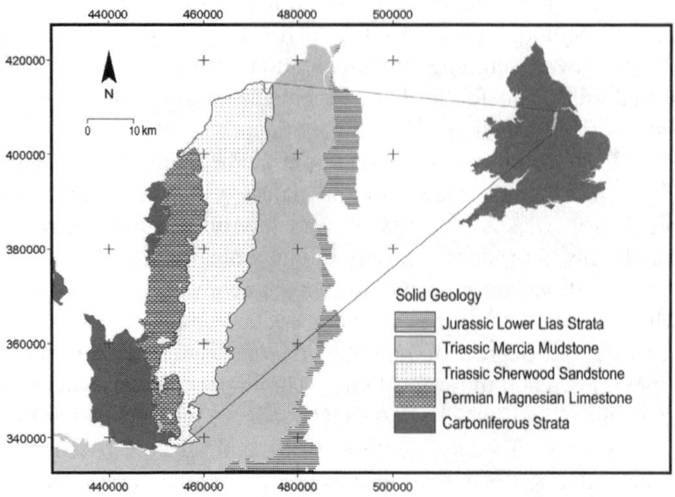

Figure 6.1 Map showing the location and extent of the unconfined Sherwood Sandstone aquifer unit used for the digital water table interpolation.

The primary reasons for choosing the Sherwood Sandstone aquifer unit for the study were: (i) the existence of a substantial data set of water level measurements contained in the Environment Agency's observation borehole network in the Midlands region; and (ii) the availability of a reliable hand-contoured paper map of the water table in the sandstone aquifer (scale 1:50,000; produced by ADAS Cartography, Gloucester) which could be used to assess the accuracy of the different interpolation methods. The hand-contoured map was based on data for a high water level period obtained during January to April 1994.

The Sherwood Sandstone Group comprises undifferentiated sandstones that are poorly cemented. The average hydraulic conductivity of the sandstones is 3.4 m day^{-1} and higher values are associated with locally-enhanced fissures induced by coal workings which produce high groundwater yields of good quality[18]. Approximately 42% of public water supplies in the area are from groundwater supplied by the Sherwood Sandstone aquifer. The Sherwood Sandstone also provides water for many major industries and is used to support irrigation of arable crops in the area[16]. The surface topography is relatively flat over much of the area and most groundwater recharge occurs through rain falling directly on to the unconfined part of the aquifer in the west of the region. Annual effective rainfall can be as low as 120 mm which, combined with a deep water table and relatively high porosity of 30%, can lead to long delays in groundwater recharge[18].

6.3 METHODS

6.3.1 Conversion of the Paper Map

For greater ease of comparison between the hand-contoured groundwater level map representing the water table surface in early 1994 and the interpolated digital maps, the paper map was converted into a digital layer in ArcView® GIS 3.2 (http://www.esri.com). This was achieved by superimposing the British National Grid on to the paper map and recording the co-ordinates of points along each water table contour. These co-ordinates and their respective depths were entered into a database and imported into ArcView GIS. The imported data were used to create a triangulated irregular network (TIN) representing the water table, and the TIN was converted to raster format at 50 m grid resolution (Figure 6.2a).

6.3.2 Data Point Selection

Data points from 59 locations were selected from a subset of 110 Environment Agency observation boreholes in the Sherwood Sandstone aquifer (Figure 6.2b). The study area boundary enclosed 82,500 ha of unconfined aquifer within the sandstone unit and included three sites identified by the Environment Agency as 'key borehole' sites where water levels are not subject to major fluctuation or influenced by local abstraction. These sites provide good quality data against which the reliability of surrounding data points can be judged, thereby limiting the amount of model input error. Hydrograph plots from each site in the entire subset were examined and boreholes exhibiting irregularities in their plots, such as gaps in the time series, or erratic fluctuations in water levels, were omitted from the selection. Nine boreholes outside the study boundary, in the confined aquifer to the east, and five further sandstone boreholes located beyond the northern and southern extremities of the study area boundary, were included in the selection so that the subsequent water level interpolation would not suffer from 'edge effects'. These phenomena, manifested as a warping of the surface, can occur when there is an absence of data beyond a boundary, resulting in unrealistically high or low values[19].

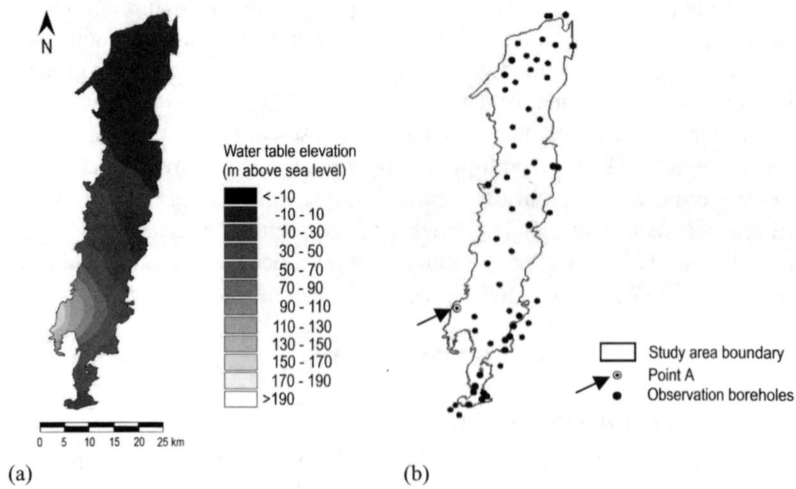

Figure 6.2 (a) Digital representation of the hand-contoured water table in the unconfined area of the Sherwood Sandstone aquifer. (b) Locations of the data points used for the water table interpolation. Point A represents a peak water level value corresponding with an elevated water table surface in the south-western corner of the study area.

6.3.3 Time Selection

To ensure consistency throughout the data set, water level data were compiled from the selected borehole records using measurements taken for a single occasion during a period of high rainfall in late 2001. All measurements fell within a three-week period during the month of October, at which time the high water levels represented minimum depth to the water table.

6.3.4 Data Interpolation

ArcView GIS was used to interpolate the water level data using spline and inverse distance weighting (IDW) methods, and kriging interpolation of the same data was executed in the GS+ program (Gamma Design Software, http://www.gammadesign.com). Although the area of interest was within the study boundary, the three interpolated surfaces extended well beyond this boundary so that differences between interpolation methods could be fully appraised. All three interpolated surfaces were expressed in grid format, with a 50 m resolution.

Spline interpolation. The spline interpolation method applies local polynomial functions to fit the smoothest possible surface through all data points, in a manner in which a closest-fit curve might be plotted through points on a graph[20]. The extent of smoothing relates to the number of points on which the polynomial curves are based; the more points, the smoother the surface produced. The value of weighting applied governs the curvature of the lines between individual data points and has little effect in areas where data points are abundant, but increased weighting leads to warping of the surface in areas where data points are sparse. The spline surface that most closely matched the hand-contoured reference map (Figure 6.2a) was achieved using a tension spline (which constrains the surface to pass through all points) based on 6-point polynomials, with a 0.1 weighting.

IDW interpolation. Inverse distance weighting is an exact local interpolation method that produces a surface whose value changes smoothly between the data points to which it is tied. The data are inversely weighted so that calculated points on the interpolated surface are more strongly influenced by nearby data points than they are by more distant points[21]. The extent of smoothing of the surface is dependent on the number of 'nearest neighbors' used for the interpolation, and on the chosen value for the decay parameter, with the sphere of influence of a data point diminishing more rapidly with higher decay values. The weight of the decay parameter is expressed as a power function[20].

In the current study, the IDW surface that most closely matched Figure 6.2a was interpolated using 12 nearest neighbors and a value of 2 for the decay parameter, giving an inverse weighting as the square of distance. This was found to give the optimum sphere of influence to most data points, producing an acceptably smooth surface without being unrealistic as to the extent to which any individual point was affecting the interpolation.

Kriging interpolation. Kriging operates in a similar manner to IDW, but uses the underlying spatial dependence of the data to calculate the most appropriate value for the decay parameter. The spatial trend of the data is described by the variogram[20,21], which shows how data values vary with distance and direction. The best-fitting variogram model can then be used to customize the kriging interpolation by calculating appropriate weights according to clustering, distance and direction of neighboring data points. In the current study, ordinary point kriging, employing a spherical variogram model, was found to produce a surface that most closely resembled the digital reference map (Figure 6.2a). The variogram parameters and associated plot are given in Table 6.1 and Figure 6.3 respectively.

Table 6.1. Kriging parameters relating to the water table interpolation

Model Parameter	Value/Type
Active lag	24,000 m
Lag class interval	3000 m
Model	Spherical
Nearest neighbors	12

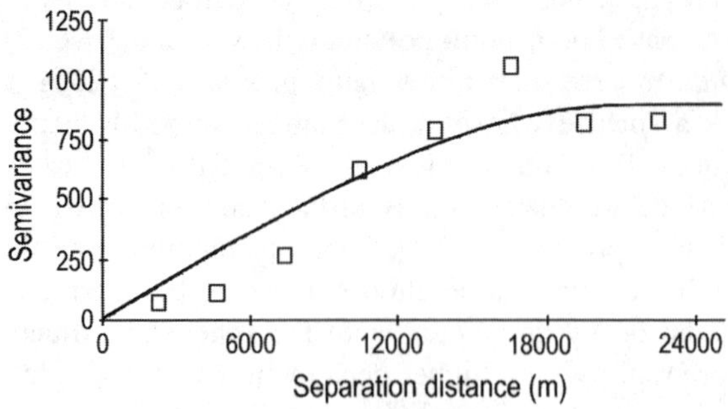

Figure 6.3 Variogram plot for the kriging interpolation of the water table.

6.3.5 Evaluation of the Surfaces

Removal of peak value. One simple test for evaluating the effectiveness of an interpolation method is to recalculate the surface after the removal of one or more significant data points[22]. This test was performed on each of the interpolated surfaces by removing a peak water level value (at Point A, Figure 6.2b) corresponding with an elevated water table surface in the south-western corner of the study area. The effects on the re-interpolated surfaces were examined for each different method.

Cross-validation. Cross-validation analysis, which removes each data point in turn and interpolates from the remaining points to estimate a value at the corresponding location[23], was performed on each of the three interpolated surfaces, using the GS+ program for the IDW and kriged surfaces and ArcGIS® 8 (http://www.esri.com) for the spline surface.

Investigation of edge effects. These were examined by comparing an interpolation that included 14 data points lying beyond the study area boundary with one that excluded these points.

6.4 RESULTS

Representations of the three different water table interpolations are given in Figure 6.4. All three surfaces exhibit a southwest to northeast decrease in water table elevation within the study area boundary, from a minimum of 0 m, to a maximum of 165 m above sea level. Point A represents a peak value in the observed groundwater level data, which corresponds to elevated surface topography in the southwestern corner of the study area. The main differences between the interpolations are evident in the curvature of the contours in the northwest and southeast corners of the map.

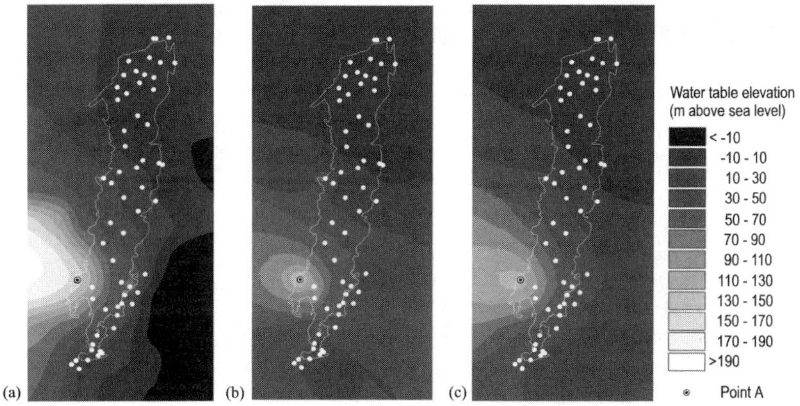

Figure 6.4 Representations of the water table in the Sherwood Sandstone aquifer, using (a) spline, (b) IDW and (c) kriging interpolation methods. The location of the peak value, Point A, is shown.

The effects of removing the peak value Point A from each interpolation are shown in Figure 6.5. Figure 6.5a indicates little change in the overall shape of the spline surface, but Figures 6.5b and 6.5c show more significant local change in the IDW and kriged surfaces, respectively. In the latter two surfaces, the 'peak contours' are shifted eastwards, closely following the change in data distribution.

Results of cross-validation analyses of the three surfaces were expressed as plots of estimated vs. observed values (Figure 6.6). The peak value Point A can be seen as the major outlier in all plots. Regression analyses on these plots indicated that the spline, IDW and kriged surfaces described 83%, 85% and 91%, respectively, of the variability in the actual water table values.

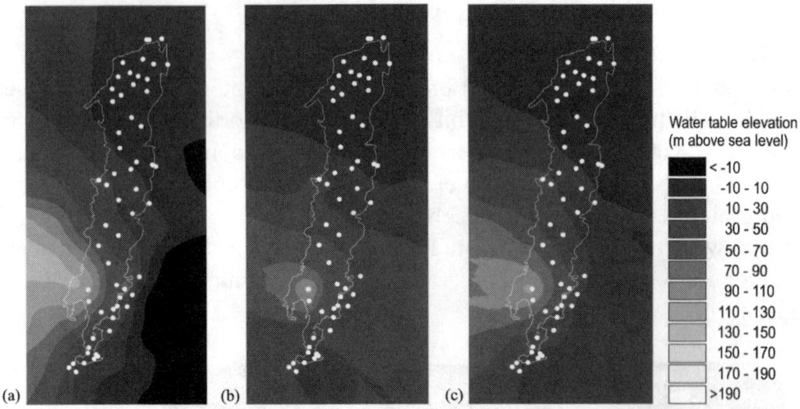

Figure 6.5 Maps showing the effect on the interpolated surfaces of removing the peak value, Point A, from the data set: (a) the spline surface, (b) the IDW surface and (c) the kriged surface.

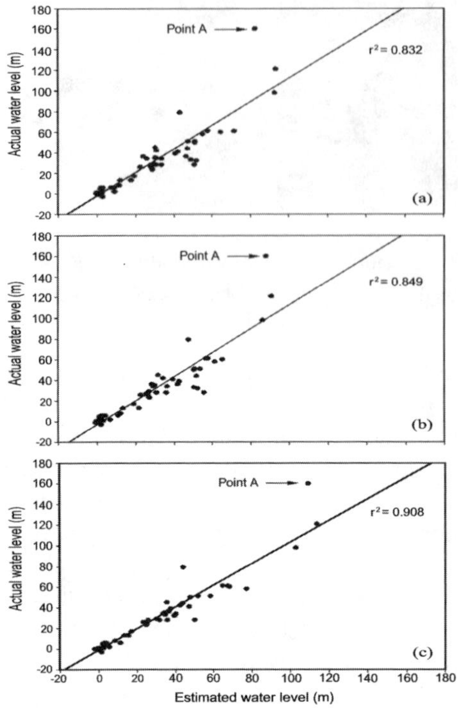

Figure 6.6 Cross-validation plots for (a) the spline surface, (b) the IDW surface and (c) the kriged surface. The peak value, Point A, is the major outlier in all plots.

Figure 6.7 shows the effect of (a) including and (b) excluding data points beyond the study boundary in the kriged interpolation. The greatest difference relates to the curvature of contours in the southeastern quarter of the map, with some lesser effects occurring around the southwestern peninsula of the study area.

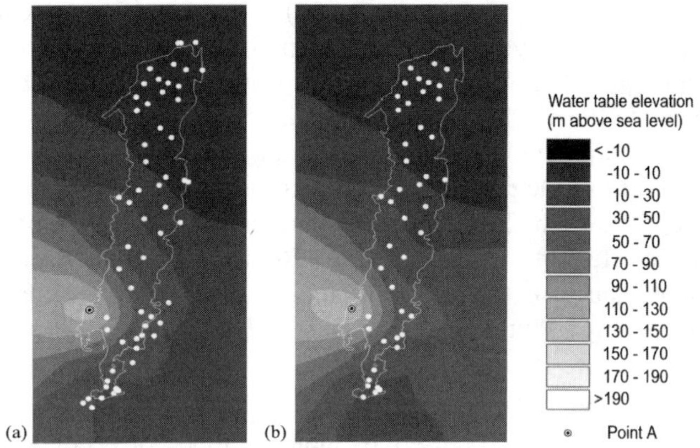

Figure 6.7 Maps showing the kriged water table interpolation: (a) using 14 data points outside the study area boundary, and (b) excluding these points.

6.5 DISCUSSION OF RESULTS

6.5.1 Visual Interpretation of the Surfaces

Although the spline interpolation method produces a credible surface in areas where data are abundant and evenly distributed (Figure 6.4a), the global nature of this method generates erratic values or warping of the surface where data are sparse, as the method attempts to produce a smooth fit through all available data points. Artifacts of this distortion are visible in the 'pinching' of the surface on each side of the southern part of the aquifer, and most particularly in the southwest, owing to the relative scarcity of data in this area.

The broadly similar surfaces produced by IDW and kriging (Figures 6.4b and 6.4c respectively) do not suffer from such effects. The close curvature of contours around certain data points (Figure 6.4b) reflects the local nature of IDW interpolation and shows the strong influence of data point values on the immediately adjacent surface. The near-circular features around some data points are not seen in the kriged surface (Figure 6.4c). Additionally, and in contrast to the IDW interpolation, the contours of the kriged surface continue to diminish in value towards the southeast corner of the map, taking the underlying spatial trend in distance and value of neighboring data points into consideration.

6.5.2 Removal of Peak Value

Comparison of Figure 6.4a with Figure 6.5a shows the very localised effect of removing the single peak value Point A from the spline interpolation. The surface values decrease in the immediate vicinity of the removed point, but the rest of the surface remains unchanged.

The revised IDW interpolation (Figure 6.5b) shows significant local change of surface shape in the vicinity of the removed point (compare with Figure 6.4b), reflecting the eastward shift of the peak surface value. This leads to a greater area of change adjacent to the southwest study boundary but, in common with the spline surface, the rest of the interpolated area remains unchanged.

The eastward shift of the peak value in the kriged surface (compare Figure 6.4c with Figure 6.5c) follows the local change in surface value, but does not exhibit the intensely localized effect of the IDW surface. Removal of Point A produces more widely distributed changes in the kriged interpolation, affecting the curvature of the contours across the entire southern area of the map.

6.5.3 Cross-Validation

The cross-validation plots (Figure 6.6) indicate good correspondence between estimated and actual water table values for all interpolation methods, with the kriged interpolation achieving the best fit, as would be expected. However, the value of the outlier Point A proved difficult to predict, resulting in underestimations of 77 m, 72 m and 52 m, in the spline, IDW and kriged interpolations, respectively.

6.5.4 Examination of Edge Effects

The effect of excluding data points beyond the study boundary is best observed in the results of two kriging interpolations (Figure 6.7). Exclusion of these points led to a marked change in curvature of the kriging contours, not only in the immediate vicinity of the excluded points, but also further afield, particularly in the southern half of the map.

6.5.5 Comparison with Hand-Contoured Data

The hand-contoured map of the sandstone water table was produced seven years prior to, and at a different time of year from the interpolated data, so direct comparisons of absolute values cannot be made, although the general shape of the interpolated and hand-contoured surfaces should be similar in the absence of major changes in the groundwater abstraction regime. It should also be remembered that the hand-contoured map is itself an approximation, the accuracy of which is not known.

Taking these issues into consideration, it was decided to subtract the digital representation of the hand-contoured surface from each of the interpolated surfaces, in order to highlight areas where interpolation might be most problematic. The resulting maps (Figure 6.8) indicated that the southern part of the aquifer was the most difficult area to model successfully. In all three interpolations, most of the estimated water table surface fell within one standard deviation of the hand-contoured surface. The isolated pockets of greater deviation from the hand-contoured map in the spline interpolation (Figure 6.8a) corresponded to the area across which greatest 'pinching' was visible in Figure 6.4a, with the least successful fit occurring near the southeastern corner. The IDW surface also showed some disagreement in the southeastern corner, though to a lesser extent than the spline surface, but produced a much less successful fit in the southwestern peninsula (Figure 6.8b). Similarly, the kriged surface produced a poor fit in and around the south-western peninsula (somewhat poorer than the IDW surface) and had some minor areas of disagreement near the western and northwestern boundaries.

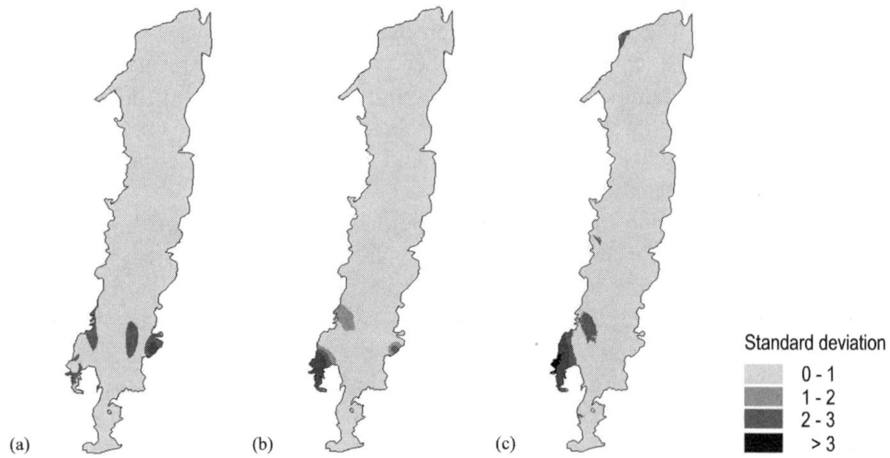

Figure 6.8 Standard deviation maps showing the areas of difference between each of the interpolated surfaces and the digital representation of the hand-contoured reference map: (a) the spline interpolation, (b) the IDW interpolation and (c) the kriging interpolation.

The main discrepancies in each case appear to relate to spatial distribution of the data. The northern part of the aquifer has a majority of fairly evenly distributed boreholes on which to base the interpolations. Therefore, although there are minor differences in the interpolated surfaces, all three methods appear to work equally well. However, in the southern part of the map, data distribution is more irregular with points clustered in the southeast corner, but fairly sparse in the southwest. Furthermore, the peak outlier (Point A) occurs in an isolated area, close to the study boundary, with few data points in the immediate vicinity.

The close correspondence between the spline and hand-contoured surfaces in the southwestern peninsula (Figure 6.8a) is likely to be coincidental, and due to the warping of the spline surface just happening to follow the curvature of the actual water table contours. The fit of the kriged surface (Figure 6.8c) in the southeastern corner is better than that of the other two interpolations because of the more accurate representation of the curvature of the contours beyond the study boundary.

6.5.6 Interpolation for Water Table Mapping Purposes

As seen in Figure 6.4a, the potential for edge effects and warping of the surface when using spline interpolation makes this method unsuitable for modelling surfaces with non-uniform distribution of data points, such as the location of observation boreholes. A good approximation of actual surface values is achieved where data points are abundant and evenly distributed, but warping of the surface where data are sparse or irregularly spaced can give rise to unreliable values, especially close to the boundary of the interpolated area.

IDW does not lead to such erratic surface values and boundary effects, and the exact nature of this method gives a very accurate representation of the water table in areas where data are abundant. Nevertheless, there is less certainty in the interpolated surface where data are sparse, and circular features can arise from the influence on the surrounding area of data point weighting.

The ability of kriging to accurately estimate and take account of the underlying data trend leads to a closer approximation of the true surface in areas where data are lacking. Of the three interpolation methods examined in this study, ordinary point kriging is therefore thought to be the most suitable for water table mapping purposes.

The change in surface shape afforded by inclusion of data beyond the confining boundary indicates that interpolation should be extended beyond such boundaries to avoid edge effects. In the Sherwood Sandstone study area, extra data points were not available beyond the western boundary, which marks the western extent of the sandstone aquifer. In such circumstances it may be appropriate to add some 'dummy' points beyond the study boundary, with values close to those of neighboring data, to improve the surface representation near the boundary.

6.5.7 Derivation of the Final Depth to Water Table Map

As a final calculation, the kriged surface (Figure 6.4c) was subtracted from a digital layer of surface topography (Ordnance Survey, Land-Form PANORAMA™ digital terrain model, 50 m grid resolution) using the Map Calculator facility in ArcView GIS. The resulting map of 'depth to water table' shown in Figure 6.9 clearly depicts channels indicative of groundwater in potential contact with the topographic surface and, therefore, likely to represent groundwater discharge areas. It was found that these channels corresponded closely with a digital overlay of the major river network where groundwater discharge would be expected to occur.

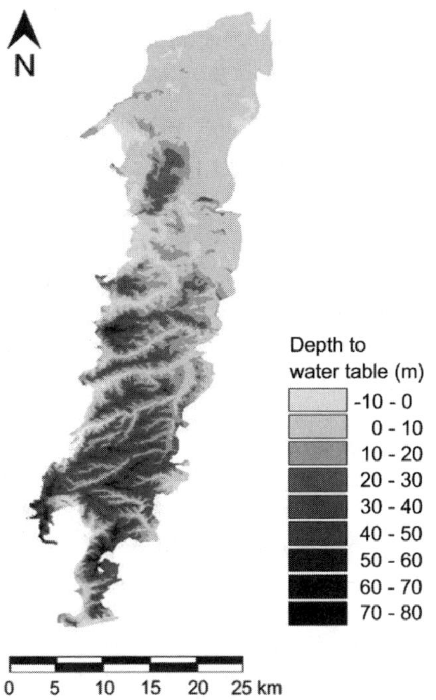

Figure 6.9 'Depth to water table' map for the unconfined area of the Sherwood Sandstone aquifer, produced by subtracting the kriged water table surface from the topographic surface. The visible channels, indicative of groundwater in potential contact with the topographic surface, correspond closely with the major river network.

The magnitude of the negative values in Figure 6.9, interpreted to represent surface flow caused by high artesian groundwater pressure, is rather high at a number of locations, due to the absence of any constraint on surface water elevations in the model. Additional analysis was therefore carried out which involved including river elevation spot heights within the water level data set. However, this did not entirely rectify the problem and also tended to distort the interpolation of groundwater levels, since the potentiometric surface does not necessarily follow surface water elevations. This approach to refining the definition of depth to water table was therefore not taken any further.

6.6 CONCLUSIONS

This paper has discussed the issues involved in interpolating groundwater level data and compared the outputs produced by three different methods (spline, IDW and kriging). The results suggest that kriging generates the most reliable digital representation of the water table surface. The methodology presented here could be applied in any region with adequate coverage of groundwater level monitoring points and topographic data, to assist in groundwater management procedures and vulnerability modelling.

6.7 ACKNOWLEDGMENTS

This study was conducted as part of a Natural Environment Research Council PhD studentship (NER/S/A/2001/06/48), with CASE partner support from the Environment Agency of England and Wales.

6.8 REFERENCES

1. Hiscock, K.M., *Hydrogeology: Principles and Practice*, Blackwell Science Ltd, Oxford, 2005.

2. Beeson, S. and Cook, M.C., Nitrate in groundwater: a water company perspective, *Quarterly Journal of Engineering Geology and Hydrogeology*, 37, 261-270, 2004.

3. Gooddy, D.C., Stuart, M.E., Lapworth, D.J., Chilton, P.J., Bishop, S., Cachandt, G., Knapp, M., and Pearson, T., Pesticide pollution of the Triassic Sandstone aquifer of South Yorkshire, *Quarterly Journal of Engineering Geology and Hydrogeology*, 38, 53-63, 2005.

4. Silgram, M., Williams, A., Waring, R., Neumann, I., Hughes, A., Mansour, M., and Besien, T., Effectiveness of the Nitrate Sensitive Areas Scheme in reducing groundwater concentrations in England, *Quarterly Journal of Engineering Geology and Hydrogeology*, 38, 117-127, 2005.

5. Sokol, G., Leibundgut, C., Schulz, K.P., and Weinzierl, W., Mapping procedures for assessing groundwater vulnerability to nitrate and pesticides, in: *HydroGIS 93: Application of Geographic Information Systems in Hydrology and Water Resources*, Vienna. IAHS Publication No. 211, 1993, 631-639.

6. Hiscock, K.M., Lovett, A.A., Brainard, J.S., and Parfitt, J.P., Groundwater vulnerability assessment: two case studies using GIS methodology, *Quarterly Journal of Engineering Geology*, 28, 179-194, 1995.

7. Burkart, M.R. and Feher, J., Regional estimation of ground water vulnerability to nonpoint sources of agricultural chemicals, *Water Science and Technology*, 33, 241-247, 1996.

8. Lasserre, F., Razack, M., and Banton, O., A GIS-linked model for the assessment of nitrate contamination in groundwater, *Journal of Hydrology*, 224, 81-90, 1999.

9. Srivastava, P., Day, R.L., Robillard, P.D., and Hamlett, J.M., AnnGIS: Integration of GIS and a continuous simulation model for non-point source pollution assessment, *Transactions in GIS*, 5, 221-234, 2001.

10. Lake, I.R., Lovett, A.A., Hiscock, K.M., Betson, M., Foley, A., Sünnenberg, G., Evers, S., and Fletcher, S., Evaluating factors influencing groundwater vulnerability to nitrate pollution: developing the potential of GIS, *Journal of Environmental Management*, 68, 315-328, 2003.

11. Foster, D., Wood, A., and Griffiths, M., *The Water Framework Directive (2000/60/EC) - an introduction*, Department of the Environment Northern Ireland (Environment & Heritage Service) Workshop, Enniskillen, 2000.

[12.] Harris, B., Groundwater management and protection in England and Wales - a backwards and forwards glance, in *Protecting Groundwater: Applying Policies and Decision Making Tools to Land-Use Planning. Environment Agency NC/00/10/Conference Proceedings*, Environment Agency, Bristol, 2001, 3-11.

[13.] Rukin, N., Boland, M., and Thurston, N., Recommendations for an improved groundwater vulnerability assessment framework, National Groundwater and Contaminated Land Centre, Report NC/99/27, Environment Agency, Olton, 2003.

[14.] Thapinta, A. and Hudak, P.F., Use of geographic information systems for assessing groundwater pollution potential by pesticides in Central Thailand, *Environment International*, 29, 87-93, 2003.

[15.] Holman, I.P., Dubus, I.G., Hollis, J.M., and Brown, C.D, Using a linked soil model emulator and unsaturated zone leaching model to account for preferential flow when assessing the spatially distributed risk of pesticide leaching to groundwater in England and Wales, *The Science of the Total Environment*, 318, 73-88, 2004.

[16.] NRA, *Policy and Practice for the Protection of Groundwater*, National Rivers Authority, Bristol, 1992.

[17.] Merchant, J.W., GIS-based groundwater pollution hazard assessment: a critical review of the DRASTIC model, *Photogrammetric Engineering & Remote Sensing*, 60, 1117-1127, 1994.

[18.] Edmunds, W. M. and Smedley, P. L., Residence time indicators in groundwater: the East Midlands Triassic sandstone aquifer, *Applied Geochemistry*, 15, 737-752, 2000.

[19.] Martin, D., *Geographic Information Systems: Socioeconomic Applications*, 2nd ed.. Routledge, London, 1996.

[20.] Webster, R. and Oliver, M.A., *Geostatistics for Environmental Scientists*, John Wiley & Sons, Ltd., Chichester, England, 2001.

[21.] Isaaks, E.H. and Srivastava, R.M., *An Introduction to Applied Geostatistics*, Oxford University Press Inc., New York, 1989.

[22.] Burrough, P. A. and McDonnell, R. A., *Principles of Geographical Information Systems*, Oxford University Press, Oxford, 1998.

[23.] Todini, E., Influence of parameter estimation uncertainty in Kriging: Part 1 - Theoretical development, *Hydrology and Earth System Sciences*, 5, 215-223, 2001.

CHAPTER 7

GIS and Predictive Modelling: A Comparison of Methods for Forest Management and Decision-Making

A. Felicísimo and A. Gómez-Muñoz

7.1 INTRODUCTION

GIS can be a useful tool for spatial or land-use planning, but only if several conditions are fulfilled. The key conditions are related to 1) the quality of basic spatial information, and 2) the statistical methods applied to the spatial nature of the data. Appropriate information and methods allow the generation of robust models that guarantee objective and methodologically sound decisions.

In this study we apply several multivariate statistical methods and test their usefulness to provide robust solutions in forestry planning using GIS. We must emphasize that in our Iberian study area, where forests have progressively decreased in extent over centuries, the main aims of forestry planning are the reduction of forest fragmentation, biodiversity conservation, and restoration of degraded biotopes.

The research develops a set of likelihood or suitability models for the presence of tree species that are widely distributed over a study area of 41,000 km^2. The utility of suitability models has been demonstrated in some previous studies[1], but they are still not as widely employed as might be expected.

A suitability model is a raster map in which each pixel is assigned a value reflecting suitability for a given use (e.g., presence of a tree species). Suitability models can be generated through diverse techniques, such as logistic regression or non-parametric CART (classification and regression trees) and MARS (multiple adaptive regression splines)[2-4]. All of these techniques require a vegetation map (dependent variable) and a set of environmental variables (climate, topography, geology, etc.) which potentially influence the vegetation distribution. The foundation of the method is to establish relationships between the environmental variables and the spatial distribution of the vegetation. Typically, each vegetation type will respond in a different way as a consequence of its contrasting environmental requirements.

Suitability is commonly expressed on a 0-1 scale (incompatible-ideal). The precise value depends on a set of physical and biological factors that favor or limit the growth of each type of vegetation. Once the distribution of suitability values across a region is known, decisions on land use and management can be made on the basis of objective criteria.

The set of suitability values for a region can be considered as the potential distribution model if presented as a map: the area defined as 'suitable' in a model should reflect the potential area for the vegetation type under consideration. Such a model also represents the relationships between presence/absence of each forest type and the values of the potentially influential environmental variables in a given region. Usually, current forest distributions are significantly smaller than the potential spatial extents because they have been systematically logged. Potential distribution models allow the recognition and delineation of such former distribution areas in order to direct current and future management plans, provide valuable data for restoration initiatives and highlight areas where such actions should be considered a priority.

7.2 OBJECTIVES

The main objectives of the study were to 1) use several different statistical methods to generate maps of potential distributions and suitability for each of three species of *Quercus* (oak) in the study area, and 2) identify the most appropriate method and assess its advantages and limitations. In order to fulfill these objectives, we developed a workflow that included sampling strategies, GIS implementation of statistical models and validation of results.

7.3 STUDY AREA

The study area was Extremadura, one of the 17 Autonomous Communities of Spain, covering 41,680 km^2, and located in the west of the Iberian Peninsula (Figure 7.1). It has a Mediterranean climate, somewhat softened by the relative proximity to the sea and the passage of frontal systems from the Atlantic.

The study subjects, which partially cover this area, were three species of the genus *Quercus* that grow in forests or 'dehesas'. Dehesas are artificial ecotypes derived from original forest clearings (Figure 7.2). Continuous forest cover disappeared centuries ago and currently only scattered patches remain over a large potential area. In some places deforestation was complete and not even the most open dehesas remain. Trees from the genus *Quercus* are the dominant constituents of forests in the area, the most important species (and those considered in the analysis) being *Quercus rotundifolia* Lam. (holm oak, 12,680 km^2, synonym: *Quercus ilex* L. ssp. *ballota* (Desf.) Samp.), *Quercus suber* L. (cork oak, 2,130 km^2) and *Quercus pyrenaica* Wild. (Pyrenean oak, 950 km^2). With some exceptions, Pyrenean oak appears most commonly in forests, while cork and holm oaks preferentially occur in dehesas.

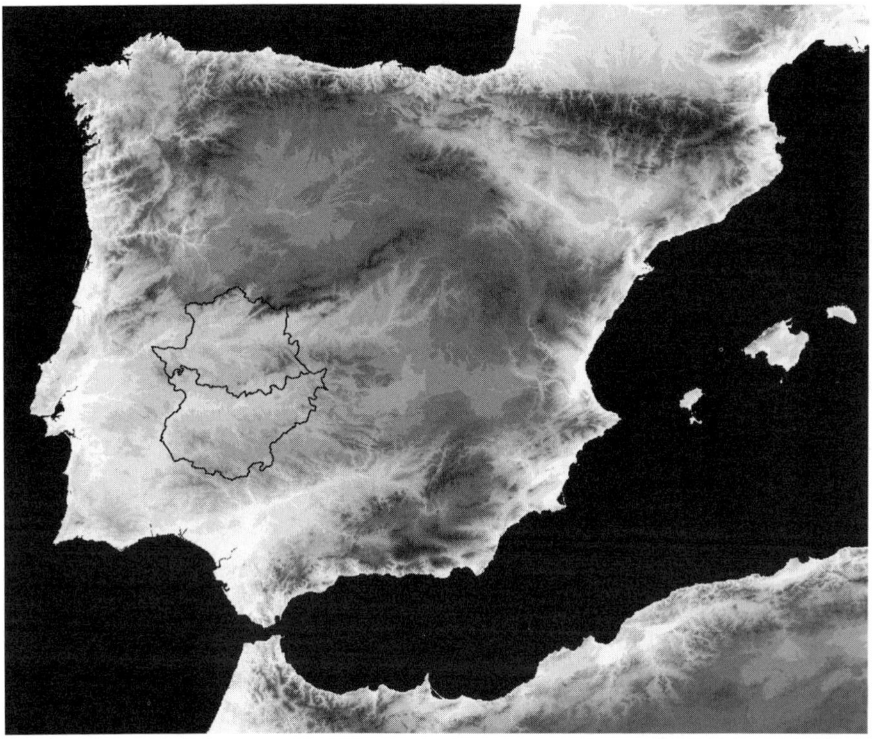

Figure 7.1 Location of Extremadura in the Iberian Peninsula.

Figure 7.2 Dehesas are artificial ecotypes comparable to savannas: a Mediterranean (seasonal) grassland containing scattered trees of the genus *Quercus*.

7.4 DATA

A set of raster maps was compiled to reflect the spatial distribution of dependent and independent (predictive) variables.

7.4.1 *Quercus* Distributions

Current *Quercus* species distribution maps were taken from the Forestry Map of Spain (scale 1:50,000), produced by the Spanish General Directorate for Nature Conservation during the period 1986-96. We used the digital version of the map to identify the main vegetation classes and the current spatial distributions (Figure 7.3).

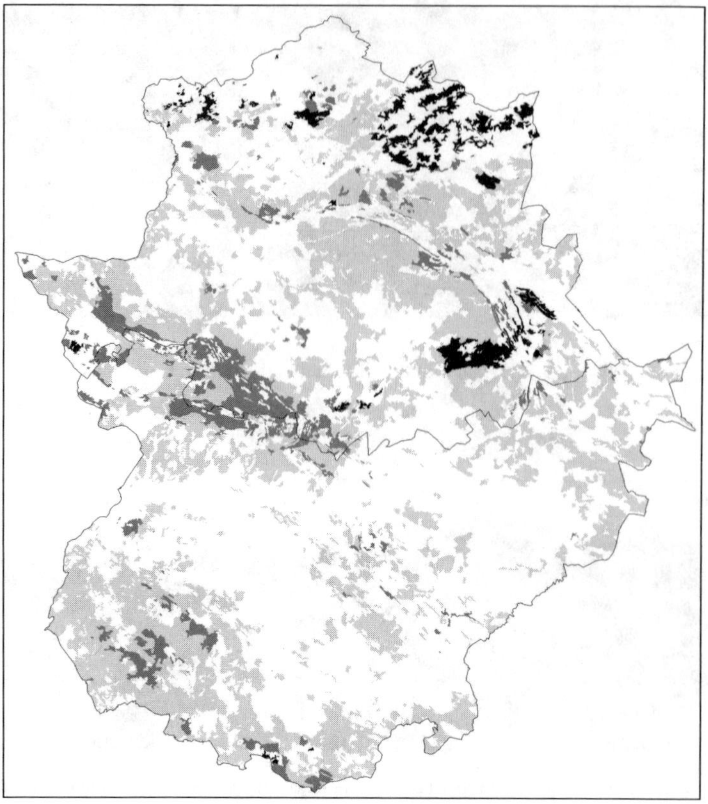

Figure 7.3 Current distribution of *Quercus* species in the study area (black represents Pyrenean oak, *Q. pyrenaica*; dark gray, cork oak, *Q. suber*; and pale gray, holm oak, *Q. rotundifolia*).

7.4.2 Predictive Variables

Raster maps were generated to represent the following independent variables:

- *Elevation.* A digital elevation model (DEM) was constructed using Delaunay triangulation of spot height and contour data from the 1:50,000 scale topographic map of the Army Geographical Service, followed by transformation to a regular 100 m resolution grid.
- *Slope* angle was calculated from the DEM by applying Sobel's algorithm[5].
- *Potential insolation.* A measure was derived following the method proposed by Fernández Cepedal and Felicísimo[6]. This used the DEM to assess the extent of topographical shading given the position of the sun at different standard date periods[7]. The result was an estimate of the time that each point on the terrain surface was directly illuminated by solar radiation. The temporal resolution was 20 minutes and the spatial resolution 100 m.
- *Temperature* maps of the annual maxima and minima were interpolated from data for 140 meteorological monitoring points (National Institute of Meteorology, Spain) using the thin-plate spline method[8,9] with a spatial resolution of 500 m.
- Quarterly *rainfall* maps were interpolated from data for 276 meteorological monitoring points (National Institute of Meteorology, Spain) using the thin-plate spline method with a 500 m spatial resolution.

These variables were selected because of their potential influence on the distribution of the vegetation and the availability of sufficient data to generate GIS digital layers. Lack of data eliminated other variables (e.g., soils) commonly used in ecological modelling.

7.5 METHODS

7.5.1 Statistical Methods

The methods used in predictive modelling are usually of two main types: global parametric and local non-parametric. Global parametric models adopt an approach where each entered predictor has a universal relationship with the response variable. An advantage of global parametric models, such as linear and logistic regression, is that they are easy and quick to compute, and their integration with a GIS is straightforward. As an example of such a model we used logistic multiple regression (LMR). This is widely employed in predictive modelling[10], but has several important limitations. For instance, ecologists frequently assume a response function which is unimodal and symmetric, yet this is often not justified[11,12].

An alternative hypothesis when modelling organism or community distributions is to assume that the response is related to predictor variables in a non-linear and local manner. Local non-parametric models are appropriate for such an approach since they use a strategy of local variable selection and reduction, and are flexible enough to allow non-linear relationships. Two examples of this type of model are CART (classification and regression trees) and MARS (multiple adaptive regression splines).

All three types of model used in this study were calculated from stratified random samples of pixels with an approximately even representation of points where each *Quercus* species was present or absent. Each random sample covered about 10-20% of the total area for each species. One sample was used to generate the models, and a second to test the reliability of the predictions.

7.5.1.1 Logistic Multiple Regression

Logistic multiple regression (LMR) has been used to generate likelihood models for forecasting in a variety of fields. It requires a dichotomous (presence/absence) dependent variable and the predicted probability of presence takes the form shown in Equation 7.1:

$$P(i) = 1 \,/\, 1+\exp[-(b_0 + b_1 \cdot x_1 + b_2 \cdot x_2 + \ldots + b_n \cdot x_n)] \quad (7.1)$$

where $P(i)$ is the probability of presence (e.g., for a tree species), $x_1 \ldots x_n$ represent the values of the independent variables, and $b_1 \ldots b_n$ the coefficients. The predicted values from the regression are probabilities which range from 0 to 1 and can be interpreted as measures of potential suitability[13]. Several studies have combined LMR with GIS tools to present such probabilities in cartographic form. For instance, Guisan et al.[14] used LMR in the ArcInfo GIS to generate a distribution model for the plant *Carex curvula* in the Swiss Alps. A similar study on aquatic vegetation was conducted by Van de Rijt et al.[15] using the GRASS GIS. In this study LMR was performed using a forward conditional stepwise method in SPSS® 11.5[16] and the results were then imported back into the ArcInfo® GIS[17] for mapping.

7.5.1.2 Classification and Regression Trees

CART is a rule-based method that generates a binary tree through 'binary recursive partitioning', a process that splits a node based on yes/no answers about the values of the predictors[2]. Each split is based on a single variable, and while some variables can be used several times in a model, others may not be used at all. The rule generated at each step minimizes the variability within each of the two resulting subsets. Applying CART often results in a complex tree of subsets based on a node purity criterion and subsequently this is usually 'pruned back' to avoid over-fitting via cross-validation.

The main drawback of CART models when used to predict organism distributions is that the generated models can be extremely complex and difficult to interpret. For example, work on Australian forests by Moore et al.[18] produced a tree with 510 nodes from just 10 predictors. In this study, the optimal tree generated from the *Quercus rotundifolia* data set had 4889 terminal nodes. Although the complexity of such a tree does not diminish its predictive power, it makes it almost impossible to interpret, which in many studies is a key requirement. Moreover, implementation of such an analysis within a GIS is difficult. Nevertheless, as part of this study we developed a method to translate the large CART reports (text files) to AML (Arc Macro Language) files that could be run with the ArcInfo GIS. Such files can be large (e.g., the text file containing the CART decision rules for constructing the *Q. rotundifolia* suitability map was 1.8 Mb in size) and execution times may be long (about 55 hours for the *Q. rotundifolia* model).

7.5.1.3 Multivariate Adaptive Regression Splines

MARS is a relatively novel technique that combines classical linear regression, mathematical construction of splines and binary recursive partitioning to produce a local model where relationships between response and predictors can be either linear or non-linear[3]. To do this, MARS approximates the underlying function through a set of adaptive piecewise linear regressions termed 'basis functions'. For example, the first four basis functions from the *Q. pyrenaica* model are:

$$\begin{aligned} BF1 &= MAX(0, PT4 - 3431) \\ BF2 &= MAX(0, 3431 - PT4) \\ BF3 &= MAX(0, MDE50 - 1181) \\ BF4 &= MAX(0, 1181 - MDE50) \end{aligned}$$

where PT4 is the mean rainfall for the period October-December ($l/m^2 * 10$) and MDE50 is elevation (m).

Changes in the slope of these basis functions occur at points called 'knots' (the values 3431 or 1181 in the above examples). Regression lines are allowed to bend at the knots, which mark the end of one region of data and the beginning of another with different functional behavior. Like the subdivisions in CART, knots are established in a forward/backward stepwise way. A model which clearly overfits is produced first and then those knots that contribute least to efficiency are discarded in a backwards-pruning step to avoid overfitting. The best model is selected via cross-validation, a process that applies a penalty to each term (i.e., a knot) added to the model in order to keep complexity as low as possible.

As in the CART analysis, we transformed the MARS text report files into AML and then generated the suitability models using the ArcInfo GIS.

7.5.2 Model Evaluation

The predictive capacity of a model can be evaluated as a function of the percentages of correct classifications, both for presences and absences (sensitivity and specificity parameters). The sensitivity and specificity of the model depend on the threshold or cut-off, which is set so as to classify each point according to its likelihood value.

To assess model performance we used the area under the Receiver Operating Characteristic (ROC) curve, particularly a measure commonly termed AUC[19]. The ROC curve is a plot of the relationship between sensitivity and specificity across all cut-off points of the model. We developed a method to construct the ROC curves by importing the databases associated with sample points into the SPSS statistical package. The ROC curve is recommended for comparing two-class classifiers, as it does not merely summarize performance at a single arbitrarily selected decision threshold, but across all possible decision thresholds[20,21]. AUC is a synthesized overall measure of model accuracy where 1 indicates a perfect fit and a value of 0.5 indicates that the model is performing no better than chance. AUC is also equivalent to the normalized Mann-Whitney two-sample statistic, which makes it comparable to the Wilcoxon statistic.

7.6 RESULTS

7.6.1 Suitability Models

All the LMR equations, MARS basis functions and CART classification rules were translated into ArcInfo GIS syntax. ArcInfo was subsequently used to generate the spatial suitability models, whose goodness-of-fit was evaluated by AUC values. Table 7.1 compares the overall results for different tree species and statistical methods, with bold text highlighting the best fitting models for each species. The AUC values indicate that the LMR models provided the poorest goodness-of-fit for each species, while the CART ones were the best performers. However, there were some differences between tree species with a relatively narrow range of AUC values for *Q. pyrenaica* (i.e., all the methods produce a good fit) and a much greater one in the *Q. rotundifolia* case. This may be related to differences in the current extent of the species (see Section 7.3) with *Q. rotundifolia* being the most common and therefore having potentially more complex environmental relationships. It is also worth noting that greater complexity (number of terminal nodes) in the CART models does not guarantee better results. This is an interesting finding that could assist in the practicalities of implementing such models within a GIS framework.

Table 7.1 Summary statistics for the suitability models

***Quercus* Species**	**Method**	**Terminal Nodes**	**AUC**	**Confidence Interval (95%)**
Q. pyrenaica, Pyrenean oak	LMR	Not Applicable	0.924	Not Available
Sample Size	MARS	Not Applicable	0.972	0.970-0.974
18,880 positive cases	CART	56	0.970	0.968-0.972
18,590 negative cases	CART	102	0.974	0.972-0.976
	CART	204	**0.979**	0.977-0.981
	CART	817	0.974	0.972-0.976
Q. suber, cork oak	RLM	Not Applicable	0.790	Not Available
Sample Size	MARS	Not Applicable	0.802	0.799-0.805
42,040 positive cases	CART	525	0.971	0.970-0.972
41,979 negative cases	CART	1016	**0.975**	0.974-0.977
	CART	2355	**0.975**	0.973-0.976
Q. rotundifolia, holm oak	RLM	Not Applicable	0.627	Not Available
Sample Size	MARS	Not Applicable	0.767	0.764-0.770
50,394 positive cases	CART	1343	0.889	0.887-0.891
50,690 negative cases	CART	2347	**0.894**	0.892-0.896
	CART	4889	**0.895**	0.893-0.897

Another feature of the CART model output became apparent when the results were converted into suitability maps. As is illustrated in Figure 7.4a the CART maps show abrupt transitions between areas of high and low suitability (darker and lighter shading respectively) which reflects the reliance on binary rules. In addition, due to the influence of climate variables, the suitability models frequently replicate the shapes of isopleths, which makes them visually less convincing. Although the backward pruning process in CART reduces the number of terminal nodes and makes the final model less complex, it does not eliminate such effects. These features are not present in the MARS-based maps (Figures 7.4b-7.4d) which show more smoothed and continuous distributions of suitability values. For this reason, we decided to use the MARS model output to generate a potential vegetation distribution.

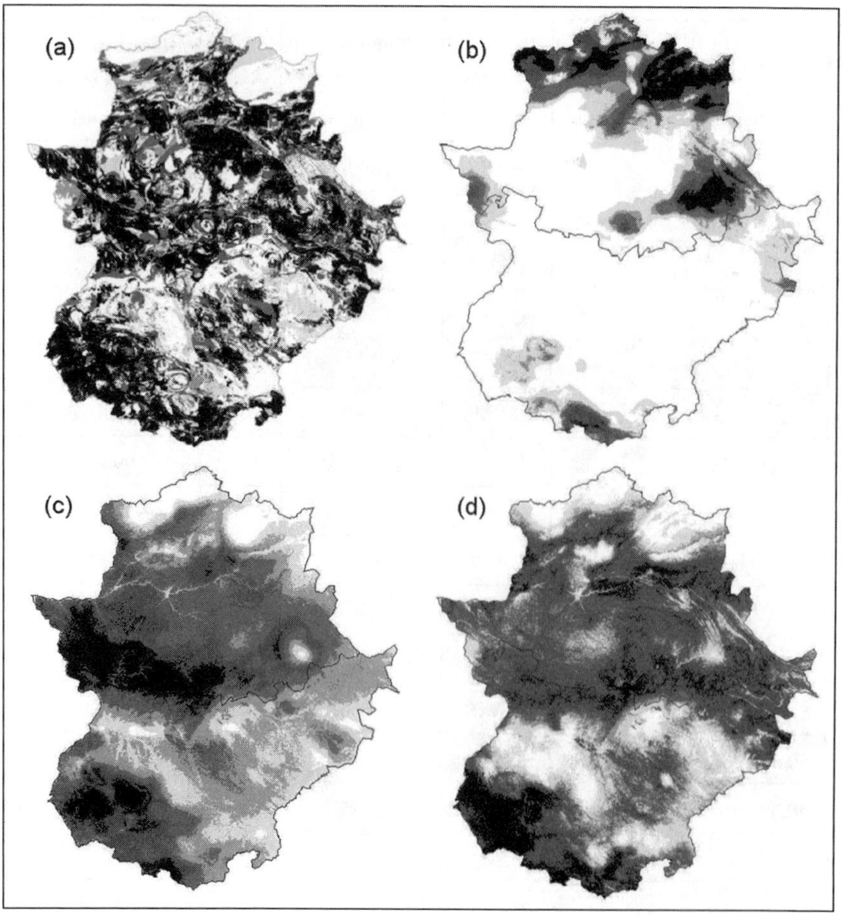

Figure 7.4 Suitability models: a) CART model for *Q. rotundifolia*, b) MARS model for *Q. pyrenaica*, c) MARS model for *Q. suber*, d) MARS model for *Q. rotundifolia*. Darker shading indicates higher suitability.

7.6.2 Potential Vegetation Model

Suitability models for the three tree species were combined to generate a potential vegetation distribution map that could be used to inform land management and decision-making. This map was generated through a decision rule that took into account both suitability values as well as proximity to the current presence of forests. We defined a function where, for each cell, the suitability value for each species was corrected by the inverse of the distance to the closest cell where the species currently grows. This correction can be considered as a coarse indicator of colonization likelihood. The result of these calculations was a model showing, for each cell, the type of forest with the highest potential value after considering colonization processes. Figure 7.5 shows the result, highlighting relatively clustered regions for *Q. pyrenaica* amidst more dispersed distributions for the other two *Quercus* species.

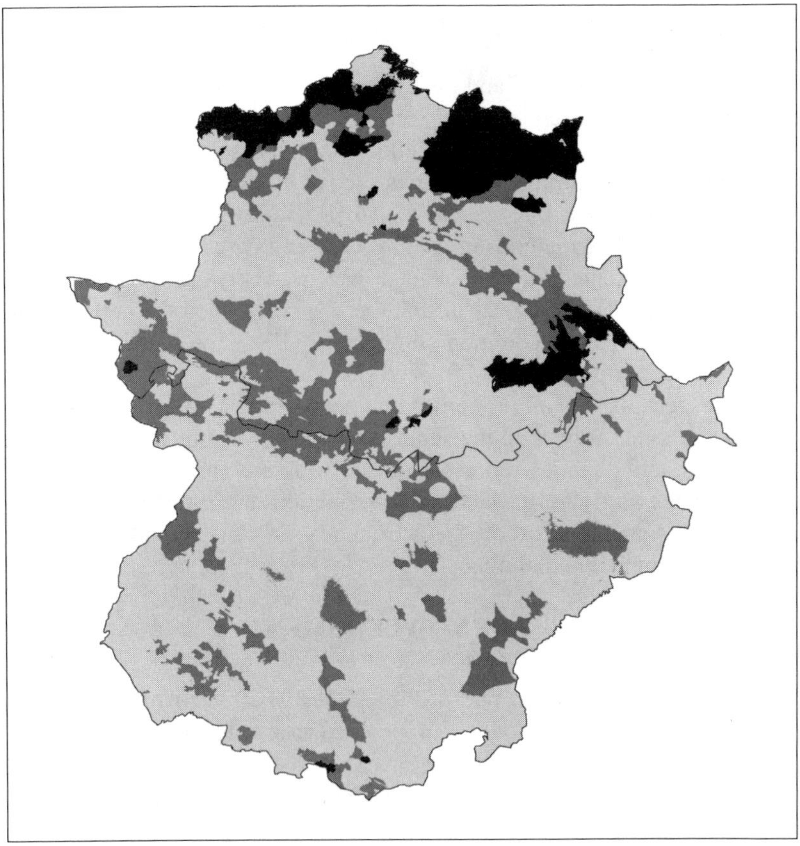

Figure 7.5 Potential distribution model of *Quercus* species in Extremadura; *Q. pyrenaica* (black), *Q. suber* (dark gray), *Q. rotundifolia* (pale gray).

7.7 CONCLUDING DISCUSSION

Suitability maps represent a useful tool for environmental management as they synthesize a wide range of knowledge which is difficult to integrate in any other way. Until recently, most potential vegetation maps were developed by largely subjective methods, usually by an 'expert'. In contrast, the approach used in this study is based on robust statistical or GIS operations, and objective cartographical information. There is an explicit procedure to produce the final result and the entire workflow of information is transparent and repeatable.

The models used are based on real data (data driven) and in our experience these methods give good results in mountainous areas because the limiting factors are mainly physical: elevation, potential insolation, slope, etc. Data on such variables are generally available (e.g., elevation) or can be derived with sufficient accuracy (e.g., potential insolation). However, the choice of statistical methods to employ can be very important.

Our results (see also Muñoz and Felicísimo[4]) show that spatial distributions can be better defined if we accept that they may follow non-linear patterns. LMR has been widely used in predictive modelling to successfully predict organism/ community distributions despite drawbacks such as an inability to deal with skewed or multi-modal responses, but we have also provided evidence that CART and MARS are very effective methods in the most difficult cases.

The analysis presented in this chapter has used various procedures to link statistical tools and GIS, but it is clear that there is still a need for better integration of such capabilities in most common commercial GIS. Transferring the potential vegetation model into practical forestry action would also require further information, especially on soil properties and economic factors. Dealing with such implementation issues is beyond the scope of this chapter, but it is evident that the generated maps and statistics represent data of obvious utility. Combining such model-based maps with current land-use information and management data could help provide decision support tools that would be extremely useful in many aspects of spatial or environmental planning.

7.8 ACKNOWLEDGMENTS

This study was conducted as part of Project 2PR01C023 co-funded by the Junta de Extremadura and FEDER (Fondo Europeo de Desarrollo Regional).

7.9 REFERENCES

1. Guisan, A. and Zimmermann, N.E., Predictive habitat distribution models in ecology, *Ecological Modelling*, 135, 147-186, 2000.

2. Breiman, L., Friedman, F., Olshen, R., and Stone, C., *Classification and Regression Trees*, Wadsworth, Monterey, 1984.

3. Friedman, J.H., Multivariate adaptive regression splines, *Annals of Statistics*, 19, 1-141, 1991.

4. Muñoz, J. and Felicísimo, A.M., Comparison of statistical methods commonly used in predictive modeling, *Journal of Vegetation Science*, 15, 285-292, 2004.

5. James, M., *Pattern Recognition*, John Wiley, New York, 1988.

6. Fernández Cepedal, G. and Felicísimo, A.M., Método de cálculo de la radiación solar incidente en áreas con apantallamiento topográfico, *Revista de Biología de la Universidad de Oviedo*, 5, 109-119, 1987.

7. Heywood, H., Standard date periods with declination limits, *Nature*, 204, 678, 1964.

8. Hutchinson, M.F., Climatic analyses in data sparse regions, in *Climate Risk in Crop Production. Models and Management for the Semiarid Tropics and Subtropics*, Chow, M.C. and Bellamy. J.A., Eds., CAB International, Wallingford, 1991, 55-71.

9. Lennon, J.J. and Turner, J.R., Predicting the spatial distribution of climate temperature in Great Britain, *Journal of Animal Ecology*, 64, 370-392, 1995.

10. Felicísimo, A.M., Francés, E., Fernández, J.M., González-Díez, A., and Varas, J., Modeling the potential distribution of forests with a GIS, *Photogrammetric Engineering & Remote Sensing*, 68, 455-461, 2002.

11. Austin, M.P. and Smith, T.M., A new model for the continuum concept, *Vegetatio*, 83, 35-47, 1989.

12. Yee, T.W. and Mitchell, N.D., Generalized additive models in plant ecology, *Journal of Vegetation Science*, 2, 587-602, 1991.

13. Jongman, R.H.G., Ter Braak, C.J.F., and van Tongeren, O.F.R., *Data Analysis in Community and Landscape Ecology*, Cambridge University Press, Cambridge, 1995.

14. Guisan, A., Theurillat, J.-P., and Kienast, F., Predicting the potential distribution of plant species in an alpine environment, *Journal of Vegetation Science*, 9, 65-74, 1998.

15. Van de Rijt, C.W.C.J., Hazelhoff, L., and Blom, C.W.P.M., Vegetation zonation in a former tidal area: A vegetation-type response model based on DCA and logistic regression using GIS, *Journal of Vegetation Science*, 7, 5005-518, 1996.

16. SPSS, SPSS 11.5, SPSS Inc, http://www.spss.com, 2004.

17. ESRI, ArcInfo Desktop, Environmental Systems Research Institute, http://www.esri.com, 2004.

18. Moore, D.M., Lee, B.G., and Davey, S.M., A new method for predicting vegetation distributions using decision tree analysis in a geographic information system, *Environmental Management*, 15, 59-71, 1991.

19. Hanley, J.A. and McNeil, B.J., The meaning and use of the area under a Receiver Operating Characteristic (ROC) curve, *Radiology*, 143, 29-36, 1982.

20. Fielding, A.H. and Bell, J.F., A review of methods for the assessment of prediction errors in conservation presence/absence models, *Environmental Conservation*, 24, 38-49, 1997

21. Ivers, R.Q., Macaskill, P., Cumming, R.G., and Mitchell, P., Sensitivity and specificity of tests to detect eye disease in an older population, *Ophthalmology*, 108, 968-975, 2001.

CHAPTER 8

A Comparison of Two Techniques for Local Land-Use Change Simulation in the Swiss Mountain Area

A. Walz, P. Bebi and R. Purves

8.1 INTRODUCTION

Mountain landscapes and ecosystems are considered highly fragile and vulnerable to environmental change[1,2]. These changes are driven by a combination of natural processes, such as ecological disturbances through avalanches or wild fires, and anthropogenic forces such as changes in economic policy or development[3]. One of the most prominent interfaces between human activities and environmental change is land-use. In the Alps, the socially, economically and politically driven transformation of land-use has exceeded natural changes both in amplitude and rate of change[4].

The Swiss Mountain Area has experienced major transitions in land-use during the last 50 years[5]. For instance, between the last two official land-use surveys (1979/85 and 1992/97) an increase in housing and infrastructure of 14% was recorded[6]. However, compared to the densely populated zone between Zurich and Geneva the Swiss Mountain Area has a low population density (for example, in the Canton of Zurich the population density is 711 people/km^2, while in the Canton of Grisons it is 26 people/km^2)[7]. Nevertheless, because of the mountainous relief and the consequent shortage of level ground, these rural areas have seen conflicts between proposed developments (often connected to economic interests in tourism) and mountain agriculture and related issues of scenic attractiveness, which are often considered as major resources for tourism. Thus on-going building development is a key political and environmental issue for the Swiss Mountain Area. Particular attention has focused on how development driven by tourism may in turn reduce future demand for tourism, for example through changes in traditional agricultural practices and corresponding reductions in the scenic quality of land or workers moving into service industries and land reverting to forest.

Future regional development scenarios[8,9] and integrated regional modelling[10] provide tools to predict future land demands in an absolute sense (i.e., what proportion of agricultural land might be built upon), but in order to answer questions where the geographical distribution is important, spatially explicit land-use simulation is required. Figure 8.1 shows schematically a set of conceptual links to model not only how much of a particular land-use class exists, but also where that land is located.

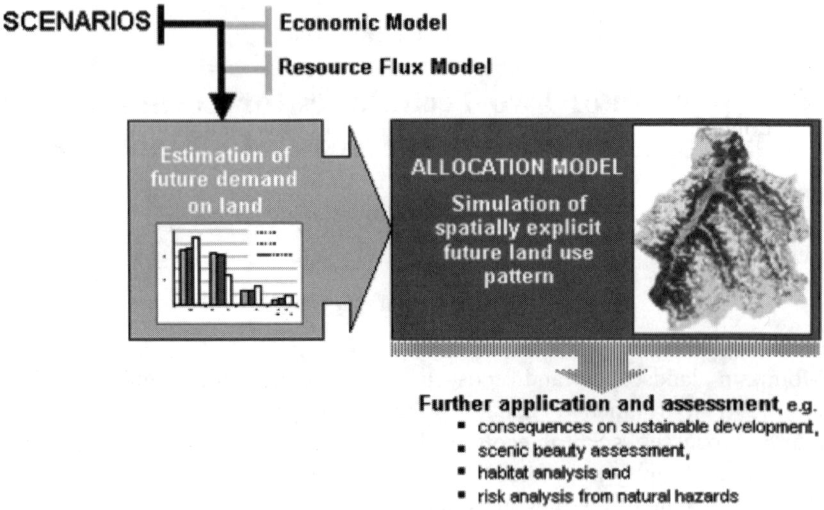

Figure 8.1 The role of spatially explicit land-use simulation within an integrated regional modelling approach. Based on future scenarios and the outcome of economic and resource flux models, demand for different land-uses is estimated. In a second step the spatially explicit land-use pattern is simulated. Simulation outcomes provide the base for further assessment and derivation of sustainability indicators.

While scenarios can represent the external drivers on a given region, a spatially explicit land-use simulation displays local consequences and further supports the decision-making process. In this chapter we focus on modelling at the local scale, because this is the level at which control of development and planning is carried out in most European countries.

8.1.1 Approaches to Spatially Explicit Land-Use Modelling

Land-use has been studied in a wide variety of contexts, including projects that have modelled future distributions for a variety of purposes[11]. In previous studies local-scale land-use modelling has mostly focused on the urban development of large cities, such as San Francisco Bay[12] , Toronto[13] or Dortmund[14], and often with a strong emphasis on traffic planning or neighborhood analysis[15-17]. These studies have generally concentrated on highly differentiated classes of urban area, while agricultural land-use and natural areas are neglected.

A variety of models have examined the interaction of different agricultural options. With exceptions[18,19], these models have usually operated on a macroeconomic level, such as at nationwide scales[20]. More recently, land-use allocation models have been developed that focus on rural areas[21-22] or the expansion of cities into the agricultural hinterland[23]. These models are based on three different approaches. While Verburg et al.[22] calculate transition probabilities for all interacting land-use classes through logistic regression, Fritsch[21] deduces 'suitability' for housing and infrastructure development using a multi-criteria evaluation and Loibl and Toetzer[23] work with an agent-based system.

There is a lack of previous land-use modelling research that focuses on the special conditions of rural communities in high mountains. In our study we concentrate on housing and infrastructure development because of its relevance to economically-orientated future scenarios and its strong impact on the fragile mountain ecosystem. Because of the limited amount of data due to the small size of area and low population density, we do not differentiate between various urban land-use types, but aggregate built-up areas, such as buildings and traffic infrastructure, and highly modified, non-agricultural open areas, such as gardens, parks or sports facilities. We also address the problem of small sample sizes in local land-use modelling and assess whether it is valid to use data from a wider area to overcome this problem. To investigate this question we present locally and regionally based strategies. While the first of these concentrates on data from only within the target region of Davos, Switzerland, the second is based on a statistical analysis from the entire Swiss Mountain Area. Firstly, we look at a modelling approach that does not require large amounts of data to deduce systematic rules, namely the use of transition probability matrixes (TPM) which can then be improved by additional rules[24]. Secondly, we use a regression model based on the entire Swiss Mountain Area, where an underlying assumption is that transition processes are similar to the target region and the resulting number of observations is sufficiently large for statistical analysis. Based on these techniques two models have been implemented and simulation results for a validation period compared.

8.2 STUDY AREA AND DATA

8.2.1 Study Area

Davos is a mountain resort town at 1560 m above sea level in the canton of Grisons, eastern Switzerland, with a population of around 11,000 permanent residents and up to 28,000 tourists during the high season in winter. The total area of Davos is about 25,450 ha, making it one of the most extensive communities in Switzerland. The central part of the main valley hosts the core settlement with a well-established urban and tourist infrastructure, while the three major side valleys and other parts of the main valley have remained relatively rural with a few small, mostly scattered settlements and a landscape still strongly dominated by mountain agriculture.

After an initial construction boom during the period of health tourism in Davos from 1880 to the late 1920s, when the core urban area developed from a number of dispersed agricultural settlements, a second peak was reached in the 1960s and 1970s when winter tourism became established. Although the proportion of built-up area is still low (~2.1% of the total) construction activity is a major concern to local stakeholders. Firstly, the establishment of higher tourist capacity that is fully utilized only in the most popular skiing season conflicts strongly with attempts to introduce more sustainable regional development. Secondly, the most attractive areas for construction often coincide with the most productive land for agriculture. Consequently the sales of this land constitute a long-term threat to the future of mountain agriculture and its associated landscape. Finally, construction work itself impacts on the appearance of the alpine landscape, which is considered a major resource for the local tourism industry. Despite these concerns the area devoted to housing and infrastructure in Davos increased from 512 ha to 566 ha (12.3%) between 1985 and 1997.

8.2.2 Data

Data were required at both the local and national scale for the methodology described in Section 8.3. The primary data set used in calculating land-use changes was the official area statistics of the Swiss Federal Statistical Office[25]. In Switzerland land-use has been surveyed since the 1940s, but only the last two survey periods from 1979-85 (ASCH79/85) and 1992-97 (ASCH92/97) are methodologically comparable. Land-cover/land-use data are extracted at 100 m intervals from a lattice overlain on aerial photographs for the whole of Switzerland. Using this sampling interval the municipality of Davos is covered by some 25,450 points, compared to about 2.5 million points for the whole Swiss Mountain Area. In the original survey 74 land-cover and land-use classes were differentiated which we aggregated into ten and five broader categories for the purposes of our study (Table 8.1).

Table 8.1 Aggregations of land-use classes in this study

<table>
<tr><th>10 Classes</th><th>5 Classes</th></tr>
<tr><td>Forest</td><td rowspan="3">Forest</td></tr>
<tr><td>Open Forest</td></tr>
<tr><td>Shrub</td></tr>
<tr><td>Meadows</td><td rowspan="3">Agriculture</td></tr>
<tr><td>Pasture</td></tr>
<tr><td>Other Agriculture</td></tr>
<tr><td>Unproductive Grassland</td><td rowspan="2">Unproductive Land</td></tr>
<tr><td>Bare Land</td></tr>
<tr><td>Housing and Infrastructure</td><td>Housing and Infrastructure</td></tr>
<tr><td>Water</td><td>Water</td></tr>
</table>

In addition to ASCH79/85 and ASCH92/97, a third dataset was reconstructed from orthorectified aerial photographs taken in 1954 by applying an identical survey methodology (ASDav54). The photos covered about 40% of the Davos area, embracing large parts of the main valley (including the core settlement) and much of the three side valleys, i.e., the most likely areas for new housing and infrastructure development. Other data used in the study allowed the derivation of a range of topographic and distance measures used in stratifying the data and developing regressions (see Section 8.3.3).

8.3 METHODOLOGY

8.3.1 Overview

Two modelling approaches were implemented based on transition probability matrices and regression. The transition probability matrix used only local land-use data from the Davos area (Figure 8.2a). For the regression-based approach a much larger sample, from the whole Swiss Mountain Area (Figure 8.2b), was used to derive regression relationships for potential changes in land-use. Figure 8.2c shows the transitions from different land-use types to housing and infrastructure between the ASCH79/85 and ASCH92/97 datasets, and underpins the assumption adopted in the regression-based approach – namely that changes in the whole Swiss Mountain Area are similar to those in Davos.

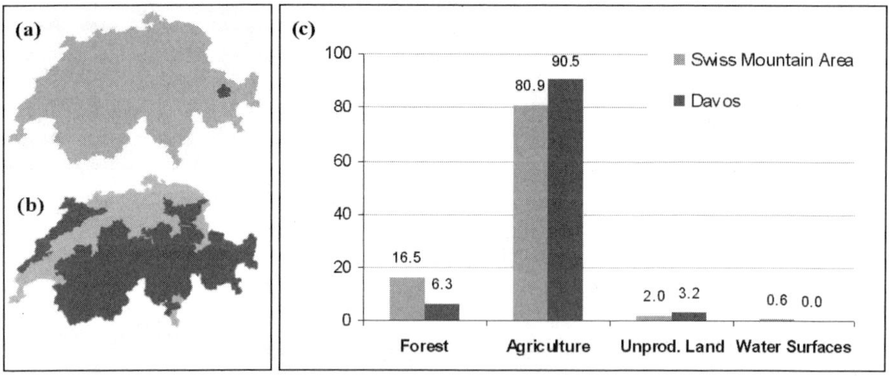

Figure 8.2 (a) Location of community of Davos within Switzerland. (b) The extent of the Swiss Mountain Area. (c) Sample points that underwent housing and infrastructure development between ASCH79/85 and ASCH92/97 classified according to their previous land-use.

8.3.2 TPM-Based Approach

Transition probability matrices (TPM) are based on rates of change from one land-use class to another during an observation period[24]. In the case of pixel-based spatial modelling, the principal idea behind TPMs is that the existing state of a pixel has a strong influence on the future state. Neither geographical position nor any further qualities of the location are taken into account in an unmodified TPM. The TPM, derived from ASCH79/85 and ASCH92/97 for Davos, is shown in Table 8.2. The matrix gives the probability that a pixel of any particular land-use class in ASCH79/85 changed to another land-use class in ASCH92/97 (or remained in the same land-use class). Thus, for instance, any pixel representing agriculture in ASCH79/85 had a probability of 0.59% of changing to housing and infrastructure in the dataset ASCH92/97. Note that only 63 pixels out of some 25,450 changed their land-use to housing and infrastructure between these two datasets.

Table 8.2 Transition matrix for Davos showing absolute numbers of changes and derived probabilities

		Land-Use Class ASCH92/97					
		Forest	Agriculture	Unproductive Land	Housing and Infrastructure	Water	Total
Land-Use Class ASCH79/85	Forest	616 99.68%	2 0.03%	14 0.23%	4 0.07%	0 0%	6181 100%
	Agriculture	190 1.96%	9427 97.34%	11 0.11%	57 0.59%	0 0%	9685 100%
	Unproductive Land	26 0.29 %	6 0.07%	8808 99.62%	2 0.02%	0 0%	8842 100%
	Housing and Infrastructure	1 0.20%	8 1.56%	0 0%	503 98.24%	0 0%	512 100%
	Water	1 0.45%	0 0%	1 0.45%	0 0%	221 99.10%	223 100%
	Total	6379	9443	8834	566	221	25443

8.3.2.1 Stratification of the Dataset by Classification Tree Analysis

Pixels which had changed between the ASCH79/85 and ASCH92/97 datasets were stratified into subsets through a classification tree analysis (Figure 8.3). The variables used to identify efficient stratification criteria were defined through analysis of the properties of individual pixels carried out with supporting spatial datasets (Table 8.3). Cross-validation indicated that the optimal classification results were achieved with four terminal nodes. The criteria identified as being most important for stratifying the dataset were based on elevation and the distance to roads. After the dataset had been stratified, transition probability matrices were calculated for each of the resulting subsets.

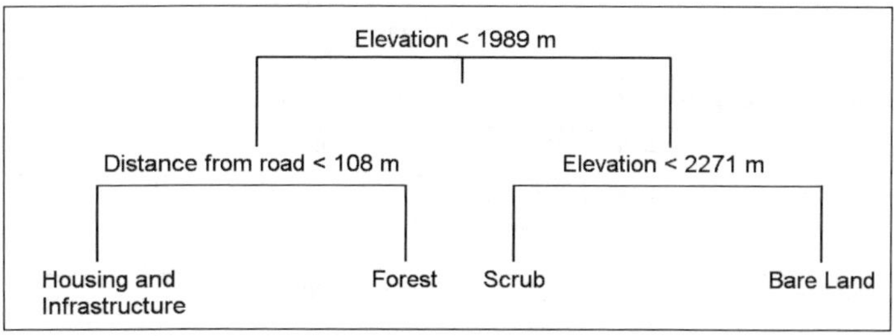

Figure 8.3 Key stratification variables and resulting subsets in the classification tree analysis of sample points in Davos which recorded land-use changes between ASCH79/85 and ASCH92/97.

8.3.2.2 Incorporation of Neighborhoods

Further modification of the basic TPM took into account the neighboring cells of sample points with changes between ASCH79/85 and ASCH92/97, and thus introduced consideration of spatial patterns into the model. This analysis demonstrated that the land-use classes of neighboring point samples had a strong influence on the future land-use at a given location. Figure 8.4 shows that highest percentages of sample sites changing to housing and infrastructure were found amongst points with 7 neighbors in Davos and 5-6 neighbors in the whole Swiss Mountain Area.

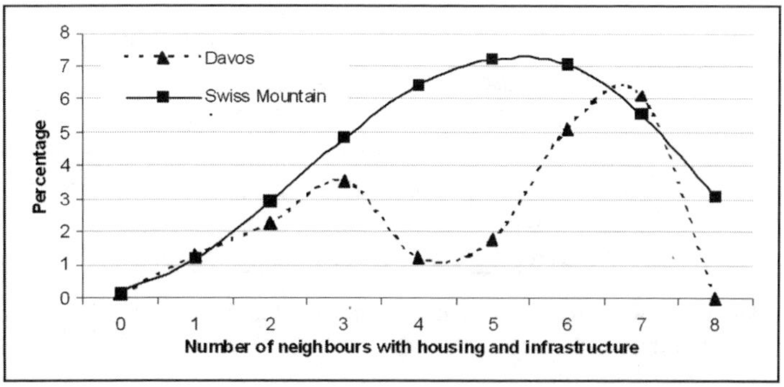

Figure 8.4 Percentage of sample points that changed to housing and infrastructure grouped by the number of neighbors that already had that classification.

Table 8.3 Variables used in the classification tree and logistic regression analyses

Variable Type	Classification Tree	Logistic Regression
Topography	Elevation	Elevation
	Slope	Slope
	Aspect	Aspect
	Curvature	Curvature
Distance measures	Distance to tarmac road	Distance to tarmac road
	Distance to river	Distance to river
	Distance to urban center	Distance to urban center
	Distance to valley floor	Distance to valley floor
Previous land-use at different levels of aggregation	1	Forest
	1	Agriculture
	1	Housing and infrastructure
	1	Unproductive land
	1	Water surface
Neighborhood in 3*3 window	2	No. of forest 3*3
	2	No. of agriculture 3*3
	2	No. of housing and infrastructure 3*3
	2	No. of unproductive land 3*3
	2	No. of water surfaces 3*3
Neighborhood in 5*5 window	2	No. of forest 5*5
	2	No. of agriculture 5*5
	2	No. of housing and infrastructure 5*5
	2	No. of unproductive land 5*5
	2	No. of water surfaces 5*5
Other	Soil characteristics	Soil characteristics
	Climatic suitability	Climatic suitability
	Geology	Geology
Dependent	Land-use class in ASCH92/97	Whether land-use changed to housing and infrastructure within the study period

Note: Variables marked with [1] were used in the basic TPM and those denoted [2] during the incorporation of a neighborhood based transition probability.

A transition probability $P_{n(L,N)}$ based on neighborhood conditions is derived from the number of points transformed to the land-use class L expressed as a proportion of the total sample points with the number of neighbors N of the land-use class L (Equation 8.1). To increase the influence of neighboring points with the relevant land-use class L, the derived probability can be transformed via an exponential function (Equation 8.2).

$$P_{n(L,N)} = A_{(L,N)} / T_{(L,N)} \quad (8.1)$$

$$P_{neigh(L,N)} = \exp(k * P_{1(L,N)}) \quad (8.2)$$

where:

$A_{(L,N)}$ = Number of points transformed to a new land-use class L and with a particular number of neighbors N of this new land-use class L

$T_{(L,N)}$ = Total number of points with N neighbors of land-use class L

L = Land-use class

N = Number of neighbors of land-use class L

k = Constant calibration factor

8.3.2.3 Simulation

For each set of possible transitions in the TPM (e.g., from agriculture to housing and infrastructure) a point with the respective land-use class (e.g., agriculture) was randomly selected. The site qualities of the point were then retrieved (e.g., distance to road, elevation, etc.) and the point associated with a specific subset (and related TPM) of the dataset. Subsequently, the properties of the neighboring pixels were examined and a transition probability derived according to Equation 8.2. A random number between 0 and 1 was then calculated, and if this number was less than $P_{neigh(L,N)}$, the transition was performed. This process was repeated until either the required number of transitions for all elements of the TPM had been performed or 300 'failed' transitions had occurred.

In order to model the impacts of a future scenario predicted transitions for all land-use classes are required. This requirement is a severe limitation, since while an economic model may provide us with input describing future demand for housing and infrastructure, it is unlikely to model the interaction between all land-use classes. Therefore, TPMs are better suited to trend-based scenarios, where it is assumed that current demand will continue for some time into the future.

8.3.3 Regression-Based Approach

A second modelling approach sought to overcome data density problems by using regression analysis to determine the probability of individual points being transformed into housing and infrastructure. Since the dataset for Davos included only 63 of these transformations within the observation period, information for the whole Swiss Mountain Area was used to perform the analysis. The increase in housing and infrastructure in this area was about 17% between ASCH79/85 and ASCH92/97, while it was 12% for Davos. However, as shown in Figure 8.2c, there were broad similarities in the profiles of previous land-uses that changed to housing and infrastructure, though agriculture was slightly more important in Davos and forest in the Swiss Mountain Area as a whole.

8.3.3.1 Sub-Sampling of Data

The ASCH encompasses 2.5 million sample points for the mountain area. Preliminary analyses identified the areas where 90% of new housing and infrastructure development occurred between ASCH79/85 and ASCH92/97. According to these assessments, sample points with elevations of < 2000 m, slopes of < 30° and a maximum distance of 1 km from the nearest road covered 90% of such changes. Subsequently an area of interest was defined on the basis of these criteria from which sub-sampling took place. This area was further sub-divided into points where new development had or had not occurred and some 700 of each category were randomly selected, reducing the impacts of spatial auto-correlation.

8.3.3.2 Variables

For each of the sub-sample points a similar set of variables was extracted from the GIS as for the classification tree analysis used in the TPM approach. Besides topographic data, distance measures and environmental parameters, previous land-use and characteristics of the sample point's neighborhood were included in the dataset (Table 8.3). A univariate assessment of power to predict development for housing and infrastructure was carried out, and only those variables with a p-value < 0.05 were included in the subsequent analysis.

8.3.3.3 Logistic Regression

Stepwise logistic regression was carried out to determine the factors that best explain whether or not a point in the sub-sample underwent housing and infrastructure development during the observation period. Assuming that site characteristics will remain similar in the future, suitability for housing and infrastructure development can be deduced for individual sites on the basis of Equation 8.3 with P_i representing the probability of transition.

$$\text{Log}\,(P_i/1\text{-}P_i) = \beta_0 + \beta_{1,i} * X_{1,i} + \beta_{2,i} * X_{2,i} + \ldots + \beta_{n,i} * X_{n,i} \qquad (8.3)$$

8.3.3.4 Simulation

To perform a simulation of future land-use change using the output of a regression model, only the desired amount of change in the class being modelled (in this case housing and infrastructure) is required. The derived annual rate of change is then distributed by initially choosing the pixel with the highest P_i-value. In a similar manner to the TPM-based model a random number is then drawn and used to decide whether a transition is accepted or not. If the random number is above the regression-derived transition probability, the transition is not accepted and the pixel with the next highest P_i is checked. When the required area of change is reached or all possible pixels have been declined, the simulation stops.

8.3.4 Implementation

The TPM-based model was implemented in VBA with ArcGIS® 8.3[26]. For the regression-based model the data were extracted from the GIS environment and processed in a bespoke C++ program.

8.3.5 Validation of the Modelling Results

The simulation models were validated with data for Davos over the time period from 1954 to 1985 (i.e., the same area but a different time slice to the observation period). This type of comparison was particularly appropriate for the TPM-based model which cannot be easily transferred to other areas. The simulations used the dataset compiled from 1954 aerial photographs as a starting point, along with rates of change extracted directly from the TPM for the first model and the observed number of transitions to housing and infrastructure in the regression-based one.

The simulation results for the validation period from 1954 to 1985 were compared with the observed data from ASCH79/85. These comparisons were undertaken on both a strict pixel-to-pixel basis and with a fuzzy assessment technique which accounted for some spatial uncertainty by matching pixels with neighbors in a 5*5 moving window. As matches were found, they were removed from the dataset in an iterative procedure, ensuring that no double-matching was permitted.

Contingency tables[27] were calculated for both validation techniques and Cohen's Kappa Index[28] used to assess the goodness of fit of the simulated housing and infrastructure allocation and to allow comparison of the modelling approaches.

8.4 RESULTS

8.4.1 Key Factors Influencing Housing and Infrastructure Development

The classification tree used to produce subsets for the TPM analysis identified 'elevation' and 'distance to road' as two key influences on land-use changes in the Davos area. As shown in Figure 8.3, three elevation categories were differentiated. The lowest of these (< 1989 m) was characterized by a mixture of housing and infrastructure, meadows and forest with an increase in housing and infrastructure and forested areas, the second (1989 < 2271 m) was dominated by grassland and increasing shrub, and the third (2271 m and above) by increasing bare land. The second classification criterion 'distance to road' introduced a distinction between the lowest areas with increasing housing and infrastructure or forested land. Some 75% of the transitions to housing and infrastructure during the observation period occurred at points within the lowest elevation band and up to 100 m from roads.

Several different sets of variables were tested in the logistic regression analysis to help avoid high inter-correlations between predictors. The final set of variables used is shown in Table 8.4. In accordance with the classification tree analysis, several measures of topography and accessibility were found to be highly significant predictors. The results also indicated that while the previous land-use class was not an important factor, the land-uses in the neighborhood of a sample point consistently were. In other words, the location of a site was more important than what existed there in determining any likely change in land-use. This finding also had parallels in the importance of neighborhood conditions in the TPM-based modelling (e.g., Figure 8.4).

Table 8.4 Predictors of built-up area in the Swiss mountain region

Predictors	**p-value**	**Coefficient (β_i)**
(Intercept)		4.6
Distance to road	< 0.001	-3.2×10^{-4}
Distance to valley floor	< 0.001	-6×10^{-5}
Elevation	< 0.001	-3.8×10^{-6}
Slope	< 0.05	-2×10^{-2}
Previous land-use: agriculture	Not significant	
Previous land-use: forest	Not significant	
Previous land-use: unproductive land	Not significant	
Neighboring cells: forest	< 0.001	-2×10^{-1}
Neighboring cells: agriculture	< 0.001	-1.8×10^{-1}
Neighboring cells: unproductive land	< 0.001	-2.6×10^{-1}
Neighboring cells : housing and infrastructure	< 0.01	2.3×10^{-1}

8.4.2 Simulation Outcomes and Validation

For validation we used the reconstructed dataset of 1954 to start both models and then ran the simulations until 1985. The distributions of housing and infrastructure land in ASDav54, ASCH79/85 and the two simulations for 1985 are displayed in Figure 8.5. Contingency tables and associated Kappa Indices for these results are given in Table 8.5. Overall there are negligible differences in the Kappa Indices for the two modelling approaches, with values for the exact comparison and moving window techniques of 0.72 and 0.83 respectively for the TPM simulation and 0.73 and 0.81 for the regression-based one.

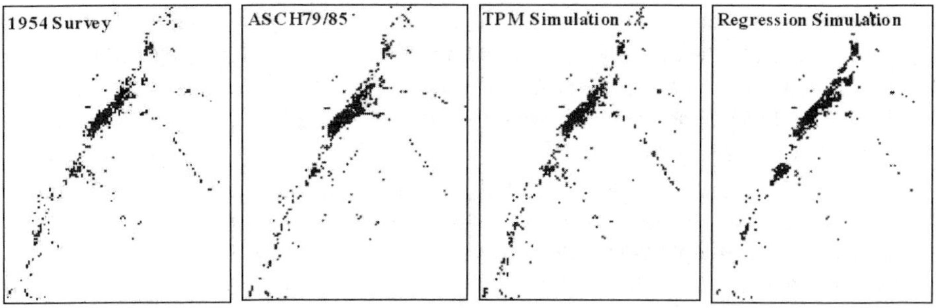

Figure 8.5 Distributions of housing and infrastructure land in different survey and simulation sources.

Table 8.5 Comparisons of simulation results and 1985 survey data: (a) and (b) show exact comparisons and (c) and (d) moving window results

(a) $\kappa_a = 0.72$

Exact Comparison		**1985 Survey**	
		H & I	Other
TPM Simulation	H & I	9901	125
	Other	127	347

(b) $\kappa_b = 0.73$

Exact Comparison		**1985 Survey**	
		H & I	Other
Regression Simulation	H & I	9900	127
	Other	128	345

(c) $\kappa_c = 0.83$

Moving Window		**1985 Survey**	
		H & I	Other
TPM Simulation	H & I	9901	69
	Other	90	403

(d) $\kappa_d = 0.81$

Moving Window		**1985 Survey**	
		H & I	Other
Regression Simulation	H & I	9900	99
	Other	84	418

8.5 DISCUSSION

8.5.1 Influences on Location of Housing and Infrastructure

Both the Davos and Swiss Mountain Area datasets suggest that similar variables (local topography and the road network) are key influences on the development of housing and infrastructure in the alpine landscape. Elevation was the most important variable for the classification of the Davos dataset. In the entire mountain area elevation also occupied a prime role in the regression, but local hypsometry determined likely areas for development. Distance to infrastructure (in the form of roads) was a strong second predictor of development both in the stratification of the TPM and the regression model.

While the TPM-based model is predicated on the idea that previous land-use is a strong determinant when modelling land-use change in alpine regions, such variables were not significant in the regression model. This suggests that development in Alpine regions is more strongly influenced by the location of a possible site, e.g., with respect to utilities and avalanche paths, than by existing land-use type. In fact, such findings emphasizing the importance of location over previous land-use are also seen in models of urban growth; for instance Clarke et al.[12] concluded that the most significant predictors for historical development in San Francisco were distance to urban area, distance to roads and slope angle.

Neighborhood variables that were strongly significant in the regression analysis were also used to enhance the probability of development in the TPM, therefore allowing some representation of the importance of location in this approach.

8.5.2 Simulation Results

A comparison between the two simulation results and the observed ASCH79/85 distribution indicates some discrepancies which can be explained by the specific characteristics of each approach, but also highlight some general limitations of land-use modelling. Global measures of fit in the form of Kappa Indices are very similar for both models. Utilizing a moving-window approach improves the Kappa value by a similar amount in both cases, suggesting that the spatial uncertainty in both models is comparable. In other words, substantial numbers of non-matched points occurred close to observed changes in land-use.

The main difference between the TPM and regression-based approaches is the intensity of housing and infrastructure development in the rural side valleys, where over-estimation occurred with the TPM-based model and under-estimation in the regression-based one (Figure 8.6). The regression-based model selects sites by assessing suitability and ranking all possible locations. The most suitable sites are developed first, and since sufficient locations are found in the main valley, little development occurs elsewhere. In contrast, the TPM randomly selects possible sites, and if such a location is found to be suitable, development may occur independently of other potentially better sites. This simulation technique is more favorable for a relatively sparse settlement pattern as exists in Davos.

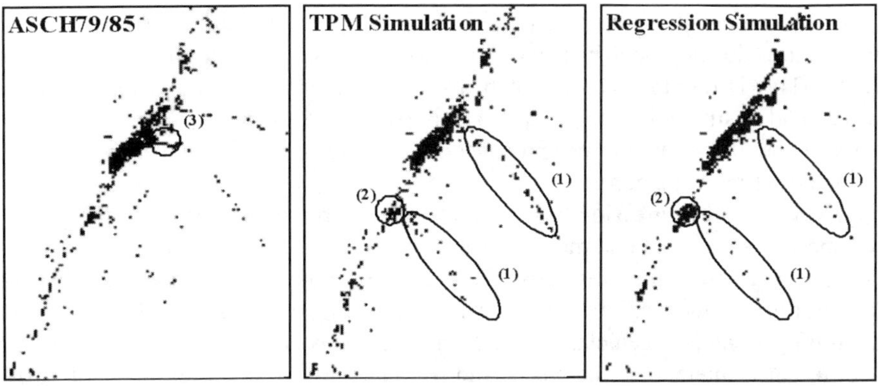

Figure 8.6 Annotated simulation results and ASCH79/85 data: (1) indicates discrepancies in the simulation outputs for side valleys, (2) marks the location of Davos-Frauenkirch and (3) the linear development at the exit of Dischma Valley.

Both simulations show the dense structure of settled areas and an enlargement of the small settlement, Davos-Frauenkirch (marked 2 in Figure 8.6), as a consequence of strong neighborhood effects. These results are partly due to a feature of the observation period (between ASCH79/85 and ASCH92/97) when strict local planning rules were introduced to limit sparse development outside existing built-up areas. However, prior to the 1970s no such policies were in place, and therefore the influence of neighboring developments may be too strong for simulations starting in 1954 when dispersed development was still accepted. Furthermore, this effect shows a limitation of all statistically-based land-use models where major changes in policy cannot be easily incorporated.

The models also failed to predict the linear development at the exit of the Dischma Valley (marked 3 in Figure 8.6). This development is in fact a 30-ha golf course which was established in 1967. Such single events cannot be systematically represented by the models, and they represent a further limitation.

8.5.3 Suitability of the Modelling Approaches for the Swiss Mountain Area

A key aim of this chapter was to investigate the suitability of two different modelling approaches to land-use change in mountainous regions and, in particular, to consider the validity of using regional data to represent local processes. Furthermore, a future aim is to extract the spatial structure of landscape change. We now consider how effective the two modelling strategies presented here are in addressing these aims.

With some important differences in spatial structure, the two modelling approaches produced broadly similar results. Thus, it is evident that the regression-based approach is valid in this case, with development in Davos following broadly similar patterns to that in the whole Swiss Mountain Area. The advantage of the

TPM-based model is in the better replication of spatial structure of housing and infrastructure development in Davos, namely the sparse development in the side valleys. This is due to its requirement for data representing all possible land-use changes and a simulation procedure that is not based on an absolute ranking of suitability for all possible development locations and so generates a more spatially distributed settlement pattern.

However, the regression-based model only requires demand for future development as an input parameter. As such input can be derived from many types of (non-spatial) planning scenarios, the regression approach is more suitable for integration in a wider regional modelling framework. In contrast, a TPM requires that all transitions between classes are calculated – this is a much more challenging exercise, since interactions between land-use changes are complex and not easily represented aspatially.

The regression-based model uses a larger sample size to produce measures of development suitability and requires less input information to run a simulation. In using a larger source area, it effectively integrates over multiple stages of land-use development and cultural regions within Switzerland which have had varying settlement structures in the past. Due to this averaging in the regression-based model and the recognition of multiple interacting land-use changes in the TPM-based approach, the latter tends to better predict the specific pattern of spatial development in a local area over a particular observed time period. However, its predictive power is limited by its requirement for input data detailing all transitions and making it hard to run future scenarios other than extrapolations. Since it is based on a broader dataset, the regression-based approach is more adaptive to possible configurations of future change. Thus, in order to replicate both the specific local pattern of land-use change and to incorporate the ability to respond to a wide range of potential future scenarios, an approach incorporating elements of the TPM, namely the integration of interactions between land-use changes, and using a regression-based engine to drive the overall pattern of change, may be a promising avenue for future work.

8.6 CONCLUSIONS

The results of this study demonstrate that the assumption of similarity in land-use change between the Swiss Mountain Area and Davos is correct over the time periods studied, and that a regionally-based regression approach is appropriate for local modelling. This type of model was as successful as a locally derived TPM-based model in reconstructing observed increases in housing and infrastructure land in the community of Davos. For the validation period, the pattern of the settlement was better reconstructed in the TPM-based approach due to model-specific simulation procedures and complex input data. However, the regression-based approach demonstrated considerable advantages for the implementation of future scenarios because of a higher adaptability to a wide range of potential land-use future configurations and simpler input requirements.

The main modelling constraints identified during the study were an inability to predict single development projects with high impact on the landscape and a neglect of changes in local spatial planning policies which can have a strong influence on future land allocations for housing and infrastructure development. Further research is therefore required with a focus on a) the improvement of the regression-based model to include interacting land-use changes and b) the integration of local spatial planning alternatives to allow the simulation of future scenarios.

8.7 ACKNOWLEDGMENTS

This project was supported by the National Research Programme NRP 48 of the Swiss National Science Foundation. We also thank Joachim Neumann for assistance with VBA programming.

8.8 REFERENCES

1. Price, M., The complex life: human land uses in mountain ecosystems, *Global Ecology and Biogeography Letters*, 6, 77-90, 1997.

2. Beniston, M., Ed., *Environmental Change in Mountains and Uplands*, Arnold, London, 2000.

3. Messerli, B., Grosjean, M., Hofer, T., Nunez, L., and Pfister, C., From nature dominated to human-dominated environmental changes, *Quaternary Sciences Reviews*, 19, 459-479, 2000.

4. Bätzing, W., *Die Alpen. Naturbearbeitung und Umweltzerstörung. Eine Ökologisch-geographische Untersuchung*, Sendler Verlag, Frankfurt a.M., 1984.

5. Egger, T., Meyre, S., Stalder, U., and Maier, T., Das Berggebiet vor neuen Herausforderungen, *LID-Dossier*, Online Article No. 392, 12.06.2002.

6. SAB, *Das Schweizer Berggebiet 2000. Fakten und Zahlen*, Schweizerische Arbeitsgemeinschaft für die Berggebiete (SAB), Brugg, 2000.

7. BFS, *GEOSTAT Benützerhandbuch*, 11th Edition, Bern, 2001.

8. Palang, H., Alumäe, H., and. Mander, Ü., Holistic aspects in landscape development: a scenario approach, *Landscape and Urban Planning*, 50, 85-94, 2000.

9. Nienhoff, D., Fritsch, U., and Bronstert, A., Land-use impacts on storm-runoff generation: scenarios of land-use change and simulation of hydrological response in a mesoscale catchment in SW-Germany. *Journal of Hydrology*, 267, 80-93, 2002.

10. Lundström, C., Walz, A., Gret-Regamey, A., Kytzia, S., and Bebi, P., ALPSCAPE - Linking models of land-use, resources and economy to simulate the development of Alpine regions, *Environmental Management*, In review.

11. Veldkamp, A. and Lambin, E.F., Predicting land-use change, *Agriculture, Ecosystems and Environment*, 85, 1-6, 2001.

12. Clarke, K.C., Hoppen, S., and Gaydos, L., A self-modifying cellular automation model of historical urbanisation in the San Francisco Bay area, *Environment and Planning B: Planning and Design*, 24, 247-261, 1997.

13. Salvini, P.A. and Miller, E.J., ILUTE: An operational prototype of a comprehensive microsimulation model of urban systems, in *Proceedings of the 10th International Conference on Travel Behavior Research*, Lucerne, Switzerland, 2003.

14. Wegener, M., *The IRPUD Model: Overview*, University of Dortmund, Dortmund, 1999.

15. Ligtenberg, A., Bregt, A.K., and van Lammeren, R., Multi-agent-based land use modelling: spatial planning using agents, *Landscape and Urban Planning*, 56, 21-33, 2001.

16. Waddell, P., UrbanSim: modelling urban development for land use, transport and environmental planning, *Journal of the American Planning Association*, 68, 297-314, 2002.

17. Miller, E.J., Hunt, J.D., Abraham, J.E., and Salvini, P.A., Microsimulating urban systems, *Computers, Environment and Urban Systems*, 28, 9-44, 2004.

18. Dabbert, S., Hermann, S., Kaule, S., and Sommer, S., Eds., *Landschaftsmodellierung für die Umweltplanung - Methodik, Anwendung und Übertragbarkeit am Beispiel von Agrarlandschaften*, Springer, Berlin, 1999.

19. Rounsevell, M.D.A., Annetts, J.E., Audsley, E., Mayr, T., and Reginster, I., Modelling the spatial distribution of agricultural land use at the regional scale, *Agriculture, Ecosystems and Environment*, 95, 465-479, 2003.

20. Kok, K. and Winograd, M., Modelling land-use change for Central America, with special reference to the impact of hurricane Mitch, *Ecological Modelling*, 149, 53-69, 2002.

21. Fritsch, U., *Entwicklung von Landnutzungsszenarien für Landschaftsökologische Fragestellungen*, Potsdam, 2002.

22. Verburg, P., Soepboer, W., Limpiada, R., Espaldon, M.V.O., Sharifa, M., and Veldkamp, A., Land use change modelling at the regional scale: the CLUE-S model, *Environmental Management*, 30, 391-405, 2002.

23. Loibl, W. and Toetzer, T., Modeling growth and densification processes in suburban regions - simulation of landscape transition with spatial agents, *Environment Modelling and Software*, 18, 553-563, 2003.

24. Turner, M.G., Wear, D.N., and Flamm, R.O., Land ownership and land-cover change in the southern Appalachian highlands and the Olympic peninsula, *Ecological Applications*, 6, 1150-1172, 1996.

25. SFSO, *GEOSTAT User Manual*, 11th Edition, Bern, 2001.

26. ESRI, ArcGIS 8.3, Environmental Systems Research Institute, http://www.esri.com, 2003.

27. Monserud, R. and Leemans, R., Comparing global vegetation maps with the Kappa statistics, *Ecological Modelling*, 62, 275-293, 1992.

28. Cohen, J., A coefficient of agreement for nominal scales, *Educ. Psychol. Meas.*, 20, 37-46, 1960.

CHAPTER 9

'Riding an Elephant to Catch a Grasshopper': Applying and Evaluating Techniques for Stakeholder Participation in Land-Use Planning within the Kae Watershed, Northern Thailand

F. Shutidamrong and A. Lovett

9.1 INTRODUCTION

Since the Food and Agriculture Organization of the United Nations (FAO) published its first forest resources assessment in early 1970s, the problem of tropical deforestation has changed from being seen as one mainly caused by timber exploitation to a situation where land use and associated forest degradation are regarded as key issues[1]. These problems are, in fact, exceedingly complex because they are connected with many other issues including population growth, poverty and the impacts of government economic policies[2-5]. Thus, in order to improve forest and land use management in the tropics, there is a clear need for effective approaches to land-use planning that can balance present and future needs for timber and other forest products (e.g., ecological services), as well as recognize the socio-cultural values of existing communities in the forests[6-9]. In addition, such approaches need to be adapted to local circumstances because of variations in forest structures, land uses, cultures, and government policies[5,10].

These concerns have been reflected in an ongoing search for better methods of land-use planning in tropical forest areas that include innovations in analytical techniques and the integration of methods from different disciplines. Nevertheless, there are still many unresolved issues in developing and applying approaches sensitive to local forest conditions and community circumstances in a manner that facilitates the resolution of land-use conflicts[8,11]. The primary aim of the research discussed in this chapter was therefore to identify, apply and evaluate a methodology that could facilitate stakeholder participation in management decisions and hopefully reduce conflicts regarding land use in a case study area - the Kae watershed of northern Thailand. Subsequent sections introduce the study area and then explain how a methodology based around spatial multi-criteria evaluation (SMCE) techniques was developed. The results of applying SMCE methods to identify desired future patterns of land use are then discussed and conclusions drawn regarding the practicalities of implementing such techniques and their potential to reduce land-use conflicts.

9.2 THE CASE STUDY

Several disastrous floods in southern Thailand during 1989 had a major impact on public attitudes and substantially boosted national campaigns regarding environmental awareness. In response to public pressures, the Royal Forestry Department (RFD) introduced a ban on all commercial logging concessions later that year. Since 1989, forest management schemes in Thailand have focused on water catchment conservation and headwater preservation[12-14]. However, a tendency for government agencies such as the RFD to consider only the physical and ecological features of areas while ignoring the views of local communities has contributed to a series of conflicts regarding land use in forested areas[14-19].

Due to a history of such problems, the Kae watershed (close to the north-eastern border of Thailand with Laos) was selected as a study site (Figure 9.1). This area represented a microcosm of the wider forest and land use management challenges that exist in northern Thailand, but was sufficiently small in size (16.5 km^2) to allow a detailed investigation of the issues involved. In addition, since 1999 there has been a small demonstration site (0.72 km^2) for integrated agricultural methods to support forest, soil and water conservation run by the Royal King's Project (RKP). The presence of this project proved very useful as a source of information and local support for fieldwork.

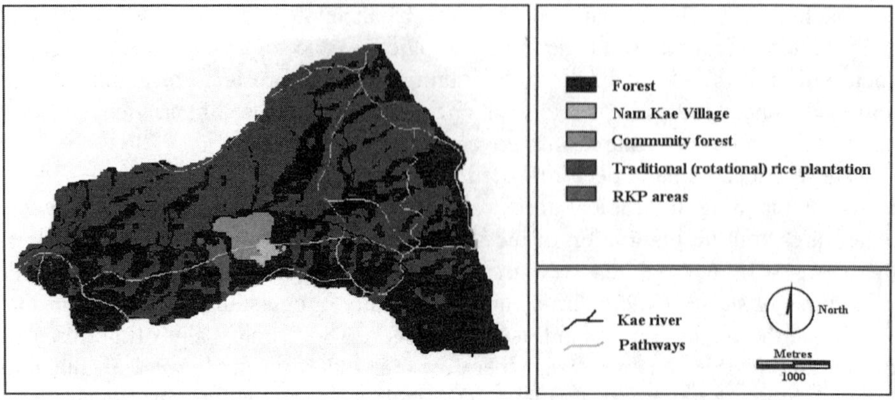

Figure 9.1 Land use characteristics of Kae watershed, northern Thailand.

Most of the Kae watershed consists of steep mountains (up to 1400 m above sea level) and valleys, with a small amount of flat land. Legally, the area is a strictly preserved forest, but in reality there is a permanent settlement (Nam Kae) of some 50 families and significant areas are under traditional rotational rice plantation for subsistence purposes (Figure 9.1). Typically rice is grown in a particular area for a year or two, then moved on to another site that has been previously used and so on, within a total cycle of five-six years. Other land uses in the watershed include a community forest around the village and a headwater zone which is strictly preserved for ecological reasons. In addition, there are 'village laws' that involve forest fire protection and wildlife hunting regulations[20].

9.3 RESEARCH METHODOLOGY

Due to the need for methods that were sensitive to the particular nature of the problem being investigated, it was decided to ask a number of experts in Thailand for their views on the most appropriate analytical techniques for the research. A questionnaire was administered during a series of meetings in August 2000 and four possible methods were discussed: a qualitative and participatory approach, spatial analysis and GIS, cost-benefit analysis and spatial multi-criteria evaluation (SMCE). The latter was rated as suitable by six of the nine experts surveyed and was selected as the primary analytical technique for the research. However, several other techniques, especially participatory approaches, were also included in the overall research methodology (Figure 9.2).

The fundamental principles of SMCE have been reviewed by Carver[21], Eastman[22], Jankowski[23] and Malczewski[24]. More recently there have been discussions about the advantages of applying an MCE decision rule called Order Weighted Average (OWA) compared to conventional Weighted Linear Combination (WLC), particularly in terms of allowing better consideration of levels of risk and tradeoff in the decision-making process[25-29]. The application of OWA is still quite experimental and so it was decided that the research should also investigate the potential of this decision rule compared to WLC as a means of resolving land-use problems.

9.3.1 Defining the Research Problem

An important principle in participatory processes is that stakeholders should be involved as early as possible[30]. Different stakeholders were therefore asked to define the main dimensions of the research problem. This included consideration of the goals for land-use planning, possible alternative land uses and the criteria that could be used to assess the suitability of areas for particular purposes. Stakeholders were first identified and classified into two groups depending on their prior experience of, and contact with, the study area. Directly-concerned stakeholders included the villagers, local foresters, RKP officers and the researcher (F. Shutidamrong). The indirectly-concerned group included national experts and samples of students and the general public.

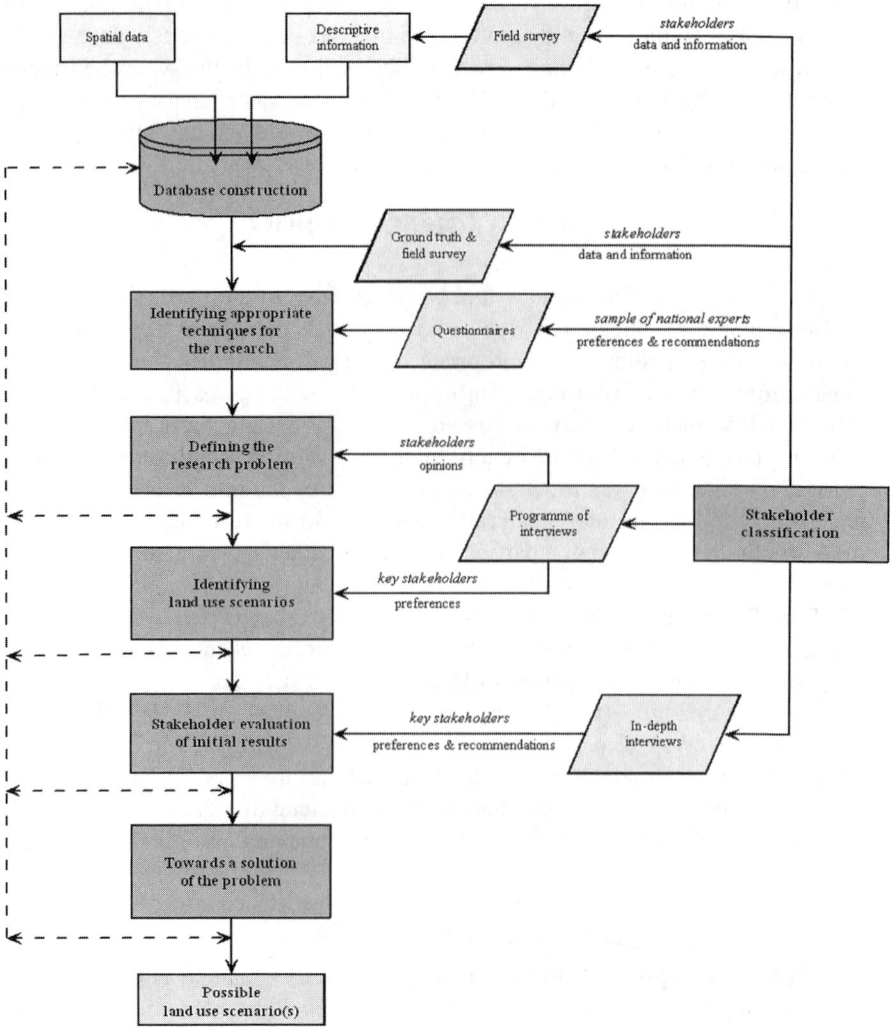

Figure 9.2 A summary of the research methodology.

A program of interviews with different stakeholders was conducted in August, 2000. It was decided to undertake informal (and focus group) interviews with the villagers and questionnaire-based interviews with other stakeholders. The results indicated a variety of opinions, but it was possible to synthesize these into the model of the research problem shown in Figure 9.3. Two main goals for land use planning were identified, namely developing the quality of life for the local community, and forest conservation. Suggested land uses included both existing (i.e., preserved forest, community forest, rice plantation and cattle grazing) and new activities (i.e., integrated agriculture and field cropping). There was general agreement on maintaining the settlement of Nam Kae village and so this was defined as a constraint that should not be altered in future land use zoning. A similar approach was taken with the land occupied by the Royal King's Project. Criteria that influenced the suitability of areas for other activities included slope angle, proximity to the river, proximity to the village, proximity to the road or pathways, and percentage of forest cover.

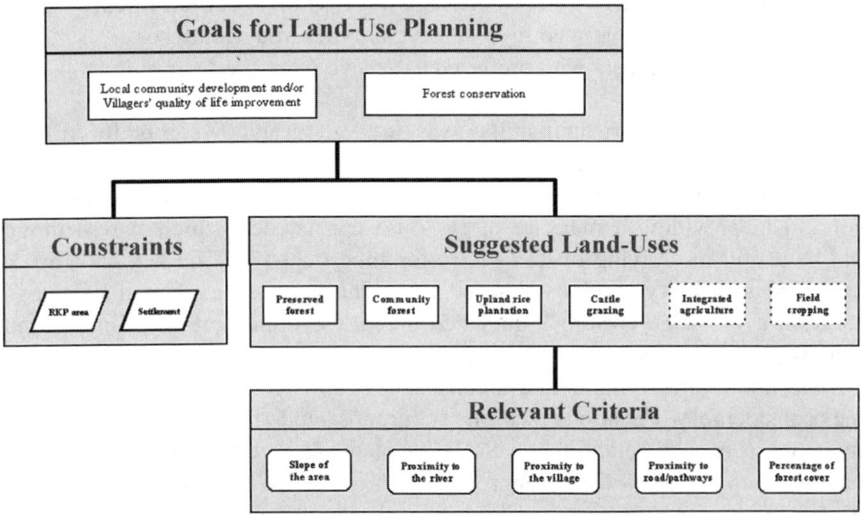

Figure 9.3 A model of the research problem.

For subsequent phases of the research it was decided to focus only on the views of 'key stakeholders' defined in terms of their importance and/or influence[31] with respect to the study area. Although it was clear that the Nam Kae residents, local foresters and local RKP officers are key stakeholders, the history of deforestation and land use problems in Thailand suggested the potential for strongly conflicting opinions between these groups. Other RKP officers and foresters working in the Northern region, environmental experts and the researcher were therefore also included as key stakeholders to provide an element of balance.

9.3.2 Stakeholder Input to Land-Use Scenarios

In a second program of interviews during March, 2001 the key stakeholders were asked to specify their preferred land uses, the sizes of areas these should occupy, and the suitable characteristics, degree of importance (weights) and tradeoff characteristics for relevant criteria. As anticipated, there were varied opinions regarding the area that should be occupied by different land uses, so it was decided to use this information to classify the stakeholders into subgroups and then derive overall criteria weights and tradeoffs for each of these. The seven subgroups of key stakeholders were:

1. Nam Kae villagers
2. Nam Kae teacher
3. Local RKP officers
4. Researcher
5. Other stakeholders who preferred an increase in preserved forest
6. Other stakeholders who favored several types of agriculture
7. Other stakeholders who preferred extensive integrated agriculture

The pairwise comparison method was used to derive weights for different criteria. This technique has been widely used in SMCE applications and has particular advantages in terms of ease of use (only two criteria are compared at a time) and in providing a measure of response consistency which it was thought could be useful in deriving overall views for subgroups[32-34]. The key stakeholders were asked to specify a relative preference value between each pair of relevant criteria on a 1-9 scale (where 1 indicated much less important and 9 much more important).

When the criteria weights and inconsistency ratios were calculated there were some considerable variations within subgroups and many responses lacked consistency (i.e., had ratios above Saaty's[32] standard threshold of 0.10). In such situations a consistency-driven approach is often the best method of deriving a consensus view for a group[35-37]. The approach adopted was therefore to exclude stakeholders in a subgroup with an inconsistency ratio above 0.15 and average the responses of the remainder according to the formula shown in Equation 9.1. This approach had the effect of giving greater emphasis to the more consistent responses. Once the average weights were calculated through this process they were then rescaled so that they summed to 1.0.

$$\begin{array}{c}\text{average weight}\\ \text{for criteria}\end{array} = \frac{\Sigma \left\{ \left(1 - \begin{array}{c}\text{inconsistency ratio}\\ \text{for stakeholder}\end{array}\right) \times \begin{array}{c}\text{stakeholder weight}\\ \text{for criteria}\end{array} \right\}}{\text{number of stakeholders involved}} \qquad (9.1)$$

For each criterion it was also necessary to identify the characteristics that made them most suitable for a particular land use (i.e., steep slope or flat). In most cases this was straightforward and where there was a divergence of opinion with a subgroup the most frequent response was used.

Another issue concerned the preferred degree of tradeoff between criteria with respect to the suitability of an area for a particular land use (i.e., the extent to which low suitability on one criterion could be compensated by better ratings on others).

Where possible, the most common response in a subgroup was taken as their representative view. In cases where such a single outcome could not be easily identified, judgements were made taking into account the subgroups' broader views regarding land use planning (e.g., the weights for particular criteria and the relative sizes of areas allocated to different land uses). For instance, it was decided to allow less tradeoff for a highly-weighted criterion because this factor would have more influence on the overall suitability of an area for a particular use. The results indicated that some subgroups (e.g., the village teacher) placed considerable restrictions on tradeoff while moderate tradeoff was favored by many of the other stakeholders, so providing another reason for investigating the merits of OWA techniques compared with a more conventional WLC method.

Order weights control the manner in which criteria are aggregated in an OWA approach. Defining order weights allows the desired levels of overall tradeoff and risk to be specified. A very risk-averse attitude can be envisaged as akin to an AND operator or minimum score (pessimistic) approach since the least favorable criterion will determine overall suitability. At the other extreme, a risk-taking perspective is similar to an OR operator or maximum score (optimistic) approach since the rating on the single most favorable criterion determines suitability. It is also important to note that order weights are different from criteria weights in that they are assigned on a case-by-case basis to criteria scores as determined by their ranking across criteria at each location being evaluated. Order weight 1 is assigned to the lowest ranked criteria for a location (i.e., the one with the lowest suitability score), order weight 2 to the next highest ranked factor, and so on[25,29].

Since OWA is still a relatively experimental technique, there is no recognized standard method of deriving order weights. Several rules were therefore developed to assign levels of tradeoff and risk to the seven subgroups of key stakeholders. For example, it was decided to adopt a risk-taking perspective to preserved forest as this would maximize the suitable area for this land use and be in accordance with the national watershed classification scheme that required strict preservation of the study area. Other judgements involved such factors as preferred levels of tradeoff, the sizes of proposed areas for particular land uses and the characteristics of the land uses or stakeholder subgroups. A risk-averse solution was taken for those land uses where a specific location or small occupied area was proposed because this would help to guarantee that any site selected was indeed suitable. With respect to the influence of stakeholder characteristics, the villagers and teacher were regarded as having particularly good knowledge about the area and confidence in their judgments. Thus, where the previously discussed considerations did not apply, a high-risk solution was adopted.

9.3.3 Generating Suitability Maps and Land-Use Scenarios

The Idrisi 32 Release 2 GIS software[38] was used to conduct the SMCE analysis. Information on the stakeholder preferences was combined with digital spatial data

(20 m cell resolution) in a GIS to create evaluation criteria and constraint maps. Each of the five main criteria (slope of the area, proximities to river, village and pathway, and percentage of crown cover) was converted to standardized score maps. The areas of Nam Kae village, the RKP, the stream network and existing pedestrian pathways were defined as constraints that should not be altered in any future land-use changes. Subsequently, the constraint and criteria maps were combined using both WLC and OWA decision rules to produce suitability maps for the alternative land uses. Considerable contrasts in some of the pairs of WLC and OWA maps were apparent, reinforcing the point that using order weights can substantially influence the outcome of suitability calculations[39].

Sets of suitability maps were aggregated to create preferred land-use scenarios for each subgroup of key stakeholders. As there were seven subgroups of key stakeholders and two approaches to suitability assessment (WLC and OWA), there were fourteen separate scenario maps. Initially, the multi-objective land allocation (MOLA) module in Idrisi 32 was investigated as a means of producing the scenario maps. This technique has the advantages of taking into account both suitability scores and the desired total areas to be occupied by different land uses[34,40], but some of the results were not especially satisfactory from practical agricultural and planning perspectives. Two particularly important limitations were that the method took no explicit account of the existing land-use pattern and sometimes produced fragmented patterns with many small patches of individual land uses. As an alternative, it was decided to begin with a relatively simple method (which involved allocating each raster cell to the land use that had the highest suitability score) and then refine the outcomes through discussion with the stakeholders.

9.4 COMPARING AND REFINING LAND-USE SCENARIOS

Table 9.1 compares the land use areas desired by different stakeholder subgroups with the initial allocations generated from the WLC and OWA suitability maps. These results show some considerable differences in totals, particularly a tendency for the area allocated to upland rice plantation to be much less than that desired (i.e., this land use rarely had the highest suitability score for a grid cell). The totals also illustrate how the variations in risk and tradeoff permitted by OWA can influence the outcomes of land use assessments. For example, the areas allocated to preserved forest in the WLC and OWA results are greater in the latter for all subgroups except the villagers. This reflects the decision to adopt a high-risk approach for preserved forest in all subgroups except the villagers, since such an assumption makes it easier to achieve higher suitability scores compared to the intermediate tradeoff implicit in the WLC method.

Table 9.1 Comparison of existing land-use areas, stakeholder preferences and initial results

Stakeholder Subgroup	Alternative Uses	Existing Areas (km^2)	Desired Areas (km^2)	Scenarios (km^2)	
				WLC	OWA
1	Preserved forest	7.37	7.37	12.27	10.65
	Community forest	0.34	0.34	2.97	2.28
	Upland rice plantation	7.85	7.85	0.39	2.70
2	Preserved forest	7.37	3.12	6.09	8.06
	Community forest	0.34	3.12	5.04	4.52
	Upland rice plantation	7.85	7.81	0.66	3.05
	Integrated agriculture	-	1.56	3.84	0.006
3	Preserved forest	7.37	6.25	6.85	11.13
	Community forest	0.34	3.12	6.07	4.43
	Upland rice plantation	7.85	-	-	-
	Integrated agriculture	-	6.25	2.70	0.07
4	Preserved forest	7.37	7.81	9.62	12.47
	Community forest	0.34	1.56	5.75	2.95
	Upland rice plantation	7.85	6.25	0.26	0.21
5	Preserved forest	7.37	9.22	10.12	13.66
	Community forest	0.34	2.58	0.45	0.67
	Upland rice plantation	7.85	1.09	1.34	0.85
	Integrated agriculture	-	2.72	3.72	0.44
6	Preserved forest	7.37	5.66	10.14	15.09
	Community forest	0.34	1.17	1.30	0.32
	Upland rice plantation	7.85	3.71	0.63	0.20
	Integrated agriculture	-	3.71	3.54	0.007
	Field cropping	-	1.37	0.02	0.007
7	Preserved forest	7.37	0.78	10.52	14.31
	Community forest	0.34	3.12	5.09	1.28
	Upland rice plantation	7.85	3.52	0.008	0.02
	Integrated agriculture	-	8.2	0.01	0.01

9.4.1 Evaluating Initial Results and Refining Scenarios

A program of in-depth interviews with the key stakeholders was carried out in August 2002 during which they were first requested to evaluate both their initial results and those of other groups and then asked for their input on how revisions should be made. Subsequently, the stakeholders' preferences and recommendations were examined to assess the extent of consensus. This exercise suggested that opinions on the scenarios essentially split into two groups, with a majority of stakeholders preferring the WLC outcome for Subgroup 6, while the villagers and researcher favored the OWA results for the directly-affected people. It was therefore decided to create further land use scenarios for the study area based on the views of these two sets of stakeholders. Revised sets of target areas to be allocated to different activities were therefore defined. In addition, the interview responses were used to specify minimum patch sizes for different land uses and helped identify the types of areas (e.g., rice plantations that had not been cultivated for at least five years) that had the greatest potential for reforestation or conversion to integrated agriculture and field cropping.

Two main techniques were used to produce revised land-use scenarios for the two sets of stakeholders. These were MOLA and a hierarchical approach that combined cellular automata (CA) and Markovian techniques with a number of rules to determine the priority with which land was allocated to different uses and eliminate small land use patches[39]. Table 9.2 indicates that the latter did not exactly match the revised preferences regarding the allocation of land to different uses, but it was much better than MOLA in producing outcomes that could form the basis of practical zoning solutions. Such an outcome accords with the findings of other studies which have noted the merits of a CA-Markov technique for land-use change modelling[28,41-43].

9.4.2 Evaluating the Two Revised Scenarios

Table 9.3 crosstabulates the land use allocations from the hierarchical approach for the two sets of stakeholders. The results indicate some substantial agreements with respect to the zoning of preserved forest, community forest and rice plantation. These areas cover a total of 11.28 km^2 (72.5% of the watershed excluding constraints). The map in Figure 9.4 also highlights the main conflicts, such as where the majority set of stakeholders would like to see existing rice plantation converted to integrated agriculture (areas primarily to the north and east of the existing community forest). In addition, disagreements exist where the majority would prefer community forest, but the villagers and researcher favor rice plantation (a zone to the west of the existing community forest).

The similarities between the land-use distribution existing in 2001 and the outcomes of the two stakeholder scenarios are summarized in Table 9.4. This table indicates that there are areas of preserved forest, community forest and rice plantation totalling 10.52 km^2 (67.61% of the watershed excluding constraints) where the 2001 land use matches that proposed in both scenarios. Disagreements regarding allocations focus mainly on existing areas of rice plantation where the majority of stakeholders would favor change to integrated agriculture or community forest (mainly areas around the northern half of the existing community forest perimeter, see Figure 9.5).

Table 9.2 Results of different land-use allocations

		Areas (km^2)				
		Preserved forest	**Community forest**	**Rice plantation**	**Integrated agriculture**	**Field Cropping**
Existing land uses		7.37	0.34	7.85	-	-
Majority of Stakeholders	**Preferred scenario**	10.14	1.30	0.63	3.54	0.02
	Revised targets	7.37	1.29	3.35	3.52	0.02
	MOLA	7.37	1.29	3.35	3.52	0.02
	Hierarchical method	7.91	1.2884	3.2664	3.0692	0.0212
Villagers and Researcher	**Researcher's preferred scenario**	10.65	2.27	2.70	-	-
	Villagers' preferred scenario	8.06	4.52	3.05	-	-
	Revised targets	7.54	1.01	7.00	-	-
	MOLA	7.54	1.01	7.00	-	-
	Hierarchical method	7.59	1.01	6.96	-	-

Table 9.3 Crosstabulation of land allocations in the two final scenarios

		Majority Set of Stakeholders Scenario (km^2)				
		Preserved forest (PrF)	**Community forest (CoF)**	**Rice plantation (RiP)**	**Integrated agriculture (InA)**	**Field Cropping (FiC)**
Villager/Researcher Scenario (km^2)	**Preserved forest (PrF)**	7.43	0.14	<0.01	0.02	<0.01
	Community forest (CoF)	0.13	0.65	0.07	0.15	0
	Rice plantation (RiP)	0.35	0.50	3.19	2.90	0.02

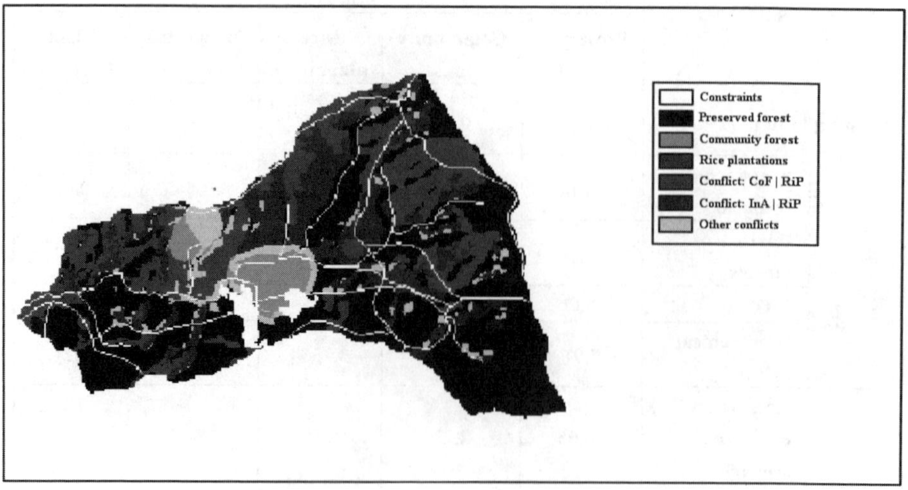

Figure 9.4 A comparison of land-use allocations in the two scenarios. The land-use codes in the legend are explained in Table 9.3. Where a category refers to two codes, the first is the allocation for the majority set of stakeholders and the second that for the villagers and researcher.

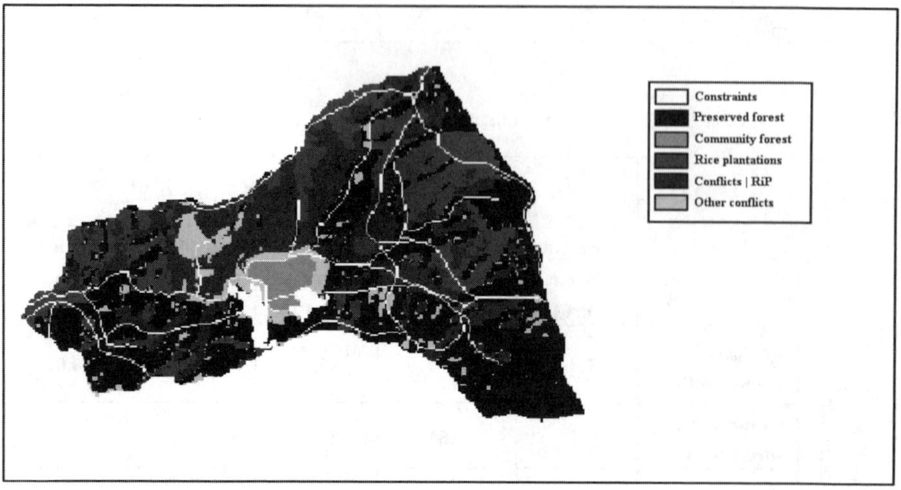

Figure 9.5 A comparison of 2001 land use and allocations in the two scenarios.

Table 9.4 Cross-tabulation of 2001 land use and scenario outcomes

<table>
<tr><th colspan="2" rowspan="2"></th><th colspan="3">Areas where both scenario allocations agree (km^2)</th><th rowspan="2">Areas with conflicting scenario allocations (km^2)</th></tr>
<tr><th>Preserved forest (PrF)</th><th>Community forest (CoF)</th><th>Rice plantation (RiP)</th></tr>
<tr><td rowspan="3">Land-use in 2001 (km^2)</td><td>Preserved forest (PrF)</td><td>6.98</td><td>0.13</td><td>0</td><td>0.25</td></tr>
<tr><td>Community forest (CoF)</td><td>0</td><td>0.34</td><td>0</td><td>0</td></tr>
<tr><td>Rice plantation (RiP)</td><td>0.45</td><td>0.18</td><td>3.19</td><td>4.03</td></tr>
</table>

Implementing either of the possible scenarios would therefore require further initiatives to resolve the remaining land use conflicts and a significant level of knowledge and experience on the part of the facilitators involved. Moreover, the ecological and socio-economic effects of the proposed land-use conversions must be seriously evaluated. For instance, are they sufficient to meet the requirements of protecting a crucial head watershed? The potential socio-economic and cultural implications for the villagers of reducing the amount of upland rice plantation also need to be investigated and much could depend on the progress of the RKP (since if the villagers perceive this project as a success they may become interested in undertaking this introduced activity themselves or incorporating it within their usual cultivation practices).

9.5 CONCLUSIONS

The research discussed in this chapter did not succeed in identifying a single compromise scenario for future land use planning in the study area, but it did demonstrate how stakeholder involvement in land management decisions could be facilitated and highlighted both the extent of current agreement and the problems that still require attention. More broadly, the study also indicated the potential of SMCE to help tackle the types of land-use conflicts that are common in tropical forest regions such as Northern Thailand. On a technical level, the research illustrated how the suitability maps (and land-use allocations) produced by WLC and OWA decision rules can be rather different, but the greater sophistication of the latter was not reflected in obviously wider acceptance by stakeholders. A clearer outcome was the manner in which CA-Markov techniques can generate improved modelling results compared to MOLA-type methods.

A number of key issues and challenges arose during the study. Several problems were encountered due to limitations in the available data. This reinforces the point that the availability of suitable base-mapping and information from stakeholders at a consistent level of detail are key requirements for any serious implementation of SMCE[44]. With respect to stakeholder involvement, it is very important to recognise that the necessary processes of engagement can be very time consuming and require considerable personal communication skills (including questionnaire design and the conduct of interviews). It is also almost impossible to avoid an element of interpretation or judgement on the part of the researcher in some circumstances, e.g., when combining individual stakeholder preferences into those for subgroups.

In conclusion, as part of the ongoing search for better management methods in tropical forest areas, this study identified, applied and evaluated a methodology to facilitate stakeholder involvement and help improve land use planning. However, there are still many unresolved issues in applying SMCE techniques and implementing them can be very demanding, akin to *'riding an elephant to catch a grasshopper'*. Nevertheless, the potential of such methods has been demonstrated and in future their full benefits may be best achieved when they are used by a group of professionals with a range of skills in GIS and remote sensing, qualitative methods and communication techniques.

9.6 ACKNOWLDGEMENTS

We are very grateful to the Royal Thai Government for sponsoring a PhD studentship and to the Royal King's Project (both in the study area and at head office) for advice and support.

9.7 REFERENCES

1. Poore, D., *Changing Landscapes*, Earthscan Publications, London, 2003.

2. Hallsworth, E., *Socio-Economic Effects and Constraints in Tropical Forest Management*, Wiley, New York, 1982.

3. Panayotou, T. and Ashton, P., *Not by Timber Alone: Economics and Ecology for Sustaining Tropical Forests*, Island Press, Washington D.C, 1992.

4. Myers, N., The world's forests: problems and potentials, *Environmental Conservation*, 23, 156-168, 1996.

5. Bawa, K. and Dayanandan, S., Causes of tropical deforestation and institutional constraints to conservation, in Goldsmith, F., Ed., *Tropical Rain Forest: A Wider Perspective*, Chapman and Hall, London, 1998, 175-198.

6. Jepma, C., *Tropical Deforestation: A Socio-Economic Approach*, Earthscan Publications, London, 1995.

7. Mackinnon, K., Sustainable use as a conservation tool in the forests of South-East Asia, in Milner-Gulland, E. and Mace, R., Eds., *Conservation of Biological Resources*, Cambridge University Press, Cambridge, 1998, 174-192.

8. Higman, S., Bass, S., Judd, N., Mayers, J., and Nussbaum, R., *The Sustainable Forestry Handbook*, Earthscan Publications, London, 1999.

9. Peuhkuri, T. and Jokinen, P., The role of knowledge and spatial contexts in biodiversity policies: a sociological perspective, *Biodiversity and Conservation*, 8, 133-147, 1999.

10. Poore, D., Burgess, P., Palmer, J., Rietbergen, S., and Synnott, T., *No Timber Without Trees: Sustainability in the Tropical Forest*, Earthscan Publications, London, 1989.

11. Food and Agriculture Organisation of the United Nations, *Monitoring and Evaluation of Participatory Forestry Projects*, FAO, Rome, Italy. 1985.

12. Flaherty, M. and Filipchuk, V., Forest management in Northern Thailand – a rural Thai perspective, *Geoforum*, 24, 263-275, 1993.

13. Ganjanapan, A., Dynamics of Local Natural Resources Management: The Situation of Thailand, (Thai document), Thai Research Fund, Bangkok, 2000.

14. Ganjanapan, A., Dynamics of Local Natural Resources Management: Policies, (Thai document), Thai Research Fund, Bangkok, 2000.

15. Lohmann, L., Forestry in Thailand: the logging ban and its consequences, *The Ecologist*, 19, 76-77, 1989.

16. Lohmann, L., The future of Thai forest conservation, *Environmental Conservation*, 19, 362-364, 1989.

17. Hirsch, P., Forests, forest reserves, and forest land in Thailand, *Geographical Journal*, 156, 166-174, 1990.

18. Pragtong, K. and Thomas, D., Evolving management system in Thailand, in Poffenberger, M., Ed., *Keepers of the Forest: Land Management Alternatives in Southeast Asia*, Kumarian Press, West Hartford, Connecticut,1990, 167-186.

19. Bhusal, Y., Thapa, G., and Weber, K., Thailand's disappearing forests: the challenge to tropical forest conservation, *International Journal of Environment and Pollution*, 9, 198-212, 1998.

20. Forestry Office of Nan Province, The Royal King's Project: A Demonstration of Forest, Soil and Water Conservation in Kae Watershed, (Thai document), Royal Forestry Department, Nan Province, Thailand, 2000.

21. Carver, S., Integrating multi-criteria evaluation with Geographical Information Systems, *International Journal of Geographical Information Systems*, 5, 321-339, 1991.

22. Eastman, J.R., Multi-criteria evaluation and GIS, in Longley, P.A., Goodchild, M.F., Maguire, D.J., and Rhind, D.W., Eds., *Geographical Information Systems: Volume 1 Principles and Technical Issues*, Wiley, Chichester, 1999, 493-502.

23. Jankowski, P., Integrating geographical information systems and multiple criteria decision-making methods, *International Journal of Geographical Information Systems*, 9, 251-273, 1995.

24. Malczewski, J., *GIS and Multicriteria Decision Analysis*, Wiley, New York, 1999.

25. Jiang, H. and Eastman, J.R., Application of fuzzy measures in multi-criteria evaluation in GIS, *International Journal of Geographical Information Science*, 14, 173-184, 2000.

[26.] Rinner, C. and Malczewski, J., Web-enabled spatial decision analysis using Ordered Weighted Averaging (OWA), *Journal of Geographical Systems*, 4, 385-403, 2002.

[27.] Malczewski, J., Chapman, T., Flegel, C., Walters, D., Shrubsole, D., and Healy, M., GIS-multicriteria evaluation with ordered weighted averaging (OWA): case study of developing watershed management strategies, *Environment and Planning A*, 35, 1769-1784, 2003.

[28.] Paegelow, M. and Olmedu,, M.T.C, Possibilities and limits of prospective land cover modelling – a compared case study: Garrotxes (France) and Alya Alpujarra Granadina (Spain), *International Journal of Geographical Information Science*, 19, 697-722, 2005.

[29.] Malczewski, J., GIS-based multicriteria decision analysis: a survey of the literature, *International Journal of Geographical Information Science*, 20, 703-726, 2006.

[30.] Hemmati, M., Dodds, F., Enayati, J., and McHatty, J., *Multi-Stakeholder Processes for Governance and Sustainability: Beyond Deadlock and Conflict*, Earthscan Publications, London, 2002.

[31.] Grimble, R. and Wellard, K., Stakeholder methodologies in natural resource management: a review of principles, contexts, experiences and opportunities, *Agricultural Systems*, 55, 173-193, 1997.

[32.] Saaty, T., *The Analytical Hierarchy Process*, McGraw-Hill, NewYork, 1980.

[33.] Basak, I. and Saaty, T., Group decision-making using the analytic hierarchy process, *Mathematical and Computer Modelling*, 17, 101-109, 1993.

[34.] Eastman, J., Jin, W., Kyem, P., and Toledano, J., Raster procedures for multi-criteria/multi-objective decisions, *Photogrammetric Engineering and Remote Sensing*, 61, 539-547, 1995.

[35.] Koczkodaj, W., A new definition of consistency of pairwise comparisons, *Mathematical and Computer Modelling*, 18, 79-84, 1993.

[36.] Finan, J. and Hurley, W., The analytic hierarchy process: does adjusting a pairwise comparison matrix to improve the consistency ratio help?, *Computers and Operations Research*, 24, 749-755, 1997.

[37.] Poyhonen, M. and Hamalainen, R., On the convergence of multiattribute weighting methods, *European Journal of Operational Research*, 129, 569-585, 2001.

[38.] Eastman, J.R., Idrisi 32 Release 2, http://www.clarklabs.org, Clark University, Massachusetts, 2001.

[39.] Shutidamrong, F., Applying and Evaluating Techniques for Stakeholder Participation in Land Use Planning in the Kae Watershed, Northern Thailand, PhD Thesis, University of East Anglia, Norwich, 2004.

[40.] Kyem, P., An application of a choice heuristic algorithm for managing land resource allocation problems involving multiple parties and conflicting interests, *Transactions in GIS*, 5, 111-129, 2001.

[41.] Brookes, C., A parameterized region-growing programme for site allocation on raster suitability maps, *International Journal of Geographical Information Science*, 11, 375-396, 1997.

[42.] Wu, F. and Webster, C., Simulation of land development through the integration of cellular automata and multicriteria evaluation, *Environment and Planning B: Planning and Design*, 25, 103-126, 1998.

[43.] Pontius, G.R., and Malanson, J., Comparison of the structure and accuracy of two land change models, *International Journal of Geographical Information Science*, 19, 243-265, 2005.

[44.] Malczewski, J., GIS-based land-use suitability analysis: a critical overview, *Progress in Planning*, 62, 3-65, 2004.

CHAPTER 10

Grid-Enabled GIS: Opportunities and Challenges

C. Jarvis

10.1 INTRODUCTION

An early definition of e-science was '*the large scale science increasingly carried out through distributed global collaborations enabled by the Internet*' (www.rcuk.ac.uk/escience/), with a stress on the Grid as an infrastructure for sharing computing resources and large collections of data. This is expressed visually within Figure 10.1, where resources might include for example computing power, databases and geographical services. Here, the Grid is a flexible cross-organizational network where communications occur on a machine-machine basis as opposed to the human-machine world of the Internet. Different 'virtual organizations' form and dissipate with each use of the network, and the choice of appropriate resources may itself be selected, as well as accessed, by machine rather than human.

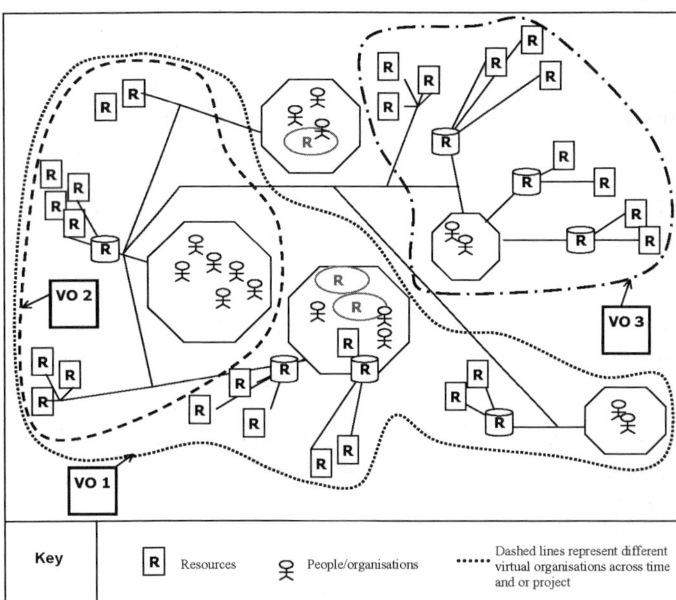

Figure 10.1 The general nature of virtual organizations in the 'Grid' (after Foster et al.[1]).

More recently it has been suggested that '*The 'Grid' ... aims to provide an infrastructure that enables flexible, secure, co-ordinated resource sharing among dynamic collections of individuals, institutions and resources*'[1]. This still encompasses issues regarding computational systems and data storage, but is a broader definition stressing collaborative (scientific) enterprise and transient virtual organizations. These last points are critical. This rationale is a superset encompassing both the earlier arguments in favor of intensive computing and also a vision of the Grid's potential to encourage changes to the very practice of science itself. Adopting this wider stance, Table 10.1 highlights just a few of the areas in which a Grid-enabled GIS might offer advantages over the status quo.

Table 10.1 Potential opportunities enabled by incorporating GIScience technologies within Grid enabled systems

Data	**Virtual organizations**
• Finding appropriate data sets automatically • Access to large data sets without downloading them completely, reducing data redundancy • A potential means of linking data held at multiple organizations • Mobile and real time sensors as input ○ Providing update through new observation ○ Requiring new computation of models ○ To give information to decision makers	• A new way of carrying out integrative modelling experiments across multiple sites • A means of bringing together elements of GI applications that plays to the strengths of individual researchers who are freed by access to appropriate interfaces • A more equitable resourcing outcome, both for researchers and governments?
Models and modelling	**Visualization for control, monitoring and decision-making**
• Access to models too complex to run at the majority of locations • A means of linking multiple models without overloading one computer system • A means of linking models developed at multiple sites without the collocation of individuals or software code • Data mining for associations/associated models • Computing power to evaluate sensitivity of simulation models/evaluate uncertainties in approach	• Interactive, multi-site visualizations to allow discussions of emerging phenomena and to support multi-user decisions • Multiple views based on a similar modelling flow, for example researchers, farmers, advisors and policy makers • Visualization methods that might assist with the monitoring of GRID processing

Turning firstly to the left hand quadrants of Table 10.1, practical computational challenges in the extent to which we are able to process increasing volumes of satellite and other data and model inter-linked critical processes at global and regional scales are perennial issues. The pooling of available computer resources across international and institutional boundaries has the potential to allow us to pursue previously intractable questions, reduce redundancy in data archives, process uncertainty bounds on simulation runs and explore geographically localized models[2]. The use of computational Grids for the processing of remotely sensed data for example has seen early progress[3,4]. Alternatively, Grid services could be used to speed up applied models to provide more responsive 'real-time' risk assessments[5]. E-science technologies also offer the possibility of drawing on expertise, data, knowledge and models in-situ in different parts of the world, opening opportunities for increased interdisciplinary collaboration and a richer set of research and socio-political perspectives. This may be deductive, or inductive through the further facilitation of data mining opportunities that the Grid presents. Grid services of the future for example should be able to find appropriate GIS models, functions and data dynamically, a considerable step forward from the currently used Web Services model.

Putting some context to these possibilities, consider a Grid approach for management and research regarding the causes and effects of urban atmospheric pollution from traffic. Figure 10.2 identifies just some of the databases, automated sensors, computing power, models (geographical and non-geographical) and expertise that are associated with these tasks. Many of these resources are currently unconnected, either in terms of easy human access or web services, let alone via a Grid; the bold lines in Figure 10.2 illustrate potential new connections across a Grid network. At present, an efficient flow of digital information to support, for example, management of risk to asthmatics from localized extreme episodes or responses to the threat of an impending critical episode is hampered by cross-institutional and cross-disciplinary barriers. The types of entity that might form a virtual organization in this case vary considerably in nature; the hospital expertise and patient data of Figure 10.2 require strong controls on the access to personal data to be in place[6], while sensor and meteorological data have less restriction. Work on Grid accessibility to this second type of data set is consequently more advanced, for example through projects such as the 'NERC DataGrid' (see http://www.bodc.ac.uk/projects/ndg.html). Similar contrasts may be identified between research and public service organizations, where progressing Grid services is understandably more in keeping with the former at this early stage. Hypothetically, advantages from all quadrants of Table 10.1 to adopting a Grid approach in this application area can be identified. This is just one very brief snapshot of the potential of cross-disciplinary and cross-institutional Grid computing in the service of an application area; more details may be found elsewhere[5]. Examples of on-going Grid work that incorporates GIS or remotely sensed data and/or functions and perspectives may be found in a diverse range of subject areas connected with environmental decision-making, such as climate modelling[7], land-use change[8] and hydrological modelling[2] among others.

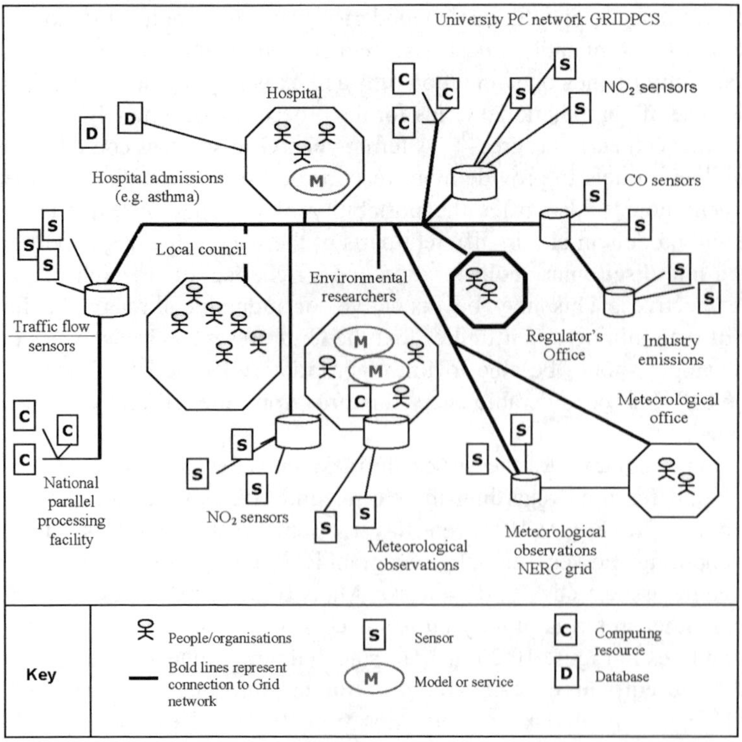

Figure 10.2 Inter-connected Grid resources for management and research regarding the causes and effects of atmospheric pollution.

Before applying Grid-enabled GIS for science and decision making however, we need to establish how close we really are to practicing GIS technologies on the Grid. The reality is that many developments in computer science will be required if data access, model integration and computing power are to be available and harnessed in a seamless and secure fashion. Figure 10.3 suggests a development profile for Grid utilization in environmental science; currently, practice is moving into the second stage but retains a data, as opposed to service, bias[7,9] that still also exists at stage one. Thus, we should not lose sight of the fact that using the Grid to support GIS applications currently requires considerable computing expertise on the part of developers; the average GIS user is a long way from logging on to the Grid in the same way that he or she logs on to a PC and searches the web.

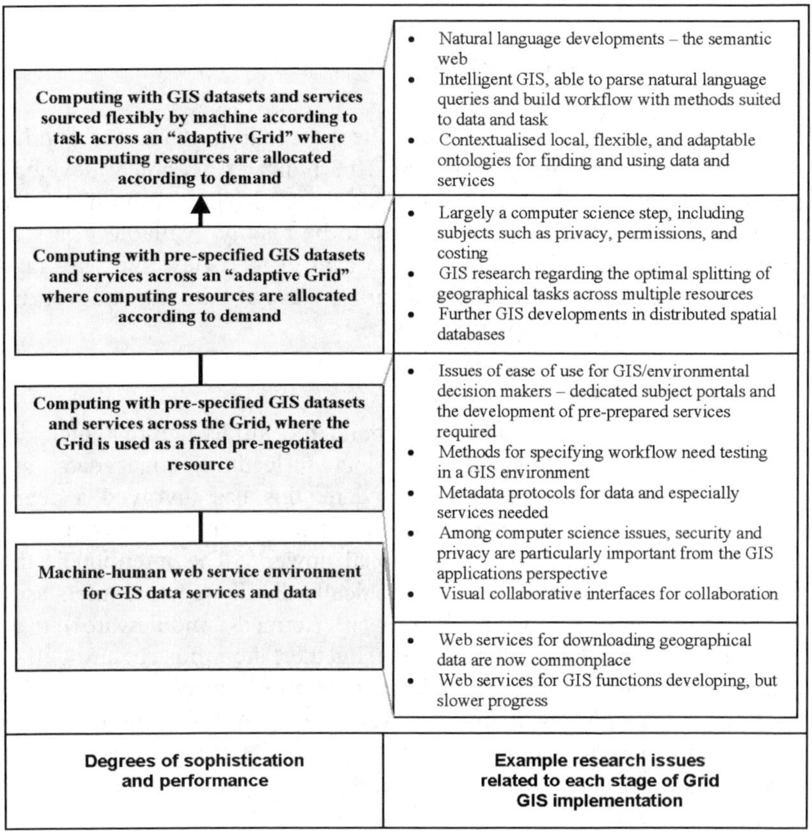

Figure 10.3 Stages of development in Grid GIS for environmental decision-making.

This chapter focuses on the technical and indeed cultural aspects of GIScience that might be further developed such that 'doing' interdisciplinary collaborative work that incorporates GIS across the Grid is both seamless and straightforward in the years to come. In other words, as Grid technologies mature, what does GIScience need to research in order that GridGIS functionality will be available to researchers and even to users who might not necessarily know that GIS technologies are serving their requests? Issues of particular current importance in meeting this goal are outlined in the right hand panel of Figure 10.3, and include further research regarding the linked themes of metadata and ontologies, distributed processing and federated databases. Work to assist users in managing remote data and processes intelligently is also relatively immature in GIS[10], while the area of collaborative analysis and spatial visualization[11] is also opening up as a dynamic research area.

10.2 PROGRESS AND CHALLENGES

The challenges involved in realising the potential of Grid for applications involving GIS are many. Firstly, as Figure 10.3 indicates, technical developments will be required both from within and outside the GIScience arena if data access, model integration and computing power are to be readily available in a seamless and secure fashion. Secondly, there will be a need to review the way in which research and data are managed, and to encourage ways of thinking and working that support collaborative interdisciplinary science.

10.2.1 Technical Issues

Part of the remit of 'e-Science' is to build the infrastructure which delivers efficient access to geographically distributed leading edge data storage, computational and network resources. To date, this has involved a change of orientation from the use of inter-connected super-computers towards a more general concept of a 'Grid' of computational power[12]. The intention is that the 'Grid' architecture will use diverse, geographically distributed computers as if they were a local resource, managed by software (termed 'middleware') that runs between the 'Grid' and the local machines. Together, the infrastructure will be one that '*enables flexible, secure, co-ordinated resource sharing among dynamic collections of individuals, institutions and resources*'[12]. A toolkit named Globus[13] is one example of emerging middleware. Emerging architectures for the Grid such as the Open Grid Service Architecture (OGSA) have incorporated the features of Web Services; the wide scale adoption of Web Services by the GIS community leaves it particularly well placed to follow a Grid pathway in this respect.

At this point in time, references to 'Grids' rather than 'the Grid' are more commonly found, and the pioneering work is being carried out on relatively small clusters of distributed machines. This is because developing 'the Grid' holds many technical challenges, from increasing network bandwidth and communication speed to security and resource scheduling; as Figure 10.3 indicates, true dynamism falls a step beyond what is currently feasible. Many of these issues fall beyond the research of GIScientists, but will impact upon the sustainability of current interest in the area. However, what emerges from the history of parallel processing and Internet usage in GIS to date is that there will be emergent GIS-specific issues relating to 'doing' GridGIS. Many of these fundamental GIScience issues can be researched with a view to their application on more robust Grids of the future, but with the expectation that smaller closed networks of resources will remain the status quo for geographical and environmental applications for some while as our understanding builds (Figure 10.3).

Fundamentally, if the opportunities presented by the Grid concept are to be maximized, then computing using the Grid as opposed to any other computing environment needs to be invisible. The applied user of GIS will not wish to grapple with scheduling and task decomposition issues, obtrusive access requirements or large seams between geographical databases. Additionally, just as how we see and label our worlds is vital when searching for data[14], any deficiencies and differences in this are expected to become even more apparent when sourcing and using networks of models and services from multiple disciplines. Furthermore, it will be important to recognize that a significant amount of interdisciplinary science is currently being carried out by researchers from one discipline stretching into the domains of another. While GridGIS potentially offers an environment within which research parties can access complementary expertise and work more fully towards their individual strengths, we should also weigh how we can incorporate expert geographical knowledge through hidden 'intelligent' infrastructures for providing assistance with services and access to resources[10]. A further, and not inconsiderable, GIScience challenge relates to how we communicate and manage data, models and results designed across multiple scales and for various purposes.

10.2.1.1 Towards the 'Invisible' Grid: Semantics

Appropriate standards for metadata have been a subject of enquiry in the GI world for some time, albeit rather focused on data[15]. This past focus on sharing data and information rather than models and service resources has led to a paucity of metadata schema and ontologies for geographical actions as opposed to objects, although recent work has begun to close this gap[16-19]. The term ontology is used here in the sense of a software engineering artifact used to describe a particular domain, '*An explicit specification of a conceptualization*'[20], as opposed to the more philosophical "science of being". Within a Grid context, further work regarding the development of metadata and ontologies for activities and objects *in combination*[16] will be a valuable contribution. We also need to consider how metadata fields might more easily be filled, for example using automated agents that mine resource-use histories to assist with this time consuming process[21]. The development of these semantic issues is important in building usable registries of services that agents may find automatically across the Grid, for developing more sophisticated data mining tools that move beyond the fixed registry approach and for making appropriate use of data and services once found.

Within this research on ontology and metadata, further work remains to be done regarding lineage, linking with Grid research on provenance. Additionally, developing methods to identify the intended meaning of words[22] will be an interesting challenge. Coming from the perspective of achieving improved interdisciplinary working, Smith and Mark[23] note the benefits of being able to account for differences in terminology for geographical processes and objects used by geographers and others. It is likely that researchers in different spheres of geography will interact with models and geographical data in different ways, as will decision makers. The question as to whether it is valuable to attempt to concatenate local ontologies into global super-sets must be opened for debate, as must the wisdom of adopting a hierarchical approach[24] to ontology building. For flexibility, given the number of permutations in ontology likely to arise when working in a global, interdisciplinary Grid context, it may be that pursuing methods to bridge ontologies through dynamic negotiation according to context will be a more fruitful avenue of research. Furthermore, incorporating changing contexts or perceptions within ontologies will be a necessary challenge, given that no ontology can ever be considered complete and immutable.

10.2.1.2 Towards the 'Invisible' Grid: Accessing and Scheduling GIS Procedures

As noted above, the average user of a GIS will not wish to grapple with many of the technical issues involved in Grid computing. The aim must rather be one of 'invisible computing', where the tools '*fit the person and tasks so well, are sufficiently unobtrusive and inter-connectivity seamless, that the technological details become virtually invisible compared to the task*'[25]. Such an aim can only be achieved by identifying and implementing appropriate Grid tools for geographical contexts. This theme links with the intelligent GIS discussed below, but also incorporates the more practical aspects of enabling and scheduling GI procedures.

Examples of geographical tools that will be desirable if we are to maximize the potential of the Grid include a comprehensive and accessible set of web services for GI functions that match those available in current GIS and beyond, and which dovetail with Grid middleware. Additionally, the creation of toolkits and frameworks that simplify model development for the Grid, such that the current extra effort in wrapping a model as a grid service is removed, might do much to make Grid computing a viable alternative for modellers[5].

A wide range of methods for specifying the processing sequence or 'workflow', that will collate and order services, is currently under investigation throughout the Grid literature[26]. Scheduling algorithms that distribute the modelling tasks specified in the workflow across multiple machines are a fundamental component of developing the Grid from a computer science perspective. This distribution will vary according to the geographical and temporal configuration of the task and resources available at any one point in time. Investigation of how these scheduling algorithms support spatial processing in particular will be useful; both previous research 'parallelizing' GI tasks[27] and more recent Grid-focused work[28,29] suggests that optimizing the way in which geographical modelling tasks are decomposed and scheduled over multiple machines may be specific to the spatial context. Indeed, understanding the changing space-time geographies of the Grid itself is likely to prove an interesting research area, since 'data "locality" can seriously affect performance'[30].

10.2.1.3 Intelligent Infrastructures

A considerable amount of interdisciplinary science is currently being carried out by those of one discipline stretching into the domains of another. While the Grid potentially offers an environment within which research parties can access complementary expertise and work more fully towards their individual strengths, we also need to consider how we can incorporate expert geographical knowledge through hidden 'intelligent' infrastructures to provide assistance with services and access to resources. At one level, this might involve metadata structures for services that encode their assumptions for use or by wrapping intelligent agents with services or data sets; at a more advanced level, the bigger challenge is to associate geographical questions expressed in natural language with appropriate workflows that are able solve the problem automatically using Grid-enabled resources.

The dangers of providing access to specialist models or resources to non-specialists have been highlighted by Anselin [31, p14-15] among others. Anselin for example suggests, in the context of spatial analysis functions, that '*with the vast power of a user friendly GIS increasingly in the hands of the non specialist, the danger that the wrong kind of spatial statistics will become the accepted practice is great*'. Seventeen years on, the creation of more 'intelligent' GIS and modelling tools to support decision makers, long identified as a priority for basic research within the environmental modelling community[32,33], remains largely unachieved. Concepts of knowledge networking, essentially a means of aggregating expertise, knowledge and information, have emerged in relation to diabetes [34] and social medicine [35] and point to possibilities for the development of 'intelligent' geographical tools for access across the Grid. However, how we encode what is often incomplete knowledge and how we evaluate versions of encoded knowledge according to their nature (e.g., prediction or interpretation) and quality are questions for further research. Case-based reasoning methods have been used to build inductive rules, models and more recently workflow procedures[36], and these complement the more formal encoding of deductively-derived cognitive knowledge[10]. Research regarding how best to link local networks and work towards global theories across spatial scales and multiple disciplines will be needed if this 'knowledge network' model is to be used as a basis for 'intelligent' collaborative support tools. Furthermore, as Zhuge[37] notes, a semantically-enabled grid is a necessary precursor to an effective knowledge grid.

10.2.2 Cultural Issues

It has been suggested that '*e-Science is about global collaboration in key areas of science, and the next generation of infrastructure that will enable it*'[38]. Gober[39] is among those who argue in a broader geographical context for the need for cultural changes and the more thorough integration of specialisms.

10.2.2.1 Interdisciplinary Practice

Issues that are likely to arise in the context of interdisciplinary modelling over the Grid, such as the need to develop theories of scale that facilitate the linking of economic and environmental models, and the better forging of links between qualitative and quantitative research, provoke questions central to geography as a discipline. It will be important for research under the banner of e-science in GIS to be more than a consideration of technical issues, and moreover it must not hinge simply on our ability to add together the 'sum of the parts' as an extension of the status quo.

10.2.2.2 Collaborative Human–Computer Interaction

If we are to use the Grid as a collaborative tool, then the design of interactive collaborative and multi-agency tools and research related to visual decision-making will be important undertakings. Developments in collaborative decision-making between different types of agency[40,41] could usefully be extended to Grid media to support smaller-scale collaborative research amongst research communities. More generic work, such as the "Access Grid" Project (http://www-fp.mcs.anl.gov/fl/Accessgrid/) which '*provide(s) a research environment for the development of distributed data and visualization corridors and for studying issues relating to collaborative work in distributed environments*', indicates starting points for such research. MacEachren[42] provides a thorough overview of work in this collaborative visualization domain from a geographical perspective, and draws up a conceptual framework for collaborative geographic visualization that explicitly incorporates scientific as well as decision-making collaborative environments. The dynamics of human, versus computer, interaction need to be pursued in a manner that explicitly considers geographical tasks by a variety of actors. While the concept of the 'collaboratory'[43], or virtual collaborative working environment, has yet to be explicitly implemented within geography, we should bear in mind that, of recent attempts at developing collaboratories, some do not succeed because '*distance still matters*'[44].

10.2.2.3 Geographically-Distributed Research

In looking to integrate research across disciplines, we must avoid the irony of geographically-distributed research issuing from a very limited number of locations and viewpoints. In this sense, it seems that technical and cultural changes need to go hand in hand if we are to succeed with GridGIS. The pursuit of technical geographies can assist in ensuring that structural and software developments in e-science are 'geographically enabled', such that they provide a supportive and potentially more democratic platform for greater collaboration, but firstly this attitudinal shift needs to take place.

10.2.2.4 Data Access, Power and Purchasing

Significantly, attitudes towards data access and variations in quality and quantity of digital data already impede global collaboration[45]. Moreover, the profile of organizational agendas, control over data, privacy issues, charges for 'public' data and political interests within geography raised by Pickles[46] are accentuated within our increasingly Internet-enabled societies. In moving to an environment of e-services, a world in which payments for the use of GIS software and data components are made on demand with each use, such contentions are likely to multiply[47]. Changes in the business model on which GI software stands will be inevitable.

Issues such as the ownership of knowledge, and the cost of encoding it, might also be expected to become important research issues given such an expected marked change in the design and use of computational and knowledge acquisition resources. For example, in relation to 'formal' knowledge, the question is being asked as to whether the days of refereed textual articles are numbered, to be replaced by entries in knowledge 'repositories'. Such viewpoints highlight the changing cultural contexts of scientific research provoked by e-science concepts. Just as the adoption of e-science is likely to affect the practice of research, new economic and social geographies are likely to emerge in its wake.

10.3 CONCLUSIONS

The adoption of an e-science framework within which to carry out GIS applications offers potential gains by providing access to both data, knowledge *and* the computer resources with which to process them in a tractable fashion. These same resources offer the potential for geographers to offer relevant, timely findings to decision and policy makers for risk management and mitigation. Arguably then, the e-science manifesto suits GIS well. However, we need to review our current progress through the stages of development required for full and easy use of the Grid by GI practitioners and users (Figure10.3); the technical and attitudinal challenges posed by the Grid at its current stage of development are many, and a long term view as to what might be possible is required.

Research regarding the ontologies of geography, and the way in which geographies are practiced and perceived by different peoples and at different locations, will be crucial if the overhead of carrying out e-science by geographers is not to be too great. So too will theoretical research regarding the 'invisible' management of geographical modelling tasks at a variety of levels across multiple resources. Culturally, fostering shifts towards greater interdisciplinary and geographical collaboration is an important goal as is work challenging some existing attitudes towards the sharing of data, knowledge and resources.

Arguably, it is the notions of collaborative scientific enterprise, semantic expression and 'invisible' computing which most stretch our current notions of

GIScience. This chapter began by noting two definitions of Grid computing. The significance of the second definition stressing the virtual organization still requires yet stronger emphasis if progress towards doing GIS over the Grid is not to be thwarted, since there may be a lack of immediately applicable 'big' compute-intensive applications. The Grid's potential to empower using remote resources and interdisciplinary communication must also be further evaluated if Grid GIS is to prosper. Linked to this theme, we also need to keep a careful watch on issues of democracy in research, security, intellectual property and privacy when moving more closely towards such a component-based, global digital research world.

The Grid has the potential to provide the technical support for exciting new developments in relevant, global geographies for the 22nd Century. However, in closing, it is important to note that empirical geography supported by Grid must be matched by developments in theory, particularly in relation to how we integrate across scales and between disciplines, if we are not simply to achieve a faster 'old' geography or a collection of small components pushed together instead of a dynamic new version.

10.4 REFERENCES

1. Foster, I., Kesselman, C., and Tueke, S., The anatomy of the Grid: Enabling scalable virtual organisations, *International Journal of Supercomputer Applications*, 15, 200-222, 2001.

2. Beven, K. J., On environmental models everywhere on the Grid, *Hydrological Processes*, 17, 171-174, 2003.

3. Aloisio, G. and Cafaro, M., A dynamic earth observation system, *Parallel Computing*, 29, 1357-1362, 2003.

4. Shen, Z., Luo, J., Zhou, C., Cai, S., Zheng, J., Chen, Q., Ming, D., and Sun, Q., Architecture design of grid GIS and its applications on image processing based on LAN, *Information Sciences*, 166, 1-17, 2004.

5. Mineter, M. J., Dowers, S., Skouloudis, A. N., and Jarvis, C. H., Towards use of grids in environmental research, management and policy, *International Journal of Environment and Pollution*, 20, 297-308, 2003.

6. Hartswood, M., Ho, K., Procter, R., Slack, R., and Voss, A., Etiquettes of data sharing in healthcare and healthcare research, in *Proceedings of the 1st International Conference of E-Social Science*, Manchester, UK, 2005.

7. Chervenak, A., Deelman, E., Kesselman, C., Allcock, B., Foster, I., Nefedova, V., Lee, J., Sim, A., Shoshani, A., and Drach, B., High-performance remote access to climate simulation data: a challenge problem for data grid technologies, *Parallel Computing*, 29, 1335-1356, 2003.

8. Edwards, P., Preece, A., Pignotti, E., Polhill, G., and Gotts, N., Lessons learnt from deployment of a social simulation tool to the semantic Grid, in *Proceedings of the 1st International Conference of E-Social Science*, Manchester, UK, 2005.

9. Ananthanarayan, A., Balachandram, R., Grossman, R., Gu, Y., Hong, X., Levera, J., and Mazzucco, M., Data webs for earth science data, *Parallel Computing*, 29, 1363-1379, 2003.

10. Jarvis, C. H., Stuart, N., and Cooper, W., Infometric and statistical diagnostics to provide artificially-intelligent support for spatial analysis: the example of interpolation, *International Journal of Geographical Information Science*, 17, 495-516, 2003.

11. MacEachren, A. M., Cartography and GIS: extending collaborative tools to support virtual teams, *Progress in Human Geography*, 25, 431-444, 2001.

12. Foster, I. and Kesselman, C., *The Grid: Blueprint for a New Computing Infrastructure*, Morgan Kaufmann, New York, 1999.

13. Foster, I. and Kesselman, C., Globus: a metacomputing infrastructure toolkit, *International Journal of Supercomputer Applications*, 11, 115-128, 1997.

14. Mark, D. M., Freksa, C., Hirtle, S. C., Lloyd, R., and Tversky, B., Cognitive models of geographic space, *International Journal of Geographical Information Science*, 13, 747-774, 1999.

15. Maguire, D. J. and Longley, P. A., The emergence of geoportals and their role in spatial data infrastructures, *Computers, Environment and Urban Systems*, 29, 3-14, 2005.

16. Kuhn, W., Ontologies in support of activities in geographical space, *International Journal of Geographical Information Science*, 15, 613-631, 2001.

17. Soon, K. and Kuhn, W., Formalizing user actions for ontologies, in *Geographic Information Science, Proceedings 2004*, University of Maryland, 2004, 299-312.

18. Duckham, M. and Worboys, M., An algebraic approach to automated geospatial information fusion, *International Journal of Geographical Information Science*, 19, 537-557, 2005.

19. Schade, S., Sahlmann, A., Lutz, M., Probst, F., and Kuhn, W., Comparing approaches for semantic service description and matchmaking, in *On the Move to Meaningful Internet Systems 2004: Coopls, Doa, and Odbase, Pt 2, Proceedings*, 2004, 1062-1079.

20. Gruber, T. R., A translation approach to portable ontologies, *Knowledge Acquisition*, 5, 199-220, 1993.

21. Gahegan, M. and Pike, W., A situated representation of GIS resources, in *Annual Meeting of the Association of American Geographers*, Denver, US, 2005.

22. Comber, A., Fisher, P., and Wadsworth, R., What is land cover?, *Environment and Planning B: Planning and Design*, 32, 199-209, 2005.

23. Smith, B. and Mark, D. M., Geographical categories: an ontological investigation, *International Journal of Geographical Information Science*, 15, 591-612, 2001.

24. Kokla, M. and Kavouras, M., Fusion of top-level and geographical domain ontologies based on context formation and complementarity, *International Journal of Geographical Information Science*, 15, 679-687, 2002.

25. Pawlikowski, K., Call for papers for InviCom 2001: International Workshop on Invisible Computing, May 16-18, Brisbane, Australia, http://coscweb2.cosc.canterbury.ac.nz/research/RG/net_sim/archive/ressim /msg00007.html, 2001.

26. Yu, H., Bai, X., and Marinescu, D. C., Workflow management and resource discovery for an intelligent grid, *Parallel Computing*, 31, 797-811, 2005.

27. Mineter, M. J., Dowers, S., and Gittings, B. M., Towards a HPC framework for integrated processing of geographical data: encapsulating the complexity of parallel algorithms, *Transactions in GIS*, 4, 245-262, 2000.

28. Hu, Y. C., Xue, Y., Wang, J. Q., Sun, X. S., Cai, G. Y., Tang, J. K., Luo, Y. W., Zhong, S. B., Wang, Y. G., and Zhang, A. J., Feasibility study of geo-spatial analysis using Grid computing, *Computational Science. Lecture Notes in Computer Science*, 3039, 956-963, 2004.

29. Wang, S. and Armstrong, M. P., A quadtree approach to domain decomposition for spatial interpolation in Grid computing environments, *Parallel Computing*, 29, 1481-1504, 2003.

30. Armstrong, M. P., Geography and computational science, *Annals of the Association of American Geographers*, 90, 146-156, 2000.

31. Anselin, L., *What is Special About Spatial Data? Alternative Perspectives on Spatial Data Analysis*, NCGIA Technical Report 89/4, NCGIA, Santa Barbara, 1989.

32. Burrough, P. A., Development of intelligent geographical information systems, *International Journal of Geographical Information Systems*, 6, 1-11, 1992.

33. Densham, P. J. and Goodchild, M. F., Spatial decision support systems: a research agenda, in *Proceedings of GIS/LIS'89*, ACSM/ASPRS/AAG, Virginia, 1989, 707-716.

34. Uelpenich, S. and Bodendorf, F., Knowledge organisation and knowledge modelling in consulting companies, *Wirtschaftsinformatik*, 43, 469, 2001.

35. Cravey, A. J., Washburn, S. A., Gesler, W. M., Arcury, T. A., and Skelly, A. H., Developing socio-spatial knowledge networks: a qualitative methodology for chronic disease prevention, *Social Science and Medicine*, 52, 1763-1775, 2001.

36. Kaster, D. S., Medeiros, C. B., and Rocha, H. V., Supporting modeling and problem solving from precedent experiences: the role of workflows and case-based reasoning, *Environmental Modelling and Software*, 20, 689-704, 2005.

37. Zhuge, H., Semantics, Resource and Grid, *Future Generation Computer Systems*, 20, 1-5, 2004.

38. Taylor, J., Director-General of the Research Councils, OST, http://www.e-science.clrc.ac.uk, 2002.

39. Gober, P., In search of synthesis, *Annals of the Association of American Geographers*, 90, 1-11, 2000.

40. Jankowski, P. and Nyerges, T., GIS-supported collaborative decision making: results of an experiment, *Annals of the Association of American Geographers*, 91, 48-70, 2000.

41. Churcher, C. and Churcher, N., Realtime conferencing in GIS, *Transactions in GIS*, 3, 23-30, 1999.

42. MacEachren, A. M., Cartography and GIS: extending collaborative tools to support virtual teams, *Progress in Human Geography*, 25, 431-444, 2001.

43. Kouzes, R., Myers, J., and Wulf, W., Collaboratories: doing science on the Internet, *Computer*, 29, 40-46, 1996.

44. Olson, G. M. and Olson, J. S., Distance matters, *Human-Computer Interaction*, 15, 139-178, 2000.

45. Hulme, M., Global warming, *Progress in Physical Geography*, 24, 591-599, 2000.

46. Pickles, J., *Ground Truth,* Guildford Press, New York, 1995.

47. Braynov, S. and Sandholm, T., Contracting with uncertain level of trust, *Computational Intelligence*, 18, 501-514, 2002.

Part III

Participation in Decision-Making

CHAPTER 11

Developments in Public Participation and Collaborative Environmental Decision-Making

I. Bishop

11.1 INTRODUCTION

Human actions have consequences. In happy circumstances everyone benefits from the actions; the win-win cliché. This is, however, seldom the case; usually there are winners and losers. The classical basis of decision-making, cost-benefit analysis, suggests that provided the benefits outweigh the costs by a reasonable margin (to account for error and uncertainty) the action should proceed. Nevertheless, this apparently sensible approach is constantly running into protests from those who bear the costs, those who rate the costs higher than the analysts, or politicians and others who appoint themselves as guardians of people or environments that will bear the cost in the future.

The cost-benefit paradigm as traditionally applied does pay some heed to the future; however the typical discount rates used (e.g., 3% or 6%) mean that any consequences beyond about a decade have little influence on the analysis. On the other hand, a key aspect of the analysis which is often wholly ignored, by the analysts if not the public, is the spatial distribution of costs and benefits.

Frequently costs are quite local – e.g., under the flight path, affected by pollutants or in the viewshed – whereas the benefits are regional or national. This has given rise to the NIMBY syndrome in which people recognize the national benefit but ask why they should carry the cost. This is a perfectly reasonable question. Sometimes governments or corporations will seek to nullify the perceived cost by offering some form of compensation – a new community swimming pool, jobs or even direct payments.

While there will always be some who perceive disadvantage and will fight for their rights, a substantial part of the contention can be eliminated by more explicit upfront analysis and communication of spatial and temporal aspects of the consequences of actions and, in particular, cost and benefit estimation[1]. In order for people to accept a decision, there appear to be certain specific aspects of the process or the outcome which must be partly or wholly satisfied. For example, from an individual perspective the criteria might be:

- My views have been recognized and taken into account
- The decision leaves me minimally worse off

- The costs and benefits are transparent
- Anyone who benefits more than me should be deserving (i.e., not already better off than me)
- The outcome is valid into the future (sustainable).

Although NIMBYism is a recent phrase, the phenomenon of local project opposition has been around for many years and spatial scientists have been arguing that there are better ways of making decisions which will be more transparent and hopefully fairer and more sustainable. As spatial scientists we have argued that *good decision-making demands good information.* This argument has not changed but now it is increasingly recognized that in addition: decision-making must carry those affected along with it. Consequently, *process is as important as information.*

There are two aspects to achieving this improved condition: (a) analysis (i.e., the base knowledge of where/when costs or benefits will accrue) and (b) communication (i.e., allowing the people affected – on both sides – to understand these distributions). The question for spatial analysts was (and remains): how can we put our tools and skills to work to improve decision-making and public confidence in decisions?

For many people the answer has been to try to improve the models: the technical process of distributing costs and benefits. Other researchers have focused on public engagement, tools for the presentation of information, the design of stakeholder processes etc. This chapter concentrates on this second aspect and reflects a personal perspective on where we have been, where we are now and where we might be going in the specific context of changes in the landscape.

First, however, we need some sort of framework for public participation. One attempt at classifying the extent of public involvement comes from Arnstein[2]. Figure 11.1 shows Arnstein's ladder of citizen participation. The terminology is somewhat judgmental but it provides a starting point for further analysis.

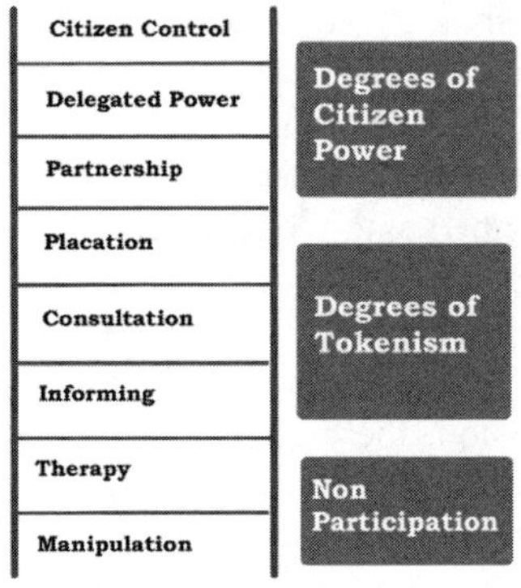

Figure 11.1 The ladder of citizen engagement (after Arnstein[2]).

11.2 SEPARATE DEVELOPMENT: GIS AND VISUALIZATION

In the 1970s several groups began working with GIS-like programs with a view to generating frameworks for more rational land use planning. Prominent among these were the group under Carl Steinitz at the Harvard Graduate School of Design, assisted by software from the Harvard Computer Graphics Lab from which sprang many of the leaders of early, and contemporary, GIS development (e.g., Jack Dangermond and Dana Tomlin). Near neighbors, and to some degree competitors, was a group under Julius Fabos at the University of Massachusetts in Amherst. Their system was called METLAND[3]. Both Steinitz and Fabos are landscape architects and their software tools were essentially raster based in their analysis and mapping. In Canberra, Australia, Doug Cocks and his team had similar objectives with their SIRO-PLAN method, but took a rather different parcel based approach[4]. In all three cases scope for public involvement was an element of the procedural design. However, the procedures used and the computer power available did not really permit these groups to think in terms of interactive mapping or visualization systems. Public involvement was orientated more towards the gathering of views in the form of weightings for aspects of the landscape or for 'policies' relating to land use locations (Figure 11.2). Generally the 'public' were experts, interest groups or the planners themselves rather than the broader community. In addition, participants often had to decide for themselves if they would be affected by particular changes in land use. There was not a lot by way of spatial models to predict the outcomes of particular actions and, especially, the populations who might be impacted.

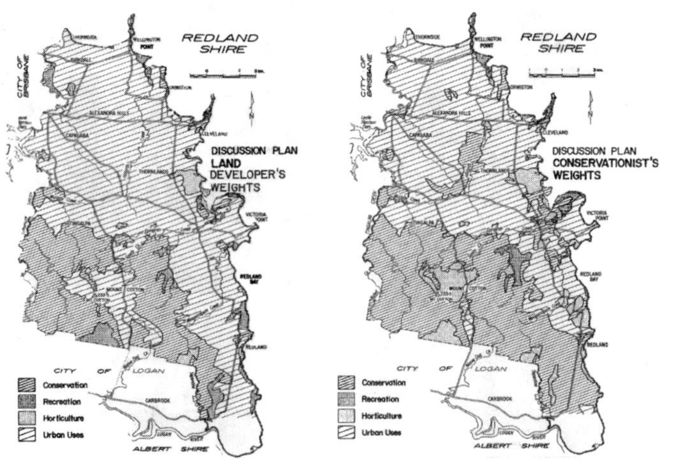

Figure 11.2 Alternative land-use plans based on different factor weightings (from McDonald and Brown[5]). Copyright Elsevier 1984 (with permission).

Among the early software products designed to determine consequences algorithmically were programs which estimated who would see or be otherwise affected by the land use changes. Landmark computer programs including VIEWIT[6] and MAP (Map Analysis Package)[7] led the way in provision of tools for landscape analysis and visual modelling. Indeed, some of their features, such as visual magnitude estimation and partial screening, are seldom found in contemporary software. These products recognized the potential of the computer to answer questions about the visual relationship between different parts of the landscape, as well as the effect of surface features on these relationships. VIEWIT was developed primarily for use in a forest management context while MAP combined the facilities of VIEWIT with a wider range of map algebra functions making it a prototypical geographic information system (GIS).

At the same time, the first examples of computer based landscape simulation were appearing. For forestry applications, wholly computer drawn images with arrows for trees were setting the standard (Figure 11.3a)[8]. In other contexts, simple perspective drawings of power stations or other industrial facilities were being superimposed onto photographs in what was then regarded as a photomosaic (e.g., Bureau of Land Management[9]) and might now be called a low-level form of augmented reality (Figure 11.3b). The purpose of these simulations was to communicate specific proposals. In certain cases alternatives were explored, but the general trend was for environmental analysis to come well after the design process was completed on functional grounds.

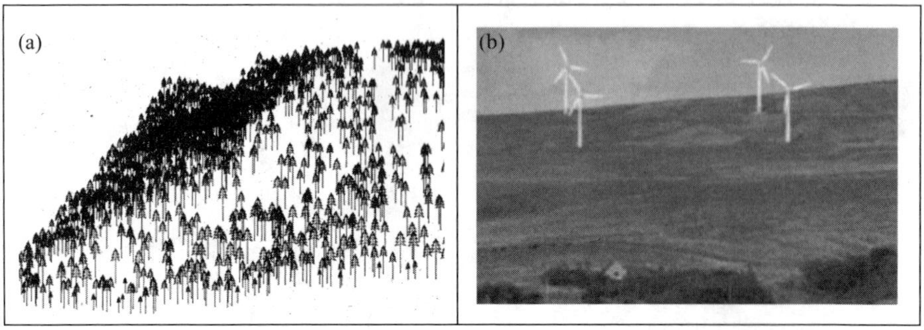

Figure 11.3 Approaches to 3D visualization for public presentation: (a) early example of forest simulation (from Myklestad and Wagar[8]), (b) modern photomontage (from Benson[10]). Copyright (a) Elsevier 1977 and (b) Taylor & Francis 2005 (with permission).

11.3 CONVERGING TECHNOLOGIES: GIS-DRIVEN VISUALIZATION

Communities are increasingly seeking opportunities to actively and deliberately manage their futures. Software products such as What if?[11] and CommunityViz[12] assist communities in exploring and envisioning possible future conditions and in assessing the consequences of planning decisions. What if? is a very clear and direct descendent of the METLAND and SIRO-PLAN systems of 20 years earlier. It is map based and relies on definition of homogenous parcels exactly as SIRO-PLAN did. The intention is to incorporate the preferences and assumptions of the user and then create a plan which is supposedly the best (or close to best) way of meeting those aspirations. This qualifies What if? as a Decision Support System (DSS) in conventional terms.

Some recent papers[13] have adopted the language of Tufte[14] and begun to use the term 'envisioning system' (EvS). An EvS differs from a DSS following the reasoning of Brail and Klosterman[15]. The goals of EvS are longer range than typical for DSS and less analytical. EvS is less directed towards identifying best solutions and more directed towards identifying achievable directions. EvS attempts to facilitate collaboration rather than enable executive decisions. This is very similar to what Michael Kwartler calls 'visioning'. In his terms: *'The quality of place, the combination of its experiential and functional attributes and group values and identity, is fundamental to visioning'*[16, p 252]. He goes on to discuss the importance of a visual representation of outcomes and the way in which this can provoke a '...that's not what I meant at all' reaction to outputs of DSS.

Bishop et al.[13] describe an EvS designed to help rural communities contemplate landscape scale changes. Simulations and models project current conditions into the future according to the constraints of scenario-based planning and available land use choices. Possible future conditions are represented visually through maps, simulations and indicator icons. The goal of an EvS is to help community members negotiate desired future conditions and implement policies which shape land use changes to produce these outcomes. Figure 11.4 shows an example of an EvS setup with back-projected screen displays and participants equipped with Personal Digital Assistants (PDAs) for input, query and response recording purposes[17].

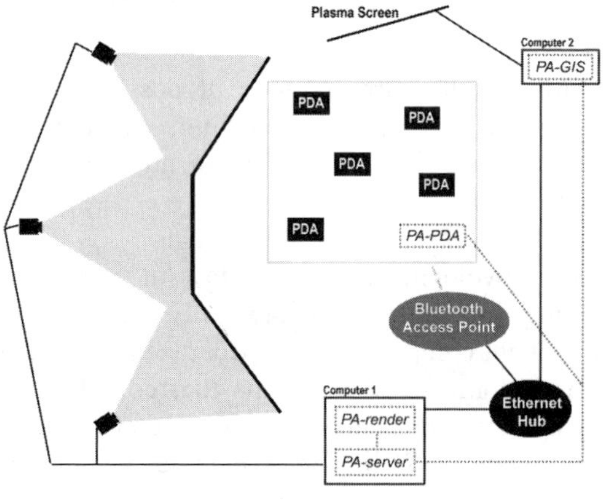

Figure 11.4 Example of a hardware setup for an envisioning system (after Stock and Bishop[17]).

Another approach to stakeholder participation is taken by Paez et al.[1]. Using a system dubbed DISCUSS they show how the spatially aggregated output of traditional cost-benefit analysis can be disaggregated using a combination of technical (process model based) and perceptual (fuzzy stakeholder input) mapping. DISCUSS works with a single user at a time who, with the aid of a trained operator, can input their perception of the distribution of costs and benefits by either:

- Agreeing with the outputs of a technical analysis
- Allocating costs and benefits to existing land parcels, or
- Drawing their own free-form polygons representing areas with different levels of impact.

In this last case the system will interpolate the mapping, using one of three different interpolation procedures, to give full spatial coverage of costs and benefits. Any output which does not fit the stakeholder perception can be adjusted iteratively. Thus, there should be no cases of '...that's not what I meant at all'. Once all the stakeholders have made their inputs, DISCUSS will map areas of consensus or dispute based on a selection of agreement metrics.

The trend towards recognition of individual preferences and behaviours is also manifest in the adoption of agent-based modelling[18] in decision-making contexts. Agent modelling is not itself a form of public participation, but the process of calibrating agent models requires close study of individuals through surveys, behavior monitoring or, eventually, observation in controlled virtual world conditions[19]. As this technology develops it provides another medium for public involvement. However the possibility exists that it could be used at either end of the Arnstein ladder – for manipulation or empowerment.

11.4 INTEGRATED TECHNOLOGIES: COLLABORATIVE WORLDS

Key factors determining the current range of possible approaches to public participation are: data availability, spatial modelling, presentation, networking and communications. Rapid changes are occurring in all these areas. Some which demand particular attention are:

Desktop graphics. Development happens fast in computer hardware – the famous Moore's law suggests a doubling of capability every 18 months. Even three years ago few people bought computers with specialized graphics cards; today they are virtually standard equipment. This means that complex 3D models can be explored interactively by most users – as they already do in computer games.

Spatial Data Infrastructures (SDI). While data has been collected digitally for sometime, and while this has increasingly been coordinated and made accessible on-line, the talk now is about adding a layer of widely accessible generic tools between the data and the user in order to allow individual value-adding to transparently available data[20]. Transparency is also aided by the development of spatial and domain specific ontologies.

Interactive linkages. Systems integration, especially using existing software packages and widely recognized standards and protocols (such as those being developed by the Open Geospatial Consortium[21]), is another trend that seems likely to accelerate in association with SDI.

Internet bandwidth. Enhanced connectivity will allow people to download complex 3D models in a reasonable time. Their graphics cards will give them the ability to move around these models in real-time. Another step forward is the process already prevalent in the world of computer gaming in which people can fight, or better collaborate, with each other through the web.

Having moved from expert-based citizen involvement in decision-making towards a more inclusive model supporting public forums and workshops, these developments will support the emergence of on-line collaborative visualization based on SDI. MacEachren and Brewer[22] and MacEachren[23] have explored this potential and developed an extensive conceptual framework for system development.

A sub-class of collaborative systems involves the use of virtual environments in which people appear as avatars and have an ability to observe and manipulate the environment in order to explore the decision space associated with a particular issue at a particular location. This scenario has a lot in common with computer games and so it is not surprising that commercial game engines are being used as development platforms for visualization[24] and also for collaborative virtual worlds[25-27]. Figure 11.5 shows example views of the system (SIEVE) which we are developing in the context of rural planning and salinity issues[26]. The initial challenges were:

- Automatic generation of virtual worlds from terrain, vegetation and built element data from the SDI
- Integration of above and below ground aspects of the salinity issue by joining hydrological modelling outcomes to realistic visualization of environmental consequences
- Development of collaborative meeting protocols and support systems

(a) Automatically generated virtual world.	(b) Linkage of procedural flood model to tree health.

Figure 11.5 Screen shots from the collaborative virtual environment system (SIEVE).

Figure 11.6 is a schematic view of our current developments and future plans which are described more fully in Bishop et al.[28] For example, the idea of providing visual representation of data to someone working in the field includes an augmented reality approach to presentation. A farmer can see a soils map draped over her paddock, can interactively plant new virtual trees into the landscape and, by sending these back through the network for server-side model processing, observe the effect of these on the water table beneath her own and surrounding properties.

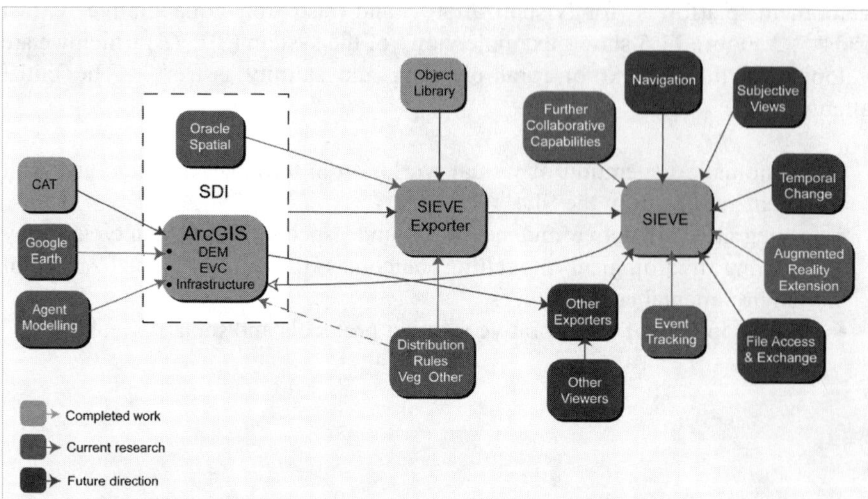

Figure 11.6 Schematic design of existing and future work towards a collaborative virtual environment.

The work to date is based on linkage of particular products: a geographic information system (ArcGIS®[29]) and a games engine (Torque[30]), but will eventually become more generic. We have developed procedures for passing data between these systems both as exported files and through a live link. These provide enormous developmental and operational flexibility.

Another key objective of our development is to support both expert users and the broader public in terms of their needs for information. The expert is typically willing to work with more abstract representations, seeks more interactivity and often works alone or with a small team. The public may be best supported by more realistic, natural modes of representation, may be content with less output or query options, but may be part of a larger group accessing the information through a planning workshop (same place) or on-line forum (different place). In addition, there are those, like our farmer above, for who the information is integral to their livelihood.

11.5 CONCLUSIONS

Technology, starting with GIS and moving into virtual worlds, has provided, and continues to provide, new opportunities for involving people in spatial decision-making. This rapid evolution has to some degree outstripped our knowledge of how the technologies may be most efficiently or appropriately applied. We also need further studies into the theory and application of technologies such as the collaborative virtual world proposed here. Do we seek to mimic face to face meeting or do we need other protocols? How does an on-line facilitator get the measure of his/her audience? These and related issues will be central to on-going research and development. As always, however, the success of systems for public involvement will depend upon freely available information and political will. Then there is a chance for win-win outcomes.

11.6 ACKNOWLEDGMENTS

Contributors to the recent work described here include Daniel Paez (DISCUSS); Christian Stock, Alice O'Connor and Alex Tao Chen (SIEVE); and Lucy Spottiswood (agent modelling). The work with SIEVE was funded by the CRC for Spatial Information. The agent modelling development is funded by the Melbourne University Research Grant Scheme (MRGS).

11.7 REFERENCES

1. Paez, D., Bishop, I.D., and Williamson, I.P., DISCUSS: A soft computing approach to spatial disaggregation in economic evaluation of public policies, *Transactions in GIS*, 10, 265-278, 2006.

2. Arnstein, S., A ladder of citizen participation, *Journal of the American Institute of Planners*, 35, 45-54, 1969.

3. Fabos, J. Gy., and Caswell, S.J., Composite Landscape Assessment: Assessment Procedures for Special Resources, Hazards and Development Suitability, Part II of the Metropolitan Landscape Planning Model, Massachusetts Agricultural Experiment Station, Amherst, 1977.

4. Cocks, K.D., Ive, J.R., Dans J.R., and Baird I.A., SIRO-PLAN and LUPLAN: An Australian approach to land-use planning. 1. The SIRO-PLAN land-use planning method, *Environment and Planning B: Planning and Design*, 10, 331-345, 1983.

5. McDonald, G.T. and Brown, A.L., The land suitability approach to strategic land-use planning in urban fringe areas, *Landscape Planning*, 11, 125-150, 1984.

6. Travis, M.R., Elsner, G.H., Iverson, W.D., and Johnson, C.G., VIEWIT: Computation of Seen Areas, Slope, and Aspect for Land-Use Planning, USDA Forest Service Gen. Tech. Rep. PSW-11/1975, Berkeley, California, 1975.

7. Tomlin, C. D. and Tomlin, S.M., An overlay mapping language, presented at the Annual Meeting of American Society of Landscape Architects, 1981.

8. Myklestad, E. and Wagar, J. A., PREVIEW: computer assistance for visual management of forested landscapes, *Landscape Planning*, 4, 313-331, 1977.

9. Bureau of Land Management, *Visual Simulation Techniques*, US Government Printing Office, US Department of the Interior, Washington, DC, 1980.

10. Benson, J.F., The visualization of windfarms, in *Visualization in Landscape and Environmental Planning*, Bishop, I.D. and Lange, E., Eds., Taylor & Francis, London, 2005, 184-192.

11. Klosterman, R.K., The What if? planning support system, in *Planning Support Systems: Integrating Geographic Information Systems and Visualization Tools*, Brail, R.K. and Klosterman, R.E., Eds., ESRI Press, Redlands, CA, 2001, 262-284.

12. Kwartler, M. and Bernard, R.N., CommunityViz: an integrated planning support system, in *Planning Support Systems: Integrating Geographic Information Systems and Visualization Tools*, Brail, R.K. and Klosterman, R.E., Eds., ESRI Press, Redlands, CA, 2001, 285-308.

13. Bishop, I. D., Hull, R. B., and Stock, C., Supporting personal world-views in an envisioning system, *Environmental Modelling & Software*, 20, 1459-1468, 2005.

14. Tufte, E.R., *Envisioning Information*, Graphics Press, Cheshire, CT, 1990.

15. Brail, R.K. and Klosterman, R.E., Eds., *Planning Support Systems: Integrating Geographic Information Systems and Visualization Tools*, ESRI Press, Redlands, CA, 2001.

16. Kwartler M., Visualization in support of public participation, in *Visualization in Landscape and Environmental Planning*, Bishop, I.D. and Lange, E., Eds., Taylor & Francis, London, 2005, 251-260.

17. Stock, C. and Bishop, I.D., 2005., Helping rural communities envision their future, in *Visualization in Landscape and Environmental Planning*, Bishop, I.D. and Lange, E., Eds., Taylor & Francis, London, 2005, 145-151.

18. Arthur, W. B., Designing economic agents that act like human agents - a behavioral approach to bounded rationality, *American Economic Review*, 81, 353-359, 1991

19. Spottiswood, L. and Bishop, I.D., An agent-driven virtual environment for the simulation of land use decision-making, in *Proceedings of the International Congress on Modelling and Simulation*, Melbourne, December 12-15, 2005, 3085-3091.

20. Williamson, I., Land administration and spatial data infrastructures: trends and developments, in *Proceedings of XXII FIG International Congress*, Washington, DC, 2002.

21. Open Geospatial Consortium, http://www.opengeospatial.org, 2006.

22. MacEachren, A.M. and Brewer, I., Developing a conceptual framework for visually-enabled geocollaboration, *International Journal of Geographical Information Science*, 18, 1-34, 2004.

23. MacEachren, A.M., Moving geovisualization toward support for group work, in *Exploring Geovisualization*, Dykes, J., MacEachren, A.M., and Kraak, M-J., Eds, Elsevier, Amsterdam, 2005, 445-461.

24. Herwig, A., Kretzler, E., and Paar, P., Using games software for interactive landscape visualization, in *Visualization in Landscape and Environmental Planning*, Bishop, I.D. and Lange, E., Eds., Taylor & Francis, London, 2005, 62-67.

25. Bishop, I.D., Stock, C., O'Connor, A., Csaky, D., Pettit, C., and Creasey, J., Interfacing visualisation with SDI for collaborative decision-making, presented at the Conference of the Spatial Science Institute, Melbourne, September 12-16, 2005.

26. O'Connor, A., Stock, C., and Bishop, I., SIEVE: An online collaborative environment for visualising environmental model outputs, in *Proceedings of the International Congress on Modelling and Simulation*, Melbourne, December 12-15, 2005, 3078-3084.

27. Stock, C., Pettit, C., Bishop, I. D., and O'Connor, A N., Collaborative decision-making in an immersive environment built on online spatial data integrating environmental process models, in *Proceedings of the International Congress on Modelling and Simulation*, Melbourne, December 12-15, 2005, 3092-3098.

28. Bishop, I.D., Stock, C., Pettit, C., Aurambout, J-P, Chen, T., O'Connor, A., and Spottiswood, L., Prospects and plans for a fully integrated collaborative virtual environment: from SDI to AR and back, *Cartography and Geographic Information Science* (to appear in special issue on collaborative GIS), 2007.

29. ESRI, ArcGIS 9.1, Environmental Systems Research Institute, http://www.esri.com, 2005.

30. GarageGames, Torque Game Engine, http://www.garagegames.com, 2006.

CHAPTER 12

Using Virtual Reality to Simulate Coastal Erosion: A Participative Decision Tool?

I. Brown, S. Jude, S. Koukoulas, R. Nicholls, M. Dickson and M. Walkden

12.1 INTRODUCTION

Decision-making in the coastal zone requires recognition of the complex dynamic interactions occurring between each and all components of the coastal system, both natural and societal. Such awareness represents a key concept in the implementation of Integrated Coastal Zone Management (ICZM), and this in turn is leading to recognition of the need for both increased participation in the decision-making process and to develop longer planning horizons[1,2]. Nevertheless, the complexity of coastal systems means that there are difficulties involved in communicating technical issues to a lay audience as a prelude to their participation in strategic decision-making. Therefore, a critical step in the process is the development of tools that compile, synthesise and communicate information on natural hazards in order to develop effective risk mitigation strategies[3]. In this chapter, we explore a Virtual Reality (VR) approach to meet this challenge, with particular emphasis on communicating the risk associated with coastal erosion in cliff areas up to the year 2100.

The case study area is located on the eastern coast of England and features extensive sea cliffs (up to 60 m in height) that extend from Sheringham to Sea Palling in north Norfolk (Figure 12.1). The cliffs are prone to rapid erosion because they consist of thick sequences of poorly-consolidated Quaternary deposits, with typical current erosion rates of the order of 0.5-2 m/year, therefore posing a substantial threat to cliff-top human infrastructure[4].

Climate represents a major driving influence on coastal systems. Although relative sea level has been slowly rising in eastern England for centuries due to land subsidence, a warming climate will most likely lead to an acceleration in the rate of sea level rise due to thermal expansion of the oceans and melting of land ice (cf. Hulme et al.[5]). It is also probable that patterns of storm activity will become modified, which will impact on coastal areas. Amongst the anticipated impacts, cliff coastlines have been suggested as one of the coastal types that will be sensitive to increased erosion[6]. Coastal planners in the area of study have therefore requested guidance on the demarcation of the possible erosion hazard zone over the coming decades. This will inform an ongoing dialog on future cliff-top development and land use, including possible relocation of infrastructure and other assets.

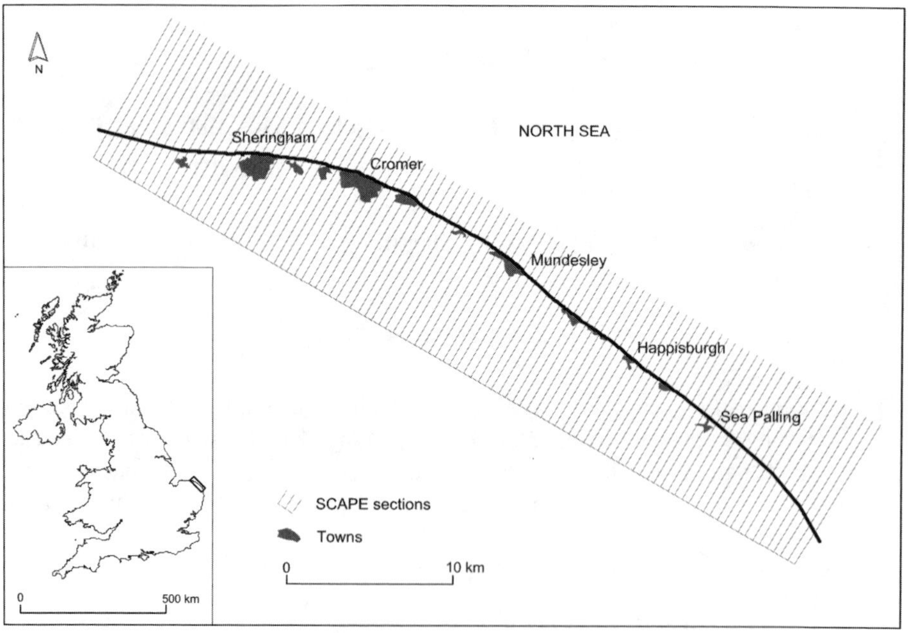

Figure 12.1 Location of the study area, showing the series of sample cross-sections along which the SCAPE model was run to predict coastal recession.

12.2 SIMULATION AND VISUALIZATION CONCEPTS

Simulation models provide, after a process of calibration and validation, composite data series that aim to synthetically replicate an event or sequence of events through time. By linking the model with a GIS, the ability to query, contextualize and communicate the raw model data can be considerably enhanced through a more complete spatial representation[7,8]. Two major difficulties arise within GIS during this coupling process: (i) the handling of temporal data relating to the modelling and analysis of change; and (ii) the inclusion of the inevitable uncertainties that accrue from data and model errors. Our approach is to explore the utility of 3D visualization in elucidating some of these issues further and to emphasise enhanced interaction in bridging the gap between scientists and two distinct sets of users: coastal decision makers and the wider general public living in coastal areas.

Visualization has been seen as an important tool in the consultation process for some time, from artistic impressions or scaled physical models, through to the development of computer images. Recent extension of this process into interactive 3D 'virtual landscapes' has been recognized as potentially providing a further useful step in engaging not only specialists, but also the wider public in participative planning issues related to coastal zone management[9]. By giving users a greater 'sense of place'[10,11], virtual landscapes can enhance the ability of participants to situate themselves within the environment, and hence by observing a process of simulated change, become more aware of its implications for the landscape[12].

In terms of visualizing future landscapes, methods have as yet tended to be based upon rather arbitrary 'what if' scenarios and an objective basis for these scenarios has been lacking. The opportunity to explore and explain scientific or decision-making issues embedded in the development of future scenarios has therefore not been taken, and the key processes driving landscape change tend not to be transparent to the user. By providing a more direct link between visualization and the scientific model used to produce the future predictions we aim to respond to this criticism, and therefore provide a more objective and rational basis for stimulating the necessary dialog on alternative management strategies.

An evaluation of VR techniques within the visualization process requires an assessment of what actually constitutes the baseline 'reality' against which it is inevitably compared. Following the concepts of Baudrillard[13], the use of VR means we are actually adopting a 'hyper-reality' approach in which we seek to intensify experience relative to the real world in order to enhance conceptual awareness of it. Scientific analysis through model-based simulation can potentially provide a replicable, rational and transparent method to explore the complex processes of the real world within a structured framework. VR visualization provides a method to replicate the key features of the landscape in an accessible and stimulating format. Both simulation and visualization involve, to varying extents, the simplification of a complex environment in order to facilitate representation and cognition. It should be apparent therefore that any resulting information cannot be an exact replication of the real world, and some discrepancy or deficiency is inevitable. However, as individual perceptions of reality vary according to societal or institutional contexts, identifying and explaining these distinctions is often dependent on the participant[14].

GIS has provided a common framework in which to link model data and visualization software through the exchange of data files (Figure 12.2). For pragmatic reasons, related to the rapid recent development of VR software, we are therefore adopting a loosely-coupled approach rather than one of full integration between components (cf. Goodchild et al.[7]). The ultimate aim, however, remains to develop a more integrated system of interoperable components, minimizing the external exchange of data, similar to that described by Bernard and Kruger[15] for atmospheric modelling.

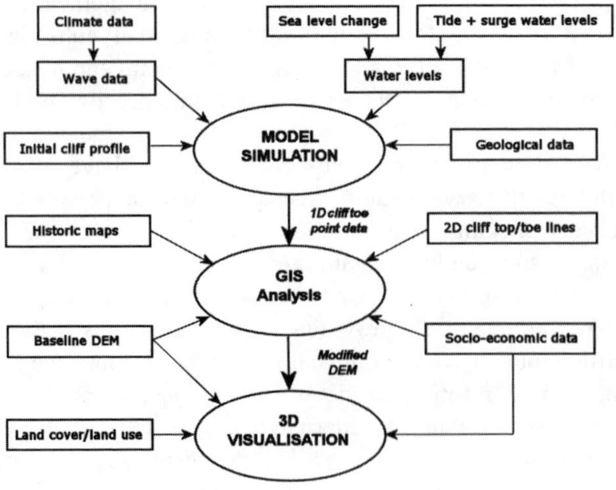

Figure 12.2 Linkages and data flows for the combined cliff erosion information system. Further details on the visualization process are shown in Figure 12.6.

12.3 SIMULATION MODELLING OF COASTAL EROSION

SCAPE (Soft Cliff and Platform Erosion) is a process-based numerical model that determines the reshaping and retreat of shore profiles along the coast[16,17]. Coastal recession proceeds through cycles of beach lowering from longshore transport, shore profile erosion, cliff toe retreat (the toe represents the base of a cliff where it meets the shore platform or beach) and the release of beach sediments from the cliff and shore platform. Erosion is driven by changes in hydrodynamic conditions; therefore key inputs for the model are waves, tides and sea level. For each of these inputs, a long time-series of historic data was compiled and generated, as a baseline for the future projections. Engineering interventions in the coastline were accounted for through the specific inclusion of seawalls (to protect cliffs) and groynes (to enhance beach volumes) at known locations.

The model operates by calculating, at every tide, the amount of sediment released from the cliff and platform, changing beach volumes, and the volume of beach material that is moved alongshore by wave action. This method has the advantage that the sediment budget is quantified in far more detail than is possible with a traditional conceptual geomorphic model. Model cross-sections of the coastline were aligned at 30 degrees from north and were spaced at 500 m intervals, as illustrated in Figure 12.1.

For the present study, the evolving shore morphology output by the model was reduced to provide only the developing position of the cliff. We also assumed that the cliff top retreated at the same rate as the cliff toe; ongoing work will extend SCAPE with a probabilistic land-sliding module to better represent the more stochastic behavior of cliff-top retreat (large events can occur here as well as more frequent smaller events). Once calibrated, the model was validated against a series of historic maps (digitized by the British Geological Survey) and this demonstrated that it could accurately represent 117 years of historic cliff toe evolution, providing a firm basis on which to predict future erosion from now until the year 2100[17,18].

Two sources of changes in future conditions were included in the simulations:

- Changes in coastal management strategies, e.g., through modifying, moving or removing protective structures;
- Climate change via two contrasting scenarios derived from Hulme et al.[5] and defined as follows:

(i) a low scenario assuming that sea level rise occurs at an average rate of 2 mm/year and the wind wave regime remains the same as present;

(ii) a high scenario in which sea level rise occurs at an average rate of 10 mm/year and wind speeds increase incrementally to reach a value 10% higher than present by 2100, with consequent effects on the wave regime.

Data flows within the analysis are shown in Figure 12.2. The SCAPE model used data on geological composition and cliff strength, waves, tides and history of engineered interventions to calculate the influence of climate change and sea level rise on the shore profile. The cliff toe positions were then imported into the GIS, where they were geo-referenced, integrated with other spatial datasets and used to edit a baseline digital elevation model (DEM). The modified DEM, representing future cliff positions, combined with other contextual data was then passed to the visualization system for specialist graphical rendering.

12.4 INTERFACE WITH GIS

The SCAPE model provides as direct output, numeric predictions of the amount of cliff recession expected to occur for a particular sample site. Expert analysis can infer rates of retreat from this data and, by comparing data from several sites, deduce the wider context such as hazard zones and infrastructure at risk. For other potential users, however, the information needs further development to facilitate interpretation of its spatial and temporal content relative to clear visual reference points. Furthermore by importing the data into a GIS, the user potentially has more flexibility in deriving relevant information based upon their own particular context and requirements.

Coupling of the simulation model and GIS required transformation of the numerical model data from the sample locations into a more complete representation of the coastline, and then integration with other contextual information. An important data requirement was the present-day cliff-line, which can generally be obtained from topographic map data in vector format. Alternatively, by using a digital elevation model (DEM), the abrupt break of slope present at both the top and base of the cliff can be used to delineate the top and toe of the cliff in digital format.

12.4.1 Interpolation from 1D to 2D

Initial import of the cliff recession simulation data was achieved through a series of Avenue scripts in the ArcView® GIS which then allowed more detailed manipulation using the map algebra capability of Arc/Info™ GRID. A spatial database containing the co-ordinates of each point, attributed with the relevant year from the SCAPE simulation, was produced by trigonometry using the recession distance along the profile line at 30° from north. This provides a series of geo-referenced points indicating future cliff positions for the sample locations. To obtain a more complete representation of the changing coastline, it was then necessary to interpolate this point data into a 2D feature. This was achieved by topologically linking together all points from a particular recession year and then importing the resulting nodes and segments as a single polyline (Figure 12.3).

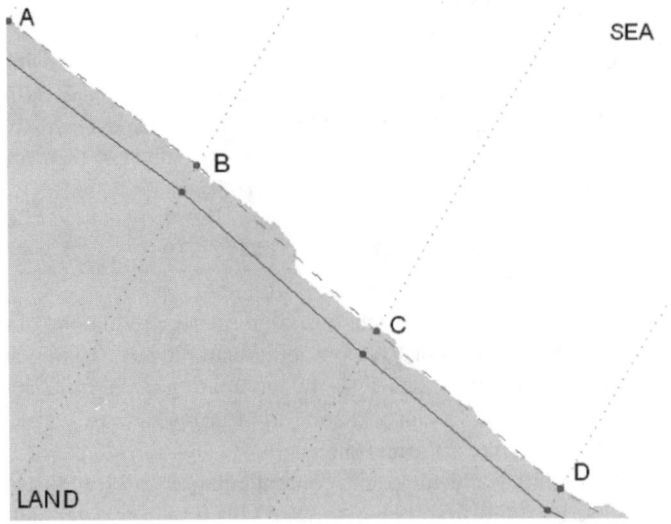

Figure 12.3 Future cliff recession data derived from the SCAPE model and imported into the GIS. The dotted line represents the SCAPE profile lines, the solid line a future 2100 scenario cliff position and the dashed line illustrates the diverge of a straight line segment from the actual cliff outline.

Production of a 2D representation of the changing coastline and integration of additional contextual information (e.g., land use data) in the GIS meant that it was now possible to visualize and calculate the area of land lost between a particular future year and the baseline year. This helped to make the simulation results more accessible to coastal managers. To facilitate areal calculations, a script was created to produce a polygon structure linking the two relevant polylines together at their end points (Figure 12.4). This could then be used, for example, to calculate the area between the present-day and 2100 cliff-line projections and in combination with a land use dataset also made it possible to estimate the proportion of different land use classes that might be lost in the future (in the absence of proactive planning responses).

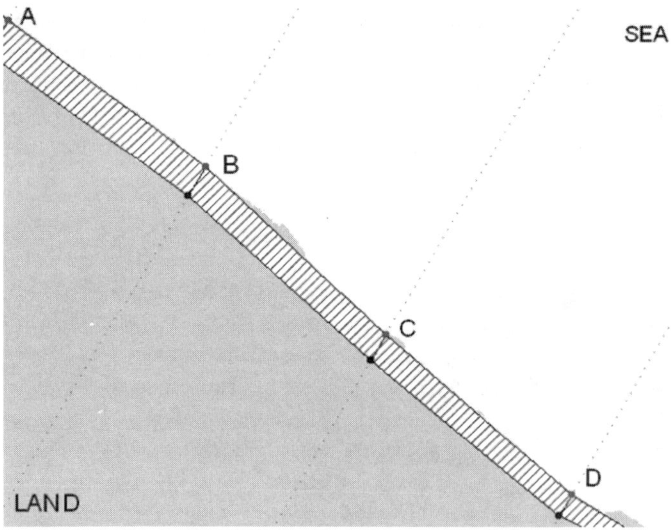

Figure 12.4 Polygons defining the area of land lost by cliff recession between a future reference year and the current position. The outline of the present position of the cliff is also shown where it differs from the straight line segment connecting SCAPE sample points.

For low resolution analysis, linking points together directly via straight-line segments is certainly adequate, but user demands for high-resolution visualization implied that further refinement of this procedure was required in order to meet their expectations. Therefore the above methodology was modified in order that future cliff-line positions could also include additional local detail in the form of inlets and headlands between the 500 m spacing of SCAPE sample points, as these are otherwise excluded by a simple straight line segment (Figure 12.3).

This high-resolution refinement used the 2D representation of the present cliff-line and projected it to a new position based upon linear interpolation of recession distances with respect to the simulation data from the two nearest SCAPE sample points (Figure 12.5). Therefore, if an intermediate point C(x_c,y_c) was closer to one of the SCAPE points A (x_a,y_a) than the other B (x_b,y_b), then the interpolated recession distance (*RDc*) was more similar to the modelled recession at A (*RDa*) than B (RD_b). This procedure can be represented by the following formula:

$$RD_c = RD_a + \frac{\sqrt{(x_c - x_a)^2 + (y_c - y_a)^2}}{\sqrt{(x_c - x_a)^2 + (y_c - y_a)^2} + \sqrt{(x_c - x_a)^2 + (y_c - y_b)^2}}(RD_b - RD_a)$$

Following the same trigonometric procedure outlined previously, intermediate nodes between SCAPE sample points were transformed from their current to future positions using the interpolated recession distance along the 30°N profile line utilized by SCAPE. Connecting together the new series of points provided the new recession line in more detailed form (Figure 12.5).

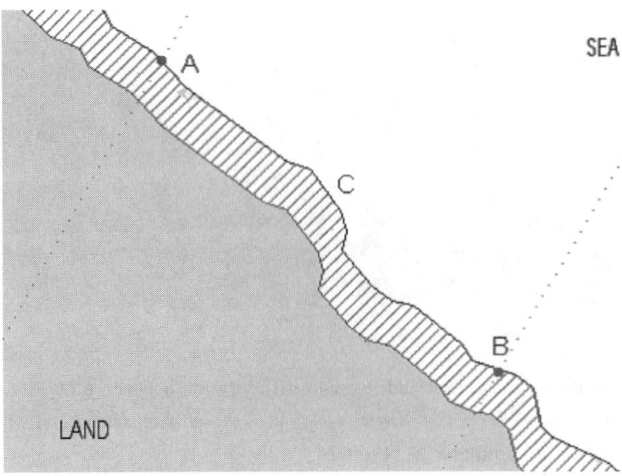

Figure 12.5 Linear interpolation of intervening points on the cliff between two SCAPE sample locations 500 m apart (A and B). The hatched area represents the land lost using this method between the present cliff position and its future location.

For some two km sections of the resulting cliff-line the full representation contained over a thousand individual nodes. Even for detailed visualization, many of these node points were superfluous and their presence substantially increased the calculation time to interpolate future cliff positions. For the purposes of the present study lines were therefore simplified by utilizing the modified version of the

Douglas-Peucker algorithm[19] as employed by Arc/Info™. The amount of line generalization applied by this algorithm can be varied by adjusting the amount of tolerance employed to weed out insignificant points: experimentation suggested that setting the tolerance value to 5 m was adequate to include all the salient features of the cliff outline.

If we zoom in to a local level, as high-resolution visualization encourages, then it can be seen that the end-result of the refined interpolation process was a future cliff outline more consistent with the present cliff outline (Figure 12.5). However, it should be noted that in terms of the amount of land or sediment lost, then the differences in value between the simple and the refined interpolation methods was generally quite small and given the other uncertainties, virtually negligible. The main advantage of the refined procedure is that it produces a visually more appealing representation, because it includes small-scale irregularities in the coastline, and therefore appears more realistic.

12.4.2 Extrapolation into 3D

The development of virtual landscapes requires extrapolation of the recession data into 3D in order to provide the basis for the additional graphical detail. This necessitates integration of the model data with a DEM. Initially, this step in the analysis utilized conventional DEM products such as the Ordnance Survey Land-Form PANORAMA® and PROFILE® datasets, but it became apparent that these could not provide an adequate baseline due to the level of accuracy required and also the rapid rate of cliff recession that has occurred since the date of survey. Problems were particularly evident where the removal of defense structures has triggered high recession rates as the cliff attempts to reach an equilibrium position with respect to the rest of the coast.

Fortunately, recent technological advances in survey instrumentation meant that the research was able to acquire high-resolution LiDAR altimetry data (from the Environment Agency). With a typical vertical ground error of ca. ±15 cm, LiDAR surveys have considerable potential in providing accurate and detailed representations of landscape features or morphology[20,21]. Each two km tile consisted of data on a 2 m grid of horizontal resolution which had been post-processed to remove objects such as houses and trees, and therefore provide values reflecting true ground elevation[22]. Further anomalies such as water bodies that can absorb the LiDAR signal and return 'no data' were corrected by interpolation based upon surrounding data values.

Once the LiDAR DEM data had been assembled, the 2D outline of the simulated cliff base for 2100 was overlaid upon the reference DEM. Using a series of map algebra and reclassification procedures in Arc/Info™ GRID, the area of the DEM between the present and future cliff base was defined, and then edited in order to reduce the elevation to values representing the shore platform and beach. This data manipulation replicated erosion of the cliff in this zone down to the beach or platform elevation. By consulting the full SCAPE simulation, elevation values for the evolving shore profile in the new DEM could be applied so that they remained consistent to the model.

The method outlined above allows for extrapolation of the cliff base, but given the readily-erodible material present in the study area it was unrealistic to assume vertical cliffs. Generally, the zone between the cliff top and cliff base is fairly wide (up to 200 m in width), which reflects the relatively low slopes (30°-50°) present. The actual detail of the cliff slope was not considered important in the present study, although in fact the LiDAR survey defines its current form with a high degree of precision providing useful proxy information on cliff stability[20]. Predicting future slope evolution with any certainty was not really possible, however, due to the sporadic stochastic processes operating there; in any case, this information is merely supplemental for any risk assessment which essentially demarcates the hazard by the cliff top. Therefore the morphology of the present-day slope area was maintained into the future by isolating it, then applying an affine transformation to that area using the set of present and future SCAPE simulation points as source and destination projections respectively. Finally, the slope data was re-integrated into the zone between the future cliff base and cliff top using the grid stitching procedures available in the Mosaic and Merge functions of Arc/Info™ GRID.

12.5 VR VISUALIZATION

The conventional 2D representation of the SCAPE simulation provided a straightforward means of displaying the raw simulation data, and may be most suitable for coastal managers who are familiar with the area of study. Conversely, it may prove less helpful to those who have less experience of detailed maps or planning documents, such as members of the public. The integration of DEM and simulation data (Section 12.4.2) introduced the possibility of extending visualization into 3D space, through the development of virtual landscapes. Potentially, this enhancement of the visual content not only increases the ability to perceive change, but through the VR technology promotes a greater level of interaction with the underlying knowledge base. In this context, the study builds on previous research which has demonstrated how techniques for visualization of future coastal states may be developed, and how they may be of potential use in coastal decision-making[9,23].

The basis for the VR visualizations was to develop landscape content based upon large-scale vector land cover data (derived from the Ordnance Survey MasterMap® dataset) and the modified LiDAR DEMs. A land cover classification was created by editing the MasterMap® data and then used to generate a series of textures that could be applied in the VR software. In addition to the present-day DEM, the SCAPE GIS interface provided the edited DEM that showed the predicted future evolution of the cliff; the VR visualization concentrated on the DEMs for the year 2100. Future land cover change coinciding with cliff recession was identified by using predictive scenarios based upon expert judgement on likely future conditions in the study area[23]. Further work is currently extending these scenarios in order to explore a wider range of possible future socio-economic changes that may impact upon the landscape.

Two approaches were used to create the visualizations:

- World Construction Set (from 3D Nature[24]) enabled the creation of a series of static images illustrating temporal changes in cliff recession for specific sites of interest, with animations also being produced to view the resulting cliff evolution.
- Interactive VR representations of the present and future coastlines were produced by importing the GIS data into real-time simulation software such as Terra Vista[25] using the process outlined in Figure 12.6.

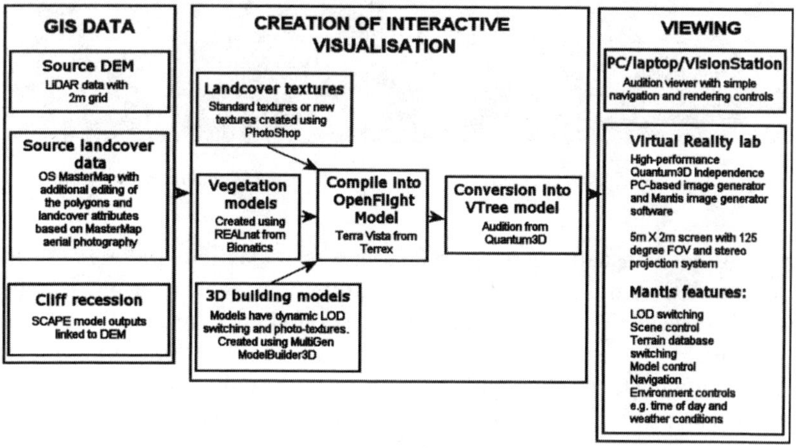

Figure 12.6 The process used to create the interactive virtual reality visualizations.

Once generated, the interactive visualizations can be explored via a series of different media, such as the following:

- Standard PCs and desktop VR packages based upon plug-in software and browsers.
- An Elumens VisionStation[26], which involves a portable hemispherical screen and provides a more immersive visual experience for a single user.
- A virtual reality laboratory, allowing a group-based immersive experience, such as that available at the University of East Anglia[27] (Figure 12.7)

Figure 12.7 Viewing an interactive visualization in the UEA virtual reality laboratory[27].

Creating such visualizations therefore makes it possible to interact with the coastal information in a range of situations, from management or public meetings, to doorstep surveys, in order to elicit opinions on alternative coastal management strategies. Future scenarios of coastal erosion are generally presented together in order to communicate the principle that no definitive prediction of the future can be generated, and that the hazard is best demarcated by a zone of risk (e.g., Figure 12.8). Ongoing work is now developing this principle further, by extending the SCAPE simulation model via a probabilistic cliff-top retreat module, with the objective of quantifying the level of risk within this zone.

12.6 DISCUSSION

Interaction between the research team and decision-makers or the wider public has generated an interesting array of responses on the simulated virtual landscapes. Undoubtedly, the extra element of 'realism' provided by the VR models has most appeal to non-experts when engaged in discussion regarding the complexities of future change. With regard to future coastal planning, and especially because of the uncertainty introduced by climate change, VR representations therefore potentially provide participants with a distinctive 'window' of insight into future trends[23]. In this context, the more abstract visualization techniques often favored by specialists engaging in data exploration tend to be rejected, even amongst experienced coastal managers, in favor of those incorporating a high level of realism. One of the factors which has influenced this, and also acted to increase user expectations, is the diffusion of multimedia technologies into wider society.

Figure 12.8 Screen-shot of one of the future VR visualizations of the cliff environment. The DEM represents the predicted situation under the low climate change scenario and gray shading defines the cliff-top area between the low and high scenarios, therefore indicating the likely risk zone.

Although we can successfully integrate temporal simulation and geo-spatial representations through a linked VR-GIS approach, the user-driven 'demand' for realistic information can lead to a tension between the processes of model-based simulation and the resulting visualization. This schism arises because it is extremely difficult to produce the quality of data that is scientifically credible, yet also provide the realistic information that end-users believe they require. The general perception amongst many users is often that the more detail that exists within a visualization, then the more accurate and believable it is[28]. This is perhaps reinforced by the increasing ability to generate detailed VR representations of the present landscape based upon high-resolution survey instruments such as LiDAR. However, to create the future visualizations in this study, at an equivalent level to the present-day, required the interpolation of significant information beyond the resolution of the SCAPE model; in effect, the model data has been interpolated from 1D to 3D. This poses a severe challenge for science-based visualization because a contradiction emerges between simulation modelling, which essentially has to generalize the real-world in order to conceptualize and replicate events, and the visualization process which is increasingly judged by its realism. One can argue that the visualization may become too good if it deceives the viewer concerning the quality of the underlying scientific knowledge that supports it. There is a risk, therefore, that VR visualizations, despite proving very popular with non-experts, potentially add new information that is not scientifically credible. At the very least, this extra information objectively represents only one credible interpolation of the model outputs, and there is a large set of other possible interpretations that remain unarticulated.

For assessing and communicating future landscape change, we therefore need to recognize two distinct sources of uncertainty within visualizations:

- The possibility of multiple states
- Multiple representations of each state

Concepts such as scenarios, which implicitly refer to uncertainty and are a common convention in the scientific world[29], can partly acknowledge the incomplete level of information on the likely future landscape state. However, it needs to be more strongly recognized that communication of both sources of uncertainty to the lay person remains a significant barrier; generally, communication media prefer crisp simple messages and uncertainty usually implies a more sustained or complicated level of information exchange.

The scope provided by VR technology to engage a new and wider audience undoubtedly represents a powerful tool for the scientific and geo-information community. Nevertheless, the present study underscores the need for more transparency within the visualization process, together with the adoption of a series of common guidelines, and most of all, for research exploring the development of new interaction techniques[30,31]. As has been similarly argued with respect to maps[32], and to GIS[33,34], visualization does not provide a neutral or objective representation. Our work strongly implies the need for further research on visualization techniques in order that they can communicate uncertainty more effectively. Significantly, the few visual communication techniques that at present include some element of uncertainty have tended not to quantify it, but rather to indicate a qualitative level of certainty:[35,36] is this the most appropriate way to interface with key users, such as decision-makers? The range of methods to be evaluated includes the use of split screens to communicate alternative multiple scenarios, visual manipulations such as screen 'flickering', animation, fuzzy boundaries, and error contours on the landscape. By analogue, animation has now become a common technological metaphor employed to communicate temporal change[37,38]. Representing uncertainty therefore seems to be essentially a cognitive challenge rather than a technological impasse[39].

12.7 CONCLUSIONS

Coupling a simulation model with GIS and VR software provided a means of generating realistic visualizations of the evolving coastal environment. This, in turn, has enabled an exploration of the role of these techniques within the decision-making process. The response from users has been that the additional visual detail and realism that the VR process provides is valuable, and that they have an important role both for communicating risk and in the local planning process. Nevertheless, to achieve this level of interest, it was necessary to extrapolate data beyond that provided by the original simulation model, which some may argue distorts the scientific rationale behind the future predictions. In this context, we conclude that geographic information science (GISc) is entering the science-policy interface characterized as 'post-normal' science by Funtowicz and Ravetz[40]. Clearly, this is an important position for GISc and one that emphasizes the need for further research on the quantification, representation and communication of uncertainty. In particular, effective risk assessment for natural hazards, such as coastal erosion, intrinsically requires an awareness of this uncertainty, as a key element of strategic decision-making[41].

12.8 ACKNOWLEDGMENTS

This research was funded by the Tyndall Centre for Climate Change Research as part of the 'Sustaining the Coastal Zone' research theme (www.tyndall.ac.uk). Wave transfer functions were developed by Professor Peter Stansby and colleagues in the School of Civil Engineering, UMIST. Wave data were supplied by HR Wallingford. The SSEVREL virtual reality facility at the University of East Anglia was funded by HEFCE. Historical mapping data were provided by the British Geological Survey, LiDAR data by the Environment Agency, and MasterMap® and Land-Form PROFILE® data by the Ordnance Survey.

12.9 REFERENCES

1. Turner, R.K., Lorenzoni, I., Beaumont, N., Langford, I., and McDonald, A., Coastal management for sustainable development: analysing environmental and socio-economic change on the UK coast, *Geographical Journal*, 164, 269-281, 1998.

2. King, G., The role of participation in the European demonstration projects in ICZM, *Coastal Management*, 31, 137-143, 2003.

3. Doody, J.P., Information required for Integrated Coastal Zone Management: conclusions from the European demonstration programme, *Coastal Management*, 31, 163-173. 2003.

4. Halcrow, *Kelling to Lowestoft Ness Shoreline Management Plan - Consultation Document*, Halcrow, Peterborough, 2004.

5. Hulme, M., Jenkins, G.J., Lu, X., Turnpenny, J.R., Mitchell, T.D., Jones, R.G., Lowe, J., Murphy, J.M., Hassell, D., Boorman, P., McDonald, R, and Hill, S., *Climate Change Scenarios for the United Kingdom: The UKCIP02 Scientific Report,*' Tyndall Centre for Climate Change Research, UEA, Norwich., 2002.

6. Nicholls, R.J., Dredge, A., and Wilson, T., Shoreline change and fine-grained sediment input: Isle of Sheppey coast, Thames estuary, UK, in *Coastal and Estuarine Environments: Sedimentology, Geomorphology and Geoarchaelogy*, Pye, K. and Allen, J.R.L., Eds., Geological Society of London Special Publication 175, Geological Society of London, London, 2000, 305-315.

7. Goodchild, M.F., Steyaert, L.T., and Parks, B.O., Eds., *GIS and Environmental Modelling: Progress and Research Issues*, GIS World Books, Fort Collins, Colorado, 1996.

8. Clarke, K.C., Parks, B.O, and Crane, M.P., Eds., *Geographic Information Systems and Environmental Modeling*, Prentice Hall, Upper Saddle River, New Jersey, 2002.

9. Jude, S.R., Jones, A.P., Andrews, J.E., and Bateman, I.J., Visualisation for participatory coastal zone management: a case study of the Norfolk coast, England, *Journal of Coastal Research*, 22, 1527-1538, 2006.

10. Brown, I.M., Developing a virtual reality user interface (VRUI) for geographic information retrieval on the Internet, *Transactions in GIS*, 3, 207-220, 1999.

11. Dykes, J.A., Moore, K.M., and Wood, J., Virtual environments for student fieldwork using networked components, *International Journal of Geographic Information Science*, 13, 397-416, 1999.

12. Brown, I.M., Kidner, D.B., Lovett, A., Mackanes, W., Miller, D.R., Purves, R., Raper, J., Ware, J.M., and Wood, J., Introduction to virtual landscapes, in *Virtual Reality in Geography*, Fisher, P. and Unwin, D., Eds., Taylor and Francis, London, 2002, 95-101.

13. Baudrillard, J., *Simulations*, Semiotext, New York, 1983.

14. Foster, D. and Meech, J.F., Social dimensions of virtual reality, in *Simulated and Virtual Realities*, Carr, K. and England, R., Eds., Taylor and Francis, London, 1995, 209-223.

15. Bernard, L. and Kruger, T., Integration of GIS and spatio-temporal simulation models: interoperable components for different simulation strategies, *Transactions in GIS*, 4, 197-215, 2000.

16. Walkden, M.J.A. and Hall, J.W., A predictive mesoscale model of the erosion and profile development of soft rock shores, *Coastal Engineering*, 52, 535-563, 2005.

17. Dickson, M.E., Walkden, M.J.A., and Hall, J.W., Systematic impacts of climate change on an eroding coastal region over the twenrty-first century, *Climatic Change*, in press, 2007.

[18.] Brown, I., Jude, S., Koukoulas, S., Nicholls, R., Dickson, M., and Walkden, M., Dynamic simulation and visualisation of coastal erosion, *Computers, Environment and Urban Systems*, 30, 840-860, 2006.

[19.] Douglas, D.H. and Peucker, T.K., Algorithms for the reduction of the number of points required to represent a digitized line or its caricature, *Canadian Geographer*, 10, 112-123, 1973.

[20.] Adams, J.C. and Chandler, J.H., Evaluation of lidar and medium scale photogrammetry for detecting soft-cliff coastal change, *Photogrammetric Record*, 17, 405-418, 2002.

[21.] French, J.R., Airborne LiDAR in support of geomorphology and hydraulic modelling, *Earth Science Processes and Landforms*, 28, 321-335, 2003.

[22.] Brovelli, M. A., Cannata, M., and Longoni, U.M., LIDAR data filtering and DTM interpolation within GRASS, *Transactions in GIS*, 8, 155-174, 2004.

[23.] Jude, S.R., The application of visualisation techniques for coastal zone management, PhD thesis, University of East Anglia, Norwich, 2003.

[24.] 3D Nature, World Construction Set software, http://3dnature.com, 2006

[25.] Terrex, Terra Vista software, http://www.terrex.com, 2006

[26.] Elumens, Elumens display solutions, http://www.elumens.com, 2006

[27.] Social Science for the Environment, Virtual Reality and Experimental Laboratories (SSEVREL), http://www.uea.ac.uk/cm/home/schools/sci/env/research/ssevrel, 2007

[28.] Appleton, K. and Lovett, A., GIS-based visualisation of rural landscapes: defining 'sufficient' realism for environmental decision making, Landscape and Urban Planning, 65, 117-131, 2003.

[29.] Shearer, A.W., Approaching scenario-based studies: three perceptions about the future and considerations for landscape planning, *Environment and Planning B: Planning and Design*, 32, 67-87, 2005.

[30.] Sheppard, S.R.J., Guidance for crystal ball gazers: developing a code of ethics for landscape visualization, *Landscape and Urban Planning*, 54, 183-199, 2001.

[31.] Lange, E., The limits of realism: perceptions of virtual landscapes, *Landscape and Urban Planning*, 54, 163-182, 2001.

[32.] Wood, D., *The Power of Maps*, Guilford Press, New York, 1992.

[33.] Pickles, J., Ed., *Ground Truth: The Social Implications of Geographic Information Systems*, Guilford Press, New York, 1995.

[34.] Pickles, J., Arguments, debates and dialogues: the GIS-social theory debate and the concern for alternatives, in *Geographical Information Systems – Volume 1: Principals and Technical Issues*, Longley, P.A., Goodchild, M.F., Maguire, D.J., and Rhind, D., Eds., Wiley, Chichester, 1999, 49-60.

[35.] Hunter, G.J. and Goodchild, M.F., Communicating uncertainty in spatial databases, *Transactions in GIS*, 1, 13-24, 1996.

[36.] Evans, B.J., Dynamic display of spatial data-reliability: does it benefit the map user?, *Computers and Geosciences*, 23, 409-422, 1997.

[37.] Raper, J.F., McCarthy, T., and Williams, N., Georeferenced 4D virtual environments: principles and applications, *Computers, Environment and Urban Systems*, 22, 1-11, 1998.

[38.] Smith, G.M., Spencer, T., and Möller, I., Visualization of coastal dynamics: Scolt Head Island, North Norfolk, England, *Estuarine, Coastal and Shelf Science*, 50, 137-142, 2000.

39. Nyerges, T.L., Cognitive issues in the evolution of GIS user knowledge, in *Cognitive Aspects of Human-Computer Interaction for Geographic Information Systems*, Nyerges, T., Mark, D.M., Laurini, R., and Egenhofer, M.J., Eds., Kluwer, Dordrecht, 1995, 61-74.

40. Funtowicz, S.O. and Ravetz, J.R., Science for the post-normal age, *Futures*, 25, 739-755, 1993.

41. Zerger, A., Examining GIS decision utility for natural hazard risk modelling, *Environmental Modelling & Software*, 17, 287-294, 2002.

CHAPTER 13

Multiple Windows on Accessibility: An Evaluation of Campus Buildings by Mobility-Impaired and Able-Bodied Participants Using PPGIS

C. Castle and C. Jarvis

13.1 INTRODUCTION

The 2001 Briefing of the Disability Rights Commission (DRC) estimated that 8.5 million people are disabled in the UK[1], of whom over 1.8 million claimed benefits for mobility difficulties[2]. The UK Government is committed to promoting rights and improving opportunities for disabled people, and argues that disabled people need to be brought into the mainstream of society in order to effect real change in their lives, and to change stereotypical attitudes entrenched in society. The implementation of the 1995 and 2005 Disability Discrimination Acts with their measures to end inequity, and the setting up of the DRC, are examples of steps being taken towards this objective[3].

One area of change that these political moves have brought has been a deliberate broadening of involvement within UK Universities, accompanied by monies from bodies such as the Higher Education Foundation for England (HEFCE) to implement urgent changes to campus environments[4]. Historically, an exceptionally small percentage of those with disabilities would have survived the educational process to reach higher education; participation rates remain low, but are increasing[5]. Nevertheless, recent papers reporting views of disabled students still point to continuing and frustrating problems in accessing their higher education[5-7]. The physical accessibility of facilities to those with ambulatory difficulties, the subject of this chapter, is one such area.

In their campus review, Shevlin et al.[8, p21] sum up their respondents' comments on the subject of access and mobility with the conclusion that *'Students expend enormous amounts of time and energy in negotiating many seemingly accessible buildings'*. This statement echoes themes common across several papers looking at UK university life for the mobility-disabled[5-7]. Two important themes of particular relevance to this chapter can be drawn from the assertion by Shevlin et al.[8] Firstly, and taking a highly pragmatic view, how can better accessibility be achieved within the immediate resource limits, whist continually increasing pressure for change; and secondly, how can we bridge the information gap between portraying buildings or routes as seemingly accessible, and reality?

The answer to this first question is clearly ongoing, with a requirement for structural and internal changes to UK campuses. In the meantime, can information regarding the accessibility of buildings be better represented so less time and energy is wasted unnecessarily? Detailed accessibility audits available via the Internet can provide some measure of information for potential and indeed current students at some Universities[7]. However, these are not a practical solution for the student seeking to orientate themselves to a new campus or unfamiliar buildings on a day-to-day basis, where a map might be more suitable. Earlier work by Vujakovic and Matthews[9], for example, demonstrates the potential of city-wide accessibility maps, while one of Chard and Couch's[7] respondents notes with disappointment the removal of mobility-related symbolism on campus mapping. Many campus maps do contain accessibility symbols, but there is little in the cartographic literature regarding the most effective style, symbolism and scale of approach to accessibility mapping. Moreover, the creation of accessibility maps has traditionally required a significant surveying investment[9], placing the ability for construction and the power of content in the hands of delegated authority rather than empowered citizen or group.

Secondly, and significantly, Shevlin et al.[8] use the term *'seemingly accessible'* to describe campus buildings. There may be pragmatic issues connected with the reliability of accessibility information produced by the able-bodied: planning rules or detailed pro-forma used in accessibility audits may be implemented accurately, yet their results miss important but individual facets of accessibility. In the context of University life with its rich potential for social contact it seems that there is a strong possibility that the UK's 'needs assessment' approach to provision, and what might be hoped for in an individual's normal development, might part company relatively easily. Neither do planning rubrics necessarily reflect the requirements of individuals, in all their diversity; as Imrie[10, p464] notes, there is *'more that divides than unites disabled people'*. These points are echoed by many poignant responses across the wider socio-cultural literature. For example, one of Imrie and Kumar's[11, p367] respondents comments about a building which frustrates him (from an access viewpoint), as it was identified as 'fully accessible' by the City Council in Newcastle (a 'fully accessible' city), *'... Now, you can imagine that other buildings which don't even match this quality are nowhere near accessible. But I think it's the perception of what's accessible and what isn't'*. Certainly Vujakovic and Matthews[9] identify significant differences in the mental maps of an area of Coventry constructed by able-bodied and disabled participants of their study, although they do not explore how different the maps of the two groups might be if they were asked to construct a map for the assistance of those with ambulatory difficulties. They do however find significant differences in choice of symbolism, albeit from a very small sample. Taken together, these varied sources suggest that accessibility information and maps may best be constructed by those who themselves understand the constraints of the subsequent users.

Many recent papers also highlight the moral imperative for partnership, particularly in a research context[10,12]. Kitchin[12, p25], following Oliver[13], observes that research should be *'both emancipatory (seeking positive societal change) & empowering (seeking positive individual change through participation)'*. Moreover, such partnerships are important if the products are to gain credibility; another of Kitchin's[12, p34] respondents suggests that *'it is important that disabled people undertake and present research because it makes more of an impact due to the fact that it is `straight from the horse's mouth``*. Whatever forms of information are produced on accessibility, *'it is only disabled people who can know what it is like to be disabled and so only disabled people can truly interpret and present data from other disabled people'*[12, p26].

Within this context, this chapter reports the findings of a pilot study exploring the potential of a Public Participation Geographical Information System (PPGIS) to build accessibility maps that draw on the local knowledge of campus respondents with ambulatory difficulties for their detailed content. PPGIS have received much attention within the GIS research community, particularly during the mid-to-late 1990s[14-18]. Critically in the context of respondent comments within the disability literature, this particular study draws on other recent studies adopting a bottom-up approach to PPGIS, where community members are able to proactively provide information rather than be restricted to responding to information supplied to them[19]. The approach taken here also follows that of Haklay and Harrison[20] and Carver et al.[16], who observe that PPGIS systems conducted via the Internet and World Wide Web (WWW) can increase public participation in planning and, by so doing, make such processes more democratic and locally relevant. Harris and Weiner's[17] work in particular stemmed from growing concerns about the potentially disadvantageous social implications of GIS, and the need to use GIS to empower disadvantaged and marginalized groups instead[21]. In the main, however, previous work with PPGIS has been levelled at community responses to planning applications, yet from recent respondents' comments within the broader disability literature PPGIS would seem an approach with high potential for accessibility mapping.

In particular, this chapter reviews the development and results of a PPGIS for accessibility mapping developed for the central campus at the University of Leicester in the summer of 2003. We investigate the ease by which the PPGIS framework was used and developed by campus participants, the perceived usefulness of the products and the potential level of empowerment that might be achieved by the approach. As an important component of the study, we also asked a sample of able-bodied students to build accessibility maps within the same setting, based on their perceptions of requirements.

13.2 METHODOLOGY

13.2.1 PPGIS: Technical Framework

The primary software package used for this study was Questionmark™'s Perception™[22]. While alternative software such as ESRI's ArcIMS® and Microsoft's MapPoint® Web Service could potentially offer greater flexibility and resources for participants developing their own accessibility maps, we were keen to avoid major obstacles in using the system for those not familiar with GIS. The Perception package offers an extensive range of question delivery options, particularly drag and drop questions, which were used to place symbols over a familiar cartographic base of the existing campus map. Thus, the software offered both the ability to create the accessibility maps in a manner that was expected to be relatively easy for the non-expert user, while simultaneously delivering the remainder of the questions for each of the cartographic trials.

The Perception mapping interface was embedded within an Internet site, which served a variety of functions. Primarily designed to collect and display the empirical data, it also served to advertise and explain the purpose of the research. Additionally, the Web site provided a schedule of the project to keep participants informed of when surveys became available online. Amongst other pages, a results archive provided the participants with the survey results and links to further reference material. This approach enabled the PPGIS to work essentially as a self-generating survey, with the potential for a cumulative rate of growth.

13.2.2 PPGIS: Stages of Development

The cartographic development of the PPGIS was divided into two trial stages, the symbolism trials and the mapping trials.

The *symbolism trials* were designed principally to investigate which accessibility symbols should be included in the final maps. Mindful of the cartographic findings of Fry[23] and Vujakovic and Matthews[9], and also the need to engage with participants at the early planning stages of the work, a secondary aim of this trial was to examine preferences regarding the symbols themselves. A questionnaire was used for this process, rather than the preferred focus group approach, owing both to data protection issues and also the timing of the study in regard to students' examinations and summer leave. Only participants with mobility difficulties were requested to complete the symbolism trial.

The *mapping trials* allowed both participants with mobility difficulties and able-bodied participants to construct an individual accessibility map of the campus. These individual products were subsequently developed into community maps. On the Internet site (available at http://www.casa.ucl.ac.uk/cjec/msc/index.htm), respondents could add predefined symbols describing accessibility levels to a familiar map of the central Leicester University campus (see screenshot in Figure 13.1). The chosen symbols from the previous symbolism trial (decided by the majority of votes received), could be 'dragged and dropped' onto a detailed map of the buildings within Leicester campus. Participants were required to 'click' on any, and as many of, the tiles which comprised the site map, for locations where they had knowledge of access to, and facilities within buildings for people with mobility difficulties. The maps were split into tiles to decrease the download time, and to make adding symbols more manageable. Questions (1-15) represented a different symbol within the cartographic trial, i.e., Q1 level access, Q2 steps, etc., but the actual symbols available for 'drag and dropping' were solid blue boxes (just visible in the bottom-left corner of Figure 13.1), representing the symbol stated at the beginning of each question. For Question 16 participants had to select accessibility classification from a pull down list which matched (as close as possible) the overall accessibility characteristics of each building the participant assessed within this tile.

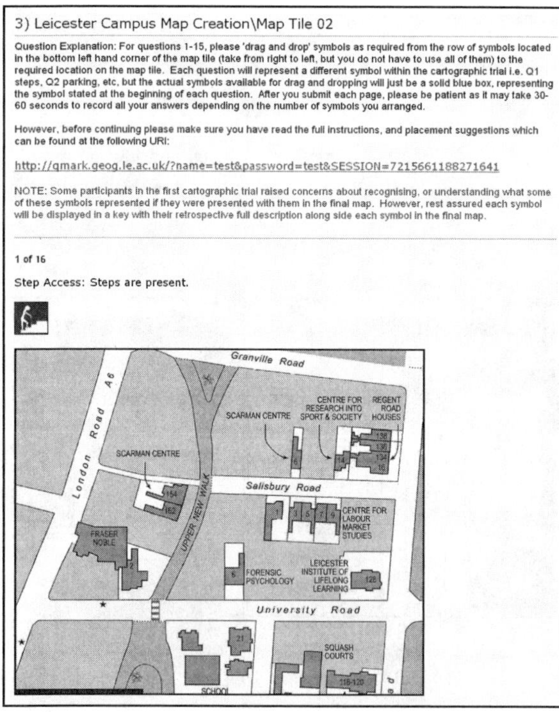

Figure 13.1 Example of the screen layout for the symbol mapping trial.

The final classification maps (see example in Figure 13.2) were created by analyzing the results of Question 16, while the individual building accessibility symbol maps were produced from Questions 1-15, as shown in Figure 13.3.

Placement of the individual symbols was determined by producing simple density and contour maps for each symbol type.

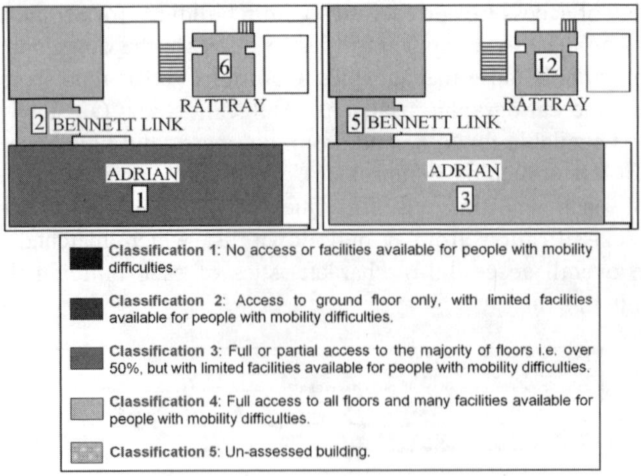

Figure 13.2 Partial view of accessibility classification maps and key produced by participants with mobility difficulties (left), and able-bodied participants (right). Numerals within each building represent the number of participants who contributed to the classification of accessibility.

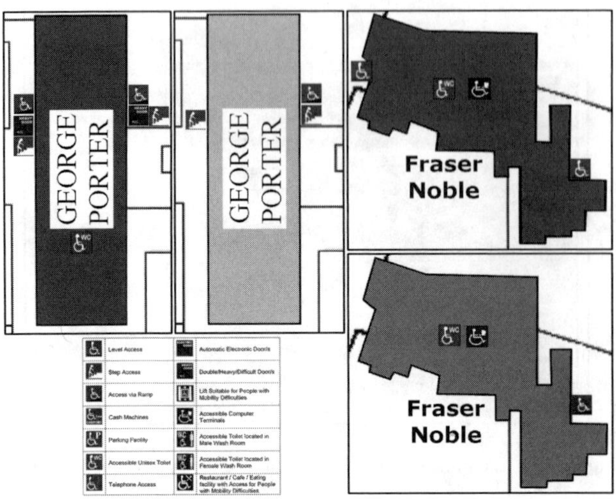

Figure 13.3 Comparison of accessibility symbol maps produced by participants with mobility difficulties (left and top), able-bodied participants (right and bottom) for the George Porter and Fraser Noble buildings. Classification of building accessibility uses the same shading as in Figure 13.2.

Respondents for both trials were recruited in several ways. An email was posted to persons known to Leicester University's AccessAbility Centre (AAC), and leaflets and posters were distributed around the campus. While the Web site was available to all, the principal targets of the initial surveys were disabled people, carers, and their families/friends. The study was open to those with disabilities of both a permanent and temporary nature, and the classification of disability was open to the respondents' interpretation. We recognize that our focus on ambulant disability covers only a proportion of disabled people, and especially neglects people with visual impairments. However, the visual nature of this work prevented people with a serious visual impairment from participating in this research.

13.2.3 PPGIS Evaluation

Participants' perceptions and attitude towards PPGIS, and their practical experience producing and using the accessibility maps, were reviewed by questionnaire, for reasons previously stated. Freeform space was however incorporated within this format, with a prompt for participants to comment regarding any matters not explored by the set questions or the relevance, use or expression of the questions. Similar to the symbolism trial, only participants with mobility difficulties completed this evaluation. Before participants started the evaluation form, a brief outline of PPGIS was provided.

Five issues were investigated within the evaluation, a list that reflects the context of the Leicester study (as a pilot for a wider town study), with an academic and computer literate populace:

- The ease and efficiency of producing the accessibility maps using the GIS system, in contrast to respondents' perceived ease of cartographic trials by hand on paper;
- Participants' general confidence in producing accessibility maps that accurately mirrored their intentions;
- Respondents' overall experience with PPGIS during the study, and suggestions for improvement;
- The value participants placed on the resultant accessibility maps;
- Respondents' perceptions of PPGIS as a means of empowering disadvantaged and marginalized groups in campaigns for public access and planning regulations.

13.3 RESULTS

Sixteen participants (11 of whom were female) with ambulatory mobility difficulties contributed to the study, representing 40% of the eligible population with a registered mobility difficulty across the university. The majority of these participants (13) were students under the age of 45, and all held a job or attended lectures that required them to use multiple locations on the campus, on a daily basis. Similar to the participants with mobility difficulties, the age profile of the 16 able-bodied participants was positively skewed, with the majority (14) of participants under the age of 35, nine of whom were female.

13.3.1 Symbolism Trial (Only Participants with Mobility Difficulties)

The first nine symbols (level access, steps, parking, cash machine, restaurant access, access via ramp, accessible computer terminal, lift access and telephone access) were accepted by the majority of respondents with mobility difficulties (more than 13 in every case) as appropriate for inclusion within the accessibility maps. Only a few respondents suggested that clearer and/or more appropriate symbols were available, but they were unable to indicate viable alternatives. However, one symbol that participants did not think would best represent the accessibility of a location (10 respondents), and should not be included in the accessibility symbol maps (11 respondents), was the minimum doorway width symbol. From the choice of a unisex symbol for toilet facilities, one for each sex, or all three of these depending on the facilities available, 10 respondents elected for all three possibilities to remain available to them. Within the open question area, additional symbols for difficult/double/heavy sets of doors (six respondents) and electronic doors (10 respondents) were requested and subsequently added to the final range.

In response to the penultimate question, asking whether participants would like to indicate areas that were particularly inaccessible using symbols struck through with a red cross, 14 participants said no.

The final question examined a possible solution if participants said that a/multiple symbol(s) offended them, or over emphasized their disability. A selection of 'abstract' and 'conceptual' symbols designed by Fry[23] were presented as alternatives to the pictorial versions. Respondents unanimously voted to use the pictorial symbols and 13 felt that these did not emphasize a/their disability.

13.3.2 Mapping Trial (All Participants)

The mapping trials generated general accessibility classification maps by participants with mobility difficulties and those who were able-bodied (see Figure 13.2, left and right respectively). Information regarding the accessibility of buildings on the campus was submitted for 28 map tiles by participants with mobility difficulties and for 47 by able-bodied participants. In total, over 95% of the spatial data created by the participants was used to create the final accessibility symbol maps (Figure 13.3). All of the accessibility symbol maps for individual buildings and the general accessibility classification maps can be viewed at http://www.casa.ucl.ac.uk/cjec/msc/index.htm.

Analysis of the building access classifications revealed several clear differences between the maps produced by the two groups of participants. Table 13.1 lists the results for individual buildings. Able-bodied students perceived 17 buildings (53%) to have full/partial access to the majority of floors, but with limited facilities available within these buildings (Category 3 in Figure 13.2). Another 12 buildings (38%) were considered fully accessible with many facilities available (Category 4).

Table 13.1 Classifications of building accessibility by study participants

Building Name	Access Category by Participant Group		Difference in Able-Bodied Participants' Perception
	Mobility Difficulties	Able-Bodied	
Fraser Noble	2	3	More Accessible
School of Education 1	1	3	More Accessible
School of Education 2	2	3	More Accessible
School of Education 3	2	3	More Accessible
School of Education 4	2	3	More Accessible
School of Social Work 1[a]	N/A	3	N/A
School of Social Work 2[a]	N/A	3	N/A
Museum Studies 1[a]	N/A	3	N/A
Museum Studies 2[a]	N/A	3	N/A
Institute of Life Long Learning	N/A	1	N/A
Regent Road House 1[a]	1	N/A	N/A
Regent Road House 2[a]	1	N/A	N/A
Regent Road House 3[a]	1	N/A	N/A
Bennett	4	4	Same
Bennett Link	4	4	Same
Adrian	3	4	More Accessible
Maurice Shock	3	N/A	N/A
Physics and Astronomy	4	4	Same
Rattray	4	4	Same
George Porter	2	4	More Accessible
Charles Wilson	3	4	More Accessible
Chemistry Research	1	N/A	N/A
Computer Centre	1	1	Same
Attenborough Tower	4	4	Same
Attenborough Seminar Block	2	3	More Accessible
Percy Gee	3	3	Same
College House	2	3	More Accessible
Maths and Computer Science	N/A	4	N/A
Ken Edwards	4	4	Same
Engineering 1[b]	N/A	3	N/A
Engineering 2[b]	N/A	3	N/A
Engineering 3[b]	N/A	3	N/A
Library	4	4	Same
Law	3	3	Same
Fielding Johnson	3	3	Same
Astley Clarke	3	4	More Accessible
Security Lodge	N/A	2	N/A

Note: N/A = Not assessed. [a] Ordered from top on map. [b] Ordered clockwise from top right.

Participants with mobility difficulties perceived the accessibility and facilities of the buildings rather differently. Six buildings (21%) were considered to have no access or facilities available for people with mobility difficulties (Category 1). A further eight buildings (29%) were considered to have access only to their ground floors (Category 2). Of these buildings, six were classed as Category 3 by the able-bodied participants, and one was rated as Category 4. Seven buildings (25%) were classed as Category 3 by participants with mobility difficulties, but of the six also assessed by the able-bodied students only three were given the same classification and the other three were graded as Category 4. Finally, another seven buildings (25%) were rated as Category 4 by the participants with mobility difficulties, all of which were assessed and given the same classification by the able-bodied participants.

Overall, 22 of the 37 campus buildings were classified by both groups of participants. In half of these cases (i.e., 11) the ratings from the two groups were the same and in the other half the buildings were perceived as more accessible by able-bodied respondents than those with mobility difficulties. None of the buildings were regarded as less accessible by the able-bodied participants.

Not only did able-bodied participants perceive campus buildings to be generally more accessible, but for the buildings that both groups assessed, able-bodied participants only contributed 77% of the accessibility symbols provided by those with mobility difficulties (excluding parking symbols). In fact, although able-bodied participants assessed 14% more buildings, they contributed 21% fewer symbols. Figure 13.3 illustrates these two trends. Able-bodied participants classified the George Porter building as being fully accessible with many facilities available for people with mobility difficulties. However, they were unable to provide any indication of these facilities. On the other hand, participants with mobility difficulties assessed the George Porter building as only having ground floor access, with limited facilities. Similarly, the Fraser Noble building (an examination hall) was rated by able-bodied participants as fully/partially accessible with limited facilities while respondents with mobility difficulties classified it as ground floor access only, with limited facilities. Several participants with mobility difficulties confirmed that their classification was a result of the building containing multiple steps and staircases to other floors, with no lift access. Other examples of such differences were also recorded and can be found at http://www.casa.ucl.ac.uk/cjec/msc/leicester_uni_ppgis_able_bodied_access_maps.htm (not navigable from the main Internet site).

Copies of the final accessibility maps, and information regarding the project Web site, were sent to Leicester University's AccessAbility Centre.

13.3.3 PPGIS Evaluation (Only Participants with Mobility Difficulties)

The majority of respondents perceived the PPGIS system an efficient means of submitting and representing their local knowledge. They were comfortable with the computer environment, and were pleased that the system allowed them to control their time on screen. For example, one participant commented that *'the ability to add information regarding individual facilities for small areas allowed, and ensured, that I kept on coming back to add information when it was convenient to me'*. Respondents were also largely confident of their ability to make submissions to the mapping system that were being captured correctly, although a number of small technical matters were highlighted for adjustment in future work.

With respect to the advantages of the PPGIS approach, most participants agreed a priori that accessibility maps were a useful idea, and which would have been desirable when they first arrived at the university. In particular, one respondent noted that *'the University has produced a rudimentary accessibility map, but it only has some detail, it doesn't indicate all of the facilities available in buildings like these maps'*. Of the 16 participants, six had actively used the final accessibly maps by the time they had completed the evaluation questionnaire. Eighty five percent found the level of information contained within the maps to be excellent, while the remainder found it sufficient. The general classification maps scored highly on reliability across the cohort, while those regarding symbolism were rated slightly lower. Nevertheless, the latter were still marked as between excellent and sufficient by all but one respondent. Fifteen of the 16 respondents found the Internet availability of the final maps to be advantageous.

Importantly in regard to the objective of empowerment, 13 study participants thought that the accessibility maps would give them considerably more confidence, two slightly more confidence, and one about the same amount of confidence when navigating around the campus. The level of confidence fell when respondents used the map to access unfamiliar buildings for the first time, with four and twelve participants stating that they had been given considerably more, and slightly more confidence (respectively) in their ability to move about the campus. More generally, after this project 15 of the 16 respondents perceived the use of PPGIS as a means of producing future accessibility projects as valuable, and endorsed their questionnaires with comments that included *'hopefully a national scheme will be developed, this would make visiting new places so much easier'*, and *'very promising for disabled people'*. The final conclusion of the respondents was almost unanimous that PPGIS had the potential to empower disadvantaged and marginalized groups, although the viability of the concept in practice was in slightly more doubt.

13.4 DISCUSSION

Although a number of technical adjustments were required to improve the PPGIS interface, the respondents were confident with its use and in submitting their contributions. This was expected of the university cohort involved, since computer literacy is a central tenet of Leicester's teaching and learning strategy, but could not necessarily be anticipated in a wider city center context. These issues will be explored further in a future Enfield Town study. In cartographic terms, the outright rejection of Fry's[23] symbolism for accessibility supports the findings of Vujakovic and Matthews'[9] smaller study. The campus area for this investigation has broad paths with smooth pavements, dropped curbs, and only a few very steep slopes, so allowing a focus on building characteristics. This contrasts considerably with Vujakovic & Matthews'[9] city study, in which the participants placed more emphasis on the external environment (cobbles, curbs, etc). The Leicester respondents also demonstrated an interest in further pan and zoom facilities, raising interesting issues of scale and generalization that will require consideration in the follow-on Enfield study, where the population is more diverse and better representative of the UK as a whole. Further research is also required to establish the best means of presenting PPGIS to those with visual impairment, an important sub-group of those with mobility difficulties that is not addressed in this study.

From a practical perspective, the PPGIS approach to accessibility mapping allowed a body of local knowledge to be built up relatively rapidly, at low cost, in the participants' own time frame and independent of authority. However, the relatively limited geographical context of this study, together with the low numbers of those with disability represented at the University, suggests that the further growth potential of this particular PPGIS is relatively limited within a short time scale. Respondents enjoyed being able to continually assess accessibility throughout the study; incorporating information about new facilities they might encounter, or modifications to the campus. Some further thought is required on the incorporation of rolling improvements to facilities with respect to building classifications, where the mapped information becomes out of date until it receives high numbers of corrective inputs. Libraries, canteens and bars are also a critical part of intellectual and social life at university and Borland and James[5], together with discussions in Leicester, highlight the important possibility of extending the coverage of mapping to frequented off-campus settings such as local pubs and cafes.

While in many senses this PPGIS is, from a technical viewpoint, an example of what has been termed a community mapping system, we have nevertheless retained the PPGIS terminology for an important reason; the need for participation and partnership is a strong theme in recent disability research. The PPGIS nomenclature echoes this parallel research more strongly. Moreover the characteristics of the accessibility map sets produced by the two groups of contributors are significantly different. This provides methodological triangulation to the body of individual comments on representations of disability, adding further weight to the need for partnership to achieve credible research and practical outcomes. There has been little recent work on accessibility mapping and one of the respondents in this study hinted at their perception of standard mappings as incomplete or unrepresentative and therefore unusable. This is despite a genuine attempt to construct a useful cartographic product in the case of the Leicester campus, a situation potentially echoed elsewhere. To cover ground in the detail achieved in this study, if organized in a standard fashion, would require resource intensive survey campaigns[7,9]; the PPGIS framework therefore offers an efficient means of achieving the required detail and reliability.

Questions regarding the extent to which the PPGIS increased confidence in mobility for existing students suggest individual empowerment at some level, the nature of which could usefully be examined further through unstructured interviews. This is important; the study aimed to address the concept of individual experience at University, not average experiences or community campaigning in isolation. We acknowledge that respondents evaluating the PPGIS in favorable terms were also those who supplied the information it contained, yet the construction of information by the community for the community and the amalgamation of individual experience by building and across campus are notions central to the study. Forty percent of the registered potential cohort at Leicester engaged with this research, suggesting an overall benefit to those for whom the research was directly applicable. Respondent comments from previous studies[12] indicate that this is an outcome relatively rare in disability research, although much desired.

With regard to the low representation of those with disabilities at University, we might also wonder whether the growth of reliable accessibility maps constructed by those with credence have the potential to encourage wider access. One of Kitchin's[12, p34] respondents suggests for example that '*it's quite easy for you to see that I can't get on the bus, I can't get into 60% of the shops, I can't get into most of the universities'*. It is clear from this study that many physical barriers remain across the Leicester campus, as others[5,7]. However, if a PPGIS approach can add confidence to those already partly familiar with an environment, then there is also the possibility that the detailed information provided may also give assurance to those considering what is largely unknown territory. While disability is an inherently individual experience[10] it would be interesting to explore whether the participatory nature of the PPGIS serves to build a sense of connection.

This chapter takes a largely pragmatic and passive approach from a political stance, looking at mitigating ambulatory disability on campus through disseminating information generated from the collective local knowledge of the community. In this sense, the PPGIS research direction presented has yet to become emancipatory. However, the collective information provides a strong evidence base when considering campaigns for change across the campus. Handley[24] points out that a rights-, rather than needs-based approach to disability provision may be overly idealistic owing to the sheer individuality of need within contexts of limited resources. This PPGIS offers individuals an independent means of defining physical barriers affecting their mobility and access requirements in the built environment, which may be presented as the collective priority for change.

The anonymous face of PPGIS will not however lever change on its own; it will need champions to shape the concept, spread the word on its value and to make political mileage with the broader community. Further research will also be needed in regard to ensuring that the views represented are those of genuine respondents without excessive 'people monitoring'.

13.5 CONCLUSIONS

Several interesting and distinct conclusions can be drawn from the results presented in this study. In particular, the research demonstrates that Internet and questionnaire strategies can be successfully and effectively adapted to evaluate, revise and contribute to disability access mapping within a designated area. Moreover, clear differences were apparent between the accessibility maps produced by participants with mobility differences and those who were able-bodied. In general, the able-bodied participants perceived the buildings within Leicester University campus to be more accessible, and to have more facilities available than those with mobility difficulties. Half of the buildings assessed by both parties were evaluated as more accessible by able-bodied students and none of them were rated as less accessible. In addition, able-bodied participants were only able to contribute 77% of the accessibility symbols that participants with mobility difficulties provided. These findings confirm the need for partnership when developing accessibility maps. Importantly, the results also identify significant improvements in the confidence of individual users who have access to what they consider to be detailed, relevant and reliable products. These initial results suggest that accessibility mapping using PPGIS has the potential for empowerment.

Although there have been, and continue to be, improvements in access to facilities for those with ambulant disabilities on campus, further enhancements are still required. In addition to questions of resource, the results of this study suggest that the perceptions of able-bodied people concerning access requirements may be why some services retain inaccessible features. While responses to individual 'rights' are resource intensive, PPGIS offers a means by which individuals may state their own requirements for access in a manner that builds a collective priority for action and provides weight to requests for change. In this emancipatory context, however, careful consideration must be applied, and further research undertaken, to prevent any constituent of a participating population being ostracized while ensuring the credibility of the knowledge sources.

13.6 REFERENCES

1. Disability Rights Commission (DRC), Disability briefing: February 2001, http://www.drc-gb.org/disabilitybriefing/feb01/Page356.asp, Accessed on October 8th, 2003.

2. Department for Work and Pensions (DWP), Disability living allowance quarterly statistics, http://www.dwp.gov.uk/asd/dla.asp, 2006.

3. Department for Work and Pensions (DWP), Information for disabled people and carers, http://www.dwp.gov.uk/lifeevent/discare/, 2006.

4. Higher Education Funding Council for England (HEFCE), *Access to Higher Education: Students with Special Needs*, HEFCE, Bristol, 1995.

5. Borland, J. and James, S., The learning experience of students with disabilities in Higher Education: a case study of a UK university, *Disability & Society*, 14, 85-101, 1999.

6. Parker, V., Personal assistance for students with disabilities in HE: the experience of the University of London, *Disability & Society*, 14, 483-504, 1999.

7. Chard, G. and Couch, R., Access to higher education for the disabled student: a building survey at the University of Liverpool, *Disability & Society*, 13, 603-623, 1998.

8. Shevlin, M., Kenny, M., and McNeela, E., Participation in higher education for students with disabilities: an Irish perspective, *Disability & Society*, 19, 15-30, 2004.

9. Vujakovic, P. and Matthews, M., Contorted, folded, torn: environmental values, cartographic representations, and the politics of disability, *Disability & Society*, 9, 359-374, 1994.

10. Imrie, R., The role of access groups in facilitating accessible environments for disabled people, *Disability & Society*, 14, 463-482, 1999.

11. Imrie, R. and Kumar, M., Focusing on disability and access in the built environment, *Disability & Society*, 13, 357-374, 1998.

12. Kitchin, R., The researched opinions on research: disabled people and disability research, *Disability & Society*, 15, 25-47, 2000.

13. Oliver, M., Changing the social relations of research production, *Disability, Handicap and Society*, 7, 101-114, 1992.

14. Armstrong, M.P., Densham, P.J., and Kemp, K.K., Collaborative spatial decision making: scientific report for the specialist meeting, National Center for Geographic Information and Analysis, Santa Barbara, 1995.

15. Ball, J., Towards a methodology for mapping 'regions for sustainability' using PPGIS, *Progress in Planning*, 58, 81-140, 2002.

16. Carver, S., Evans, A., Kingston, R., and Turton, I., Web-based public participation geographical information systems: an aid to local environmental decision-making, *Computers, Environment and Urban Systems*, 24, 109-125, 2000.

17. Harris, T. and Weiner, D., *GIS and Society: Scientific Report of the 1-19 Meeting*, NCGIA Technical Report 96-7, National Center for Geographic Information and Analysis, Santa Barbara, 1996.

18. Craig, W.J., Harris, T.M., and Weiner, D., Empowerment, marginalisation, and public participation GIS: report of the Varenius workshop, http://www.ncgia.ucsb.edu/varenius/ppgis/PPGIS98_rpt.html, 1999.

19. Craig, W.J., Harris, T.M., and Weiner, D., Community participation and geographic information systems, in *Community Participation and Geographic Information Systems*, Craig, W. J., Harris, T. M., and Weiner, D., Eds., Taylor and Francis, London, 2002, 3-16.

20. Haklay, M.E. and Harrison, C.M., The potential for public participation GIS in U.K. environmental planning: appraisals by active publics, *International Journal of Environmental Planning and Management*, 45, 841-863, 2002.

21. Pickles, J., Ed., *Ground Truth: The Social Implications of Geographic Information Systems*, Guilford Press, New York, 1995.

22. Questionmark, Perception software, http://www.questionmark.com/us/perception/index.htm, 2006

23. Fry, C., Maps for the physically disabled, *The Cartographic Journal*, 25, 20-28, 1988.

24. Handley, P., Trouble in paradise - a disabled person's right to the satisfaction of a self-defined need: some conceptual and practical problems, *Disability & Society*, 16, 313-325, 2000.

CHAPTER 14

Visualization Techniques to Support Planning of Renewable Energy Developments

D. Miller, J. Morrice, A. Coleby and P. Messager

14.1 INTRODUCTION

The European Union (EU), through Directives, sets a context for much of the implementation of environmental policies in member states in Europe. These are translated into national initiatives, and 'landscape' is one topic that cuts across a number of policy boundaries. Relevant 'horizontal' measures at the European Community scale have included the Programme of Policy and Action in relation to Environment and Sustainable Development[1] and the Sixth Environment Action Programme[2]. Both of these have involved landscape protection and management as global means of ensuring that wider environmental goals are achieved.

Public policy has also been influenced by government commitments to the Aarhus Declaration[3] on public participation and access to environmental information. Although the terms used by different organizations may vary between 'engagement', 'involvement', 'consultation' and 'awareness raising', they all echo the aspirations of greater public participation in decision-making, as outlined under the Aarhus Declaration and also built into the European Landscape Convention[4].

There are a number of techniques that can be used to involve communities in direct decision-making. These include 'Planning For Real', design days and Community Planning Weekends[5]. Planning for Real has been used since the late 1970s, giving local people a 'voice' and professionals a clear idea of local people's needs to bring about an improvement to their neighborhood or community[6]. It has also been recognied that engagement need not be undertaken only when there is a dispute to be resolved, and that raising awareness and discussing topics with a wide audience can be undertaken over a period of time to develop a relationship between stakeholders in a geographic area, or associated with a particular theme[7]. In Scotland, the Scottish Executive has published plans for increased community involvement in the planning process, with specific reference made to the potential of 3D modelling in '*... engaging communities and assisting planners and Councillors to visualise and assess the visual impact of development proposals*' [8], reflecting the importance of the prospective visual impacts of changes to public audiences and the potential of visual media for communication to different types of stakeholders.

With respect to the development of wind turbine sites, there has been a great deal of variation in the way that the visual impacts of such developments are assessed, which has led to the development of guidelines on recommended practice for agencies such as Scottish Natural Heritage[9]. According to Lange and Bishop[10], being able to visually represent the existing real world as well as potential alterations is essential for landscape planners to express and communicate their thinking to the wider public. Nevertheless, although 3D models viewed on desktop computers and 3D immersive virtual reality (VR) are increasingly used, Piekarski and Thomas[11] suggest that they lack the ability to provide the planner with a first person perspective. More broadly, Appleton and Lovett[12] and MacFarlane et al.[13] argue that there is a lack of research on audience perception and understanding of visualization tools, and that these issues require addressing if such approaches are to make significant contributions towards wider public involvement in environmental decision-making.

This chapter describes the development of one protocol for the use of VR tools to engage members of the public in issues related to the design and layout of wind turbine developments. The aim was to assess stakeholder feedback on the strengths and weaknesses of using VR tools in environmental decision-making. To this end, a hypothetical model of a wind turbine development was developed for a site in north-east Scotland, in the vicinity of the town of Huntly. The model was used in a VR facility (the 'Virtual Landscape Theatre', VLT) at an event in Huntly to explore the use of such tools in an environment where a real planning proposal was being considered, but not addressing that proposal specifically. Interviews of participants were carried out to assess the extent to which such tools could be used in practice.

14.2 METHODS

14.2.1 Virtual Reality Environment

The VLT was used as a medium for knowledge exchange between stakeholders in relation to the layout of a proposed wind turbine development. The VLT comprises a curved screen (~ 6 m x 2.2 m high), 160° in curvature, that is portable and designed for use in local community venues. It can host a maximum of 15 people, and is equipped with a handset-based polling system to enable capture of audience opinion on the landscapes and changes shown. Three high specification PCs are linked by a local area network, each one of which drives a 3DP X25i data projector, through which the geometry of the projected images is warped to fit the screen. Calibration of the imagery produces a 'seamless' display of the landscape model on the screen.

Figure 14.1 shows the VLT from a view behind the control PC and navigator. The audience are being taken through a model of the landscape, along a road from which turbines are hidden from view in this image.

Figure 14.1 Virtual reality facility, showing screen and data projectors.

14.2.2 Model Creation

14.2.2.1 Landscape Model

The prototype model was created in ERDAS IMAGINE VGIS[14] software. Input data came from the Ordnance Survey 1:10,000 Digital Terrain Model (DTM)[15], with ground textures obtained from color orthophotography (flown in 2000 at a 0.25 m resolution) supplied by the Forestry Commission. To this topographic background, two types of surface feature were added: 'billboard' images (e.g., of trees) and full 3D models[16]. The 3D models of specific features were obtained from libraries or created using suitable design packages (e.g., Creator[17], 3D Studio[18]). That for the wind turbine was based upon specifications of a Vestas V90 2MW and allowed the turbine blades to be shown as moving when the model was converted into the Openflight format for use in the VEGAPrime display environment[19].

14.2.2.2 Wind Turbine Siting

The wind turbines were located on a hill where there was some previous interest in the development of a wind farm, but no proposal had been submitted. This site was chosen on the basis that it was credible, technically feasible for development, but not that of a real proposal. Such a site was used so that the opinions expressed and choices made related as much as possible to the model presented rather than any real proposal. This distinction was also stressed to audiences in the introductions to the event.

14.2.3 Event Operation

The assessment took place as part of a Landscape Research Week, the venue for which was the public hall in the town of Huntly, north-east Scotland. In testing the role of VR tools, a number of variables could have been considered with respect to wind farm characteristics. These included the number of turbines, their height, the spatial layout and factors such as color and design. Evidence from wind farm developers and local authorities suggested that the factor which is most often changed during the period of consultation and planning is the number of turbines[20]. This factor was also identified by Bishop and Miller[21] as that which had the most significant influence on viewer opinion concerning the perceived impact of a wind farm.

The approach used in the literature on preference surveys suggested some form of conjoint analysis in which alternative images with different numbers of turbines were shown, possibly in different configurations. However, to enable a more direct input to the process of selection, a procedure was devised which gave participants the opportunity to influence the size of the hypothetical wind farm by voting on the removal of turbines. This provided:

1. A means of identifying relationships between opinions expressed and the choice of turbines (if any) for removal;
2. A mechanism for participant selection of turbine numbers;
3. A method of assessing participant reactions to the functionality of the VR environment.

Displays involved a presenter and a second person who navigated around the model. The sequence of activities was as follows:

1. Introductory slides to provide a context for the event
2. An explanation of the theater and the presentation that was to follow
3. A 'walk-through' of the model area
4. Introduction of the windfarm (containing seven turbines) to the model
5. A 'walk' to beside the windfarm
6. Change in height of the viewer to that of the turbine hub
7. A 'fly-through', away from the turbines towards a vantage point to the east
8. A 'fly-through' to a viewpoint on the agricultural land in the middle of the model, looking towards the windfarm
9. Audience selection of viewpoints
10. A change in the time of day and year
11. Alterations in the viewing distance, illustrating the effects of mist and fog
12. Changes in the number of turbines

The last of these activities used the voting handsets in which participants could select the turbine that should be removed, if any. The options provided were a number from 1 to 7 for the turbine identification and 8 for 'none'. In the results discussed here the rounds of voting took place during one afternoon with 52 participants, all initially voting in small groups on the choice of one turbine which would reduce the numbers from seven to six, or leave the *status quo*. Once the overall result of this round was known the selected turbine was removed from the visualization by the VR operator and the participants were asked to vote again on a reduction from six to five. When not involved in the voting, participants could visit other parts of the exhibition. This process continued until no turbines remained. Keywords and comments were recorded by participants to best describe the view after each round of voting. Following the formal presentation, members of the audience were invited to try navigating through the landscape themselves, or to nominate a location from which they wished to see the wind farm. Finally, feedback from the participants in the event was collected through further voting and semi-structured interviews.

14.3 RESULTS

14.3.1 Turbine Siting and Numbers

Figure 14.2a shows a view of the wind farm with seven turbines and Figures 14.2b-14.2h illustrate how this changed as the turbines were gradually removed. Table 14.1 summarizes the voting results and reveals that in no round was there a majority of participants in favor of a single course of action (e.g., removal of a particular turbine, or for no change), so the action taken was determined by the option selected by the greatest number of people.

Another feature of the results was the diversity of choices, with the maximum level of agreement among participants occurring in the final round when 41.1% voted for the removal of Turbine 3 in Round 6. The option of 'no change' in the number of turbines attracted 10-20% of votes in the first five rounds and just over 30% in the final one. There were always a larger number of participants in favor of removing a particular turbine so the 'no change' option was never the most popular one. Some rounds in Table 14.1 have fewer than 52 total votes due to either a failure to use the handsets correctly, or a decision not to vote.

Table 14.2 summarizes the keywords and comments made after each round of voting. These suggest that most of the participants felt that the initial number of turbines proposed was too great, with negative impressions of the effect of such a development on the skyline. A reduction of one or two turbines did not seem to assuage the nature of the concerns voiced (i.e., the number and the perceived level of intrusion). The removal of an additional turbine (to leave four) resulted in a reduction of their density on the horizon and, due to the choice of turbine to remove, a lessening in the visual overlap of the rotating blades.

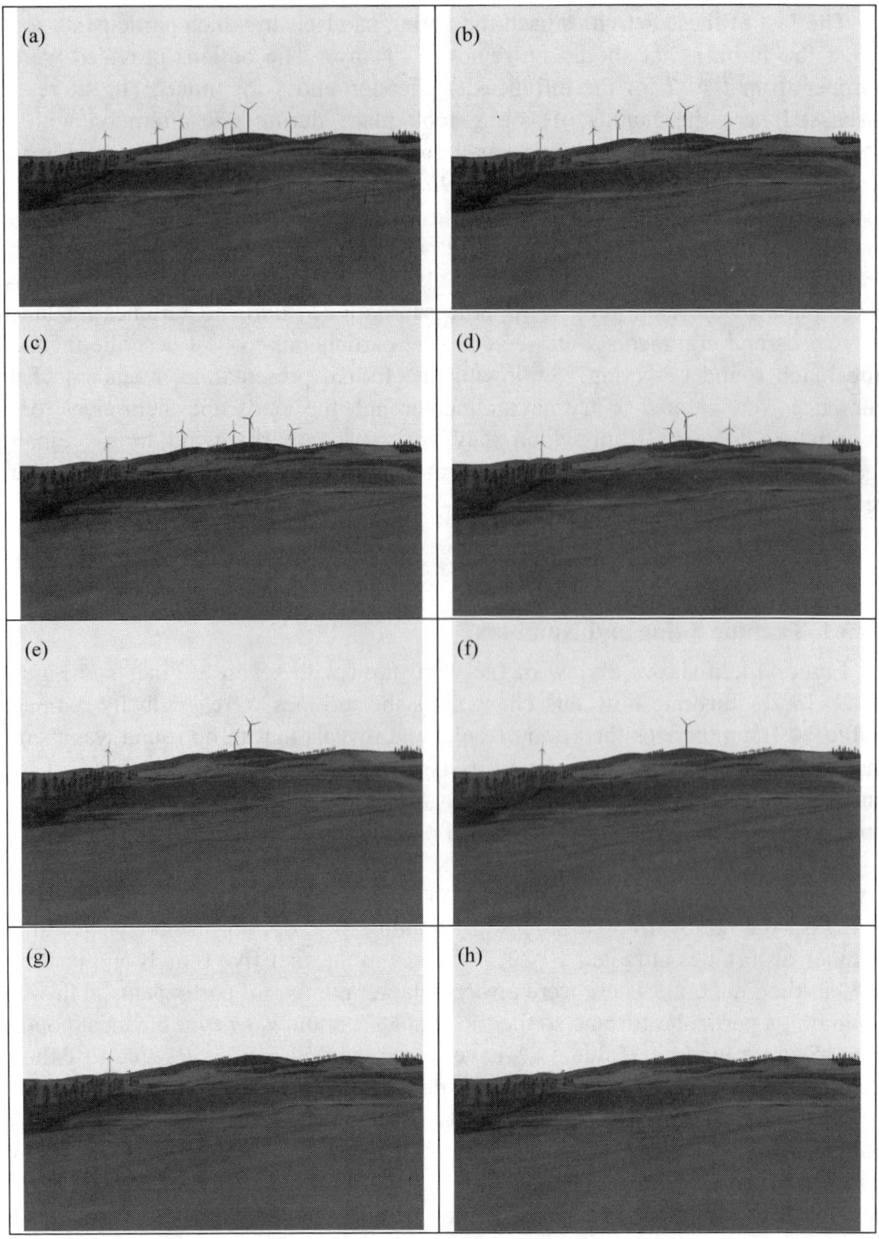

Figure 14.2 Results from the afternoon session, showing which turbines were left after the progressive removal of the wind turbines: (a) 7 turbines, (b) 6 turbines, (c) 5 turbines, (d) 4 turbines, (e) 3 turbines, (f) 2 turbines, (g) 1 turbine, (h) no turbines.

Table 14.1 Votes cast for either the removal of a particular turbine, or no change, in each round

Voting Round	Number of Turbines	Turbine Number							No Change		Total
		1	2	3	4	5	6	7	Number of Votes	% of Votes	
1	**7**	11	4	4	7	7	3	5	8	16.3	49
2	**6**	0	8	5	14	10	4	4	6	11.7	51
3	**5**	0	13	5	0	10	6	7	8	16.3	49
4	**4**	0	0	10	0	16	6	11	9	17.3	52
5	**3**	0	0	15	0	0	9	19	7	14.0	50
6	**2**	0	0	21	0	0	14	0	16	31.4	51

Note: The turbine identifier numbers above are based on map positions and do not correspond to a simple left to right sequence in Figure 14.2.

Table 14.2 Examples of keywords or comments made after each round of voting

Number of Turbines	Keyword/Comment
7	Too many, intrusive, spoils the view. Stark on skyline.
6	Six are much the same as seven. Need to remove more to change view. Could now rearrange the turbines. Don't want wind turbines of any number.
5	Start to see a change in view but only slightly, still too tall on the horizon.
4	Four is a good number visually. Less cluttered view. Would developers really want to have so few turbines?
3	Barely visible, could blend in. Will birds be able to avoid them?
2	Hardly noticeable and better for the view. Are two on their own a realistic development?
1	One is practically invisible, if turbines could be separated from the view like this who would notice them? The massive white turbines nowadays would be more visible than that.

The comments expressed following a reduction to three turbines suggest that the level of visual impact was now perceived as lower. A separate impact factor (i.e., effects on bird populations) was also mentioned. With the final two stages there was little negative comment on the number of turbines, but some questioning regarding the viability of such a proposal and one expression of skepticism regarding the visual impression being conveyed with the model.

14.3.2 Functionality of Media and Model

Additional votes and semi-structured interviews were undertaken to obtain feedback on the experience of using the VLT. Participants were asked to rate the effectiveness of the virtual environment on a scale from 1 (low) to 5 (high) with respect to different aspects of functionality and provide any comments they thought were relevant. Table 14.3 summarizes the comments and effectiveness scores (from the sample of 52 people) for six functions.

Table 14.3 Summary of comments and effectiveness scores regarding VLT functionality

Functionality	Keyword/Comment	Mean Score	Variance
Selection of viewpoint by participant	Can gain impression from different viewpoints. I like the test of view from my window. Not realistic from my viewpoint.	3.90	0.56
Movement through the model	Feeling of movement. Would react differently in the real world.	4.02	0.49
Movement within the model	Sense of turbine movement is calming. Can speed of rotation vary? Only turbines are moving. Can noise be represented in the model?	3.63	0.55
Changing time of day and season	Didn't realize effects of sun. Too dark in December view. Running through the day in a minute was excellent.	4.15	0.64
Changing atmospheric conditions	Big difference in number visible. What about snow and rain? Glad you don't assume it is always clear and sunny.	4.31	0.61
Changing number of turbines	Helpful to change layout as well as numbers. Can we add turbines? Surprising difference once 2 or 3 removed.	4.25	0.43

The rating of the ability to select viewpoints produced a mean score of 3.90. This was relatively low compared to most the other functions, although recorded comments implied that there was a desire to be able to select viewpoints. Further feedback from the questionnaires suggested that this function provided a degree of reassurance that the views were not pre-selected to give impressions of minimal visual impact.

Participants commented that the VR experience was preferable to seeing landscape futures on a flat plan such as maps, and that movement 'through' the model contributed directly to that experience, with a score of 4.02. However, views were also expressed that navigation speeds which were 'inappropriately high' to be credibly walking or driving would detract from the quality of the experience and reduce a realistic impression of the landscape.

Movement within the model was the function which scored lowest amongst the respondents (3.63). The importance of including moving turbine blades was highlighted in discussion, but the lack of movement in other elements (e.g., vehicles or animals) was also mentioned. The comparatively low score may also reflect some other topics raised in discussion, including the variability in turbine blade rotation due to changes in wind speed and associated issues of noise and bird strikes. None of these issues were represented in any way and this could have impacted upon expectations.

The most dramatic changes in view came with alterations in atmospheric conditions (levels of fog) and in the time of day or season displayed. Mean scores of 4.15 (time of day/year) and 4.31 (atmospheric conditions) suggest that a high value was attached to these aspects of functionality. By dynamically changing the scene time of day or year (season), the effects could be emphasized and this may have reinforced the strength of responses, with several discussion points relating to the differences in forelit and backlit turbines. Participants were generally surprised at how much these changes made a difference to visibility, with those that were ambivalent to the presence of turbines reacting most positively, whereas those who were against turbine construction doubted that they would disappear from view.

Changing turbine numbers was the function which received the highest mean score (4.25) and the lowest variance (0.43). Participants had a direct input into this activity, and so the score may reflect the effectiveness of those interactions. The supporting remarks also suggest that this was the most valued function. The principal criticism of the process adopted was a lack of voting on changes in layout or increasing the number of turbines. These aspects of change are being explored in other ongoing surveys not reported here.

14.4 DISCUSSION

Feedback on the opportunity for direct input to the modification of the model of the hypothetical windfarm supports the expressed enthusiasm for being provided with an opportunity and mechanism. It is recognized that there may have been an element of 'fun' involved, and that the hypothetical task of selecting turbines for removal from the landscape might not have attracted the same level of critical consideration as a real windfarm proposal could have done. However, the procedure was shown to work, and feedback from local authority representatives indicated that discussions over real windfarm cases often focused on the removal of individual turbines. The identification of such turbines was often the task of consultants to the developer or local authorities and, therefore, a means of gaining wider public input to the discussion appeared to be welcomed.

As a tool for assessing change, the approach described appears to have been received positively. Feedback from participants suggests that the opportunity for direct input to the discussion, and evidence of changes being made which could be attributed to that input, helped to enhance credibility. Anecdotal evidence also indicated that transparency in the decision-making process led to support for the outcome, and recognized the rights of others to a say. However, there were a number of limitations to the exercise, several of which were highlighted in the participant feedback. These included:

1. Layout may be as important as number of turbines;
2. Layout and number of turbines are likely to be related (i.e., for efficient power production the spacing of turbines may vary with different numbers);
3. The significance of 'no change' could have been understated as people may have felt that they were 'required' to remove a further turbine;
4. No detailed questionnaire followed each round of voting when removing turbines. As a consequence, the underlying reasons for participants identifying individual turbines were not examined and there could be an element of random choice in the results.

In general, the capability to examine the landscape from a range of viewpoints and heights allows the viewer to achieve a better understanding of landscape scale and connectivity; an understanding that maps, photographic images, drawings and even the real experience may often fail to provide. As Appleton et al.[22, p154] note *'Interactivity is the main advantage of the virtual worlds approach...this may be because it allows the user to find viewpoints which are meaningful to them and which they can relate to real life experience'*.

Participants were divided on the truth of the visualizations, with some feeling that the model was a good representation of their landscape, while others disagreed and argued that greater detail was required to show the effects of hedges, walls and existing pylons. This latter observation was also occasionally repeated alongside that of doubting the transparency of the process of model development, and a possible attempt by a turbine developer to soften the impact of a future windfarm.

The extent to which the level of realism impacts on perceptions and responses in such studies is unclear. Daniel and Meitner[23], in exploring the representational validity of landscape visualizations with varying levels of 'graphical realism', state that the appropriateness of the representation is vital in producing valid results.

They emphasize that inaccurate, poor or unrealistic representation could *'produce perceptions, interpretations and/or value judgments that are not consistent with those that would be produced by actual encounters with the environments represented'* [23, p70]. As a consequence, further empirical tests of the validity of responses obtained through visualization by comparison with those from real encounters could be a useful exercise to undertake.

Improving the means of engagement with stakeholders in issues of landscape planning potentially raises the equity with which people can participate in decisions which have a direct affect upon their local environment and lifestyle[24,25]. However, having identified the importance of engagement, and explored one approach towards enabling interactions, a number of significant issues remain, for example, including the extent of the effectiveness of the engagement. Current research is being carried out on this topic at the Macaulay Institute in Scotland.

14.5 CONCLUSIONS

The virtual landscape theater appeared most effective in the role of engaging the public, providing a means of communicating environmental information and potential change in a comprehendible manner and thus enabling them to become involved in the decision-making process. This supports previous experiences reported by Bell[26], Orland et al.[27] and Appleton and Lovett[28]. Participation was limited to an extent due to the lack of freedom for users to apply different scenarios and view a range of options for future change; and implementing such a facility would have substantial time and cost implications.

However, through observing participants, it became clear that not all were voting in time and hence their selection was not counted, and a few chose not to vote at all. In the latter situation, the use of a more discursive approach through the virtual journey proved more effective, perhaps because it provided greater freedom to expand on answers and gave more time to consider each landscape, and this is an approach which merits further consideration.

While small numbers of turbines may be acceptable in the landscape because they are perceived to be assimilated, larger numbers were often opposed because of their collective scale of imposition. Thus, the use of visualization tools could also contribute to testing thresholds for the acceptability of wind turbines in the landscape.

14.6 ACKNOWLEDGMENTS

The Scottish Executive Environment and Rural Affairs Department, and the European Commission (under project QLK5-CT-2002-01017, VisuLands) are thanked for their financial support of the research reported in this chapter.

14.7 REFERENCES

1. European Commission, *Toward Sustainability, 5th Environmental Action Programme1993-2000*, OJ C 138, 17/5/93, European Commission, Brussels, 1993.

2. European Commission, *Environment 2010: Our Future, Our Choice. The Sixth Environment Action Programme of the European Community*, OJ L 242, 10/9/2002, European Commission, Brussels, 2002.

3. European Union, *Convention on Access to Information, Public Participation in Decision-making and Access to Justice on Environmental Matters*, Aarhus, Denmark, 1998.

4. Council of Europe, *European Landscape Convention*, Committee of Ministers of the Council of Europe, Florence, 2000.

5. Kwartler, M., Visualization in support of public participation, in *Visualization in Landscape and Environmental Planning*, Bishop, I. and Lange, E. Eds., Taylor & Francis, London, 2005, 251-260.

6. Neighbourhood Initiatives Foundation, *A Practical Handbook for 'Planning For Real' Consultation Exercise*, Neighbourhood Initiatives Foundation, Telford, 1995.

7. Office of the Deputy Prime Minister (ODPM), *Participatory Planning for Sustainable Communities: International Experience in Mediation, Negotiation and Engagement in Making Plans*, ODPM Publications, Wetherby, 2003.

8. Scottish Executive, *Planning Advice Note (PAN) - Community Engagement "Planning with People": Consultation Draft*, http://www.scotland.gov.uk/ publications/2006/07/14093848/0, 2006.

9. Benson, J., Visual Analysis of Windfarms: Good Practice Guide, Report for Scottish Natural Heritage, the Scottish Renewables Forum and the Scottish Society of Directors of Planning, Scottish Natural Heritage, Edinburgh, 2005

10. Lange, E. and Bishop, I., Communication, perception and visualisation, in *Visualization in Landscape and Environmental Planning*, Bishop, I. and Lange, E. Eds., Taylor & Francis, London, 2005, 3-21.

11. Piekarski, W. and Thomas, B.H., Future use of augmented reality for environmental and landscape planners, in *Visualization in Landscape and Environmental Planning*, Bishop, I. and Lange, E. Eds., Taylor & Francis, London, 2005, 234-240.

12. Appleton, K. and Lovett, A., GIS-based visualisation of development proposals: reactions from planning and related professionals, *Computers, Environment and Urban Systems*, 29, 321-339, 2005.

13. MacFarlane, R., Stagg, H., Turner, K., and Lievesley, M., Peering through the smoke? Tensions in landscape visualisation, *Computers, Environment and Urban Systems*, 29, 341-359, 2005.

14. Leica Geosystems, ERDAS IMAGINE VirtualGIS®, Leica Geosystems, http://gi.leica-geosystems.com, 2005.

15. Ordnance Survey, Land-form PROFILE®, http://www.ordnancesurvey.co.uk/oswebsite/products/landformprofile/, 2005.

16. Discoe, B., Data sources for three-dimensional models, in *Visualization in Landscape and Environmental Planning*, Bishop, I. and Lange, E. Eds., Taylor & Francis, London, 2005, 35-49.

17. MultiGen-Paradigm, Creation products, MultiGen Paradigm Inc, http://www.multigen-paradigm.com/products/database/creator/, 2005.

18. Autodesk, 3D Studio Max, Autodesk, http://www.autodesk.co.uk, 2005.

19. MultiGen-Paradigm, Visualization products, MultiGen Paradigm Inc, http://www.multigen-paradigm.com/products/runtime/vega_prime/, 2005.

[20.] Coleby, A.M., Public Attitudes and Community Participation in Wind Farm Development, PhD Thesis, Heriot Watt University, Edinburgh, 2005.

[21.] Bishop, I.D. and Miller, D.R., Visual assessment of off-shore wind turbines: the influence of distance, contrast, movement and social variables, *Renewable Energy*, 32, 814-831, 2007.

[22.] Appleton, K., Lovett, A., Sunnenberg, G., and Dockerty, T., Rural landscape visualisation from GIS databases: a comparison of approaches, options and problems, *Computers, Environment and Urban Systems*, 26, 141-162, 2002.

[23.] Daniel, T.C. and Meitner, M.M., Representational validity of landscape visualizations: The effects of graphical realism on perceived scenic beauty of forest vistas, *Journal of Environmental Psychology*, 21, 61-72, 2001.

[24.] O'Neill, J. and Spash, C.L., Conceptions of Value in Environmental Decision-Making, Environmental Valuation in Europe, Policy Research Brief No. 4, Macaulay Land Use Research Institute, Aberdeen, 2000.

[25.] De Marchi, B. and Ravetz, J.R., Participatory Approaches to Environmental Policy, Environmental Valuation in Europe, Policy Research Brief No. 10, Macaulay Land Use Research Institute, Aberdeen, 2001.

[26.] Bell, S., *Landscape: Pattern, Perception and Process*, Spon, London, 1999.

[27.] Orland, B., Budthimedhee, K., and Uusitalo, J., Considering virtual worlds as representations of landscape realities and as tools for landscape planning, *Landscape and Urban Planning*, 54, 139-148, 2001.

[28.] Appleton, K. and Lovett, A., GIS-based visualisation of rural landscapes: defining 'sufficient' realism for environmental decision-making, *Landscape and Urban Planning*, 65, 117-131, 2003.

CHAPTER 15

The Social Implications of Developing a Web-GIS: Observations from Studies in Rural Bavaria, Germany

S. Herrmann and S. Neumeier

15.1 INTRODUCTION

Public Participation GIS (PPGIS) is now widely recognized as a potential means of empowering marginalized people and communities engaged in social change. Proponents of PPGIS argue that the GIS technology allows communities to better understand and advocate their concerns, promote the geographic visions of previously unheard people and provide for greater influence on policy-making by enabling communities to use the same tools and data as policy-makers[1-3].

In 2001 the government of Lower Bavaria and the Bavarian Ministry of Agriculture and Forestry funded a project to create a web-based GIS that would provide tourist information for visitors to the Bavarian Forest National Park Region (see location map in Figure 15.1). Agriculture and forestry are the traditional pillars of the economy in this mountainous area, but tourism has also been an important component since the late 19th century[4]. Nevertheless, the development of the region still lags behind the Bavarian average, and during recent years stagnation in the number of visitors had become noticeable. The government agencies in Bavaria therefore funded the Technical University of Munich to create a Web-GIS in order to help support the regional tourism industry. It was anticipated that the project would help stimulate regional economic growth and social well-being by transferring technical know-how, by better promoting regional attractions and by contributing to more tourist visits which, in turn, would expand the flow of money into the regional economy[5].

During the research project it became apparent that while the technical aspects of similar PPGIS are often addressed within research papers, the social implications of the process involved in developing a PPGIS, as well as those initiated by system use, have been given less attention. Given this situation, our aim in the study discussed in this chapter was to help bridge this gap in the PPGIS literature. Since the tourism project only provided a perspective into the system development process, a similar operational Web-GIS (*info-bgl*) in Berchtesgaden (see Figure 15.1) was also included in the study to help gain insights into the social implications of system use.

Figure 15.1 Location of the study area in Germany.

15.2 DEVELOPMENT OF THE 'WEB-GIS TOURISMUS TUM'

15.2.1 Conceptual Approach

Interviews with regional experts and an analysis of resources available online and in the literature allowed the specific requirements for the tourism Web-GIS to be defined. Nearly all sources suggested that it should consist of two elements. Firstly, a simple information system that provides an introduction to the destination by explaining why the visitor should come there and what he/she can do. Secondly, after interest in the destination has been awakened, specific information with a strong spatial context is needed (e.g., possible walking routes). This suggested that a fusion of classical information system and GIS capabilities would be necessary.

15.2.2 Software

Given the above requirements, the Web-GIS was created using ESRI's Internet Map Server ArcIMS®, the ArcSDE® middleware and an Oracle® 8i. database. Since it is often the case that the general public object to using Internet services that require the download and installation of additional software components[6] a thin client approach[7] was adopted and the system development was carried out with the ArcIMS HTML client. This meant that the end-user only needed a plain web browser capable of executing JavaScript to access GIS data and functions over the Internet. Nevertheless, in order to adapt and enhance the basic Web-GIS software (graphical user interface, functionality) for use in a tourism application, extensive programming was necessary using HTML, JavaScript and Perl.

15.2.3 Data Sources

In addition to various raster maps (topographic and cadastral maps, digital orthorectified imagery) used as background layers, points of interest (POI) for tourists were obtained either by digitizing (mountain summits, public transport stops etc.) or by address geocoding (hotels, restaurants etc.). At the start of the project in 2001 there was no suitable official geocoded database for properties in the Bavarian Forest National Park so an address table had to be generated out of different official datasets.

To save time in future updating, an additional stand-alone interface was developed to allow owners of tourist facilities such as hotels and restaurants to edit their own descriptive data via the Internet. Unfortunately, ESRI's feature classes - when saved in a database via ArcSDE - only allowed the insertion of data with proprietary ESRI tools. If the data were edited with other database tools, then ArcGIS did not handle the changes properly. Although it complicated the database model, it was therefore decided to separate geometric and descriptive data by using two different database tables. Subsequently, when defining the map service by generating an ArcXML-file, the two tables could be joined via a Query or Spatial-query tag, so that ArcIMS handled them as one. But as ArcIMS was not capable of handling the long data types that were used in order to save detailed descriptive or binary data for tourist POIs (e.g., pictures), a third table containing these long data types had to be introduced. So, in the final database, there were three tables to represent each POI.

Despite its complexity, this approach made attribute data handling totally autonomous from proprietary ESRI tools/formats and allowed the use of cheap and simple programs for attribute data manipulation. It also resulted in lower costs for system maintenance and allowed POI details to be kept up-to-date by a pool of users rather than requiring all edits to be made through the system operator. In addition, the database structure readily provided for the expansion of the available thematic data as new tables could easily be added and activated for use by editing only a few items in the server map configuration file.

15.2.4 Web-GIS Functionality

Alongside standard tools for zooming, buffering and querying, the developed Web-GIS (Figure 15.2) was capable of delivering dynamic database records including location maps, pictures and links to existing booking and reservation systems. This contrasted with ArcIMS's standard tabular data return and plain HTML-hyperlink functionality on the client side. Furthermore, it was possible to edit map annotations that could either be saved and reloaded in a future session or sent to other potential visitors for discussion.

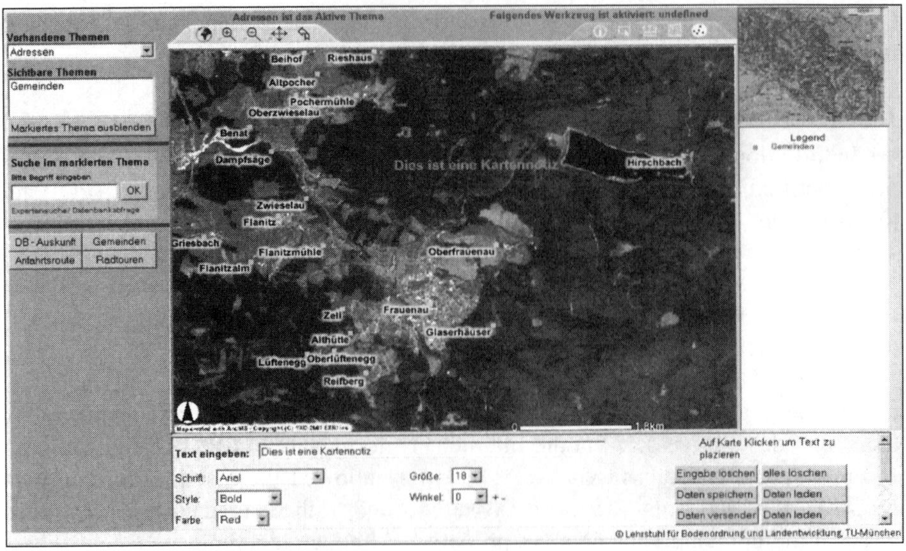

Figure 15.2 A screenshot of the Web-GIS Tourismus TUM.

Another key feature of the system was the incorporation of a routing engine (through co-operation with a company specializing in transport planning software). This routing engine could deliver the walking route to the nearest public transportation stop, the timetable of the corresponding means of transport, and the walking route from the closest stop to the final journey destination, using any marked departure and destination point on the map. The routing engine was conceptualized in such a manner that it could be extended by incorporating other means of transport and route networks (e.g., bicycle and hiking trails) in a future upgrade of the system, provided that the necessary data are available and pre-processed for GIS and routing use. The Web-GIS is currently maintained on a server at the Technical University of Munich and awaits a decision on further funding for operational use.

15.3 METHODS AND THEORETICAL FRAMEWORK

15.3.1 Research Methodology

As stated previously, the main aim of the study was to examine the social implications of Web-GIS development and system use. These issues were investigated through qualitative empirical social research methods. Qualitative methods, in comparison to quantitative methods, allow more detailed insight into the object of investigation in a real-life setting. They are especially suitable for exploratory surveys - such as the current study - where the conceptual design of the research methodology cannot build upon existing reference material[8].

The findings discussed in the remainder of the chapter are based upon information obtained during a period of three years (May 2001 - May 2004). Sources included a combination of research reports, two workshops, 15 interviews with regional politicians, tourism and nature conservation officials, direct participation in meetings and observation. In addition, the attitudes of tourists towards web-based information systems were examined prior to the project within the scope of two diploma theses[9,10].

As noted earlier, we were not only interested in the social implications of developing a Web-GIS, but also in the operational use of such a system. Since the tourism project only allowed us to gain insights in the development process, we decided to gather information about the implications of system use by analyzing another prototypical Web-GIS with similar aims. This Web-GIS for Berchtesgaden (*info-bgl*) was put into operational use at the same time the Web-GIS Tourismus TUM was being developed and was based on similar technology. In-depth interviews with selected regional stakeholders were used to explore the effects of this system.

15.3.2 Actor-Network Theory

The insights gained through the above methodology were subsequently considered through the perspective of Actor-Network Theory (ANT) in order to try to find an explanation for the observations. ANT is a research paradigm that differs from most other approaches to the investigation of socio-technical relationships because it abolishes the subject-object distinction characteristic of 'classical' sociological research[11-13]. This means that in a socio-technical system, such as a GIS, people, organizations, regulations and even inanimate objects are seen as playing active roles and influencing each other. As Harvey[14, p30] states: *'Technologies incorporate and merge different interests in bundled socio-technical relationships. In summary, the network model for actor network theories is that nodes are people, institutions, and artefacts; connections are agreements and exchanges'*. Similarly, Tatnall and Gilding[15] describe ANT as being concerned with *'studying the mechanics of power as this occurs through the construction and maintenance of networks made up of both human and non-human actors'*.

An important benefit of ANT is that it does not view technology as either a mere product of social action or a deterministic influence on society. Instead, agency is seen as multiple and distributed. This facilitates insights into political aspects of the system implementation process[16,17] and therefore helps to answer questions like 'who influenced the project, how, and why?' and 'what is the reason that it developed in this way?'[18].

In order to study and analyze these networks of actors, explain their outcomes and implications as well as subject-object relationships Latour[19] suggests to 'follow the actors' involved - through interviews and/or examination of documents produced by the actors, which means not only to examine what they do, but also investigate their motives and beliefs. The insights generated from applying such an approach to the Baviarian Web-GIS examples are discussed at the end of the following section.

15.4 OBSERVATIONS AND INTERPRETATION

15.4.1 Actor Involvement

The project was initiated by a small group of four experts from regional policy and science backgrounds. Subsequently there was a steady increase in participants, with some 64 individual actors in the system development process. These actors can be categorized into five main groups: state (Bavarian) politicians and officials (~5), regional politicians and officials (~27), scientists (~4), businesses and economic organizations (~14) and individual citizens or interest groups (~14). However, these numbers should not be interpreted too precisely since some actors represented multiple interests and certain organizations had multiple participants at meetings.

Developing the Web-GIS therefore engaged a diverse group of actors, but towards the end of the project there was a noticeable drop in the number of individuals involved with a group from administrative backgrounds taking over the helm. In the final stages, discussion about the Web-GIS was led by these actors within the scope of the Bavarian Parliament's Commission on Economy, Infrastructure, and Traffic, and a special meeting at the government offices for Lower Bavaria. This change was concurrent with a shift in the project from discussions about functionality, content, 'look and feel' etc. of the Web-GIS itself towards issues with greater political implications such as future funding and system operation.

Another feature of the whole development process was that it was not autonomous. It was initiated by a small group of actors with a special interest in the project, who then encouraged other stakeholders to actively participate. This meant that participation took place within a moderated framework and that the group that started the project maintained a supporting role throughout the process.

15.4.2 Motivation to Participate

Many of the regional actors first became involved through an invitation from the system development team to participate in an initial workshop in autumn 2001. However, other actors heard about the project through word of mouth and effectively enrolled themselves in the system development process. As the project progressed it became apparent that motives for participation were varied and included:

- Enhanced regional development and promotion of tourism (state and regional politicians, tourists boards, businesses and economic development agencies)
- An opportunity to research technical aspects of Web-GIS and the social implications of the system development process (scientists)
- Ensuring that nature conservation interests were represented in the Web-GIS and helping to secure a future for data resources whose use was under threat due to funding shortages (environmental NGOs)
- An opportunity to promote existing traffic information systems (a local software business)
- Increasing existing technical knowledge of Web-GIS (regional development association)

In many cases, therefore, participants became involved in the Web-GIS project because they could see it as a means of also fostering their own particular interests in some way.

15.4.3 Development of GIS Knowledge

Soon after the project began in 2001 an initial workshop was held with the stakeholders to discuss the desired functionality of the Web-GIS. However, the experience from this meeting was that although participants could articulate existing problems (e.g., inadequate promotion and lack of consistent standards) there was little awareness of what a Web-GIS could do and how it would differ from a standard regional Internet portal. The decision was therefore made to go ahead with the development of a prototype Web-GIS and then present this to the different parties. Such an approach certainly stimulated interest, to the extent that as the project advanced more technical ideas and visions about GIS use were introduced (e.g., scope for integration with other GIS databases or PDA-based mobile GIS) while tourism and economic development issues were less prominent.

During the project there was a general augmentation of GIS awareness and know-how among the actors involved. Initially, the GIS technology was something new for nearly all participants. As the system developed, all participating actors gained a clear idea what a GIS is, what it can do, and how it could benefit their region by contributing to marketing. Nevertheless, as questions in the interviews revealed, the further social or political implications of GIS and its possible relevance for regional development aims were not generally recognized.

15.4.4 Strengthening Regional Co-operation

At the start of the project discussions tended to take place within small, dispersed groups of interested individuals and organizations. Over time, there were more large meetings and workshops. This resulted in a changing attitude among participants. Initially, each community or interest group focused strongly on the benefits of the project for themselves, but gradually there was increasing recognition that only closer regional collaboration concerning tourism promotion and economic development would be successful. One particular example of this was agreement at a workshop in spring 2003 to promote the region under the heading of 'Bayerwald Counties' rather than more localised names. The active participation of all six chief administrative officers of the so-called 'Bayerwald-Landkreise' from the 2003 workshop until the end of the project can be seen as a further indicator of this collaboration.

15.4.5 Social Implications of System Use

Interviews regarding the impacts of the *info-bgl* Web-GIS in Berchtesgaden suggested that it had been particularly important in helping to enhance the public transportation system and promoting the region as a car-free spa and tourist destination. More specific examples of impacts included highlighting opportunities for new bus routes and a need for better coordination of the public transport timetables in Berchtesgaden and the neighboring Salzburger Land in Austria. In addition, *info-bgl* had been used for purposes other than those originally planned (e.g., providing routing information for practice rescue exercises to an auxiliary fire brigade) and had stimulated interest in other possible GIS-based applications such as a tree register.

15.4.6 Actor-Network Theory Interpretation

From an ANT perspective the development of the Web-GIS Tourismus TUM can be characterized as the formation of a 'network of aligned interests'. In ANT vocabulary such a process can be viewed as involving stages of 'problematization', 'interressement' and 'delineation and coordination'[16] as shown in Figure 15.3.

- *Problematization.* Initially there were only a few actors interested in bringing the Web-GIS into being. These actors came almost exclusively from political or science backgrounds and saw the project as a means of also fostering their own particular interests. Both sets of parties were also influenced by information about regional development objectives and the potential of GIS technology to contribute to such aims.

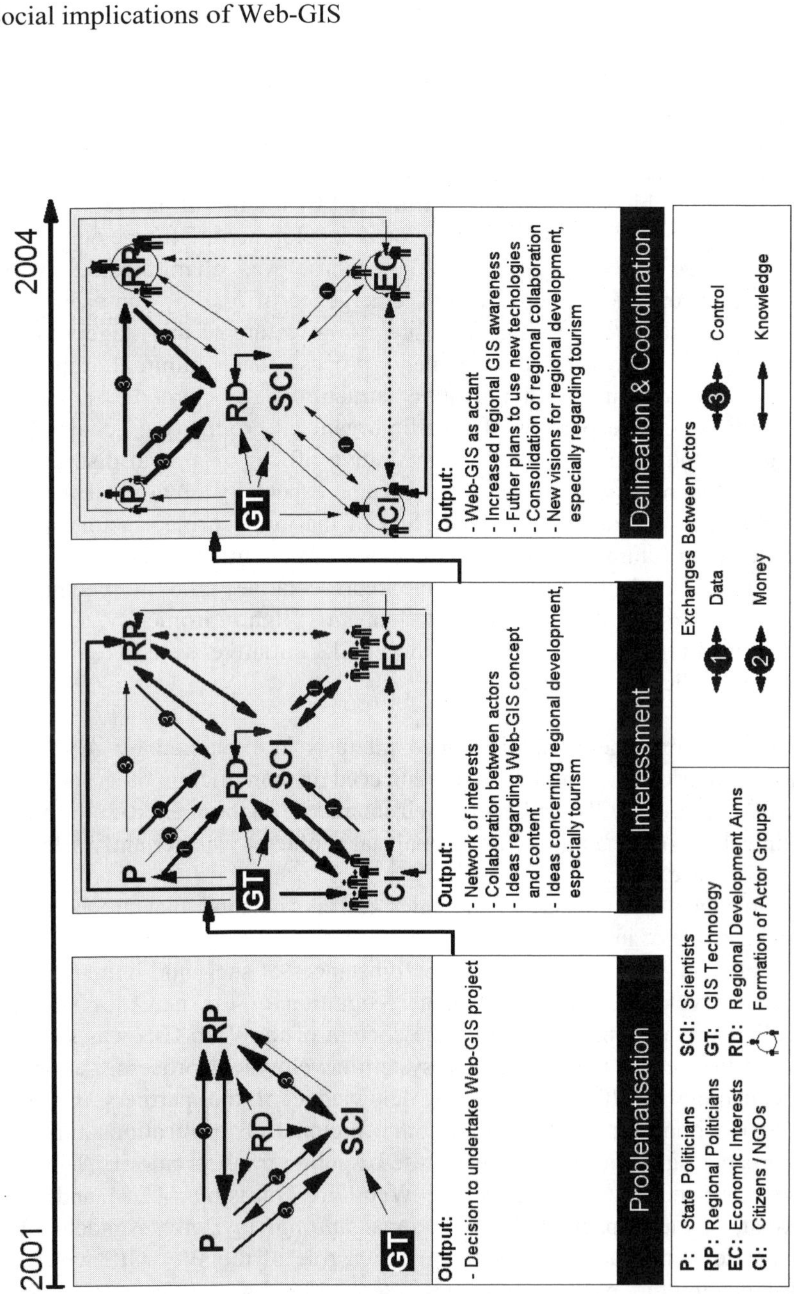

Figure 15.3 Stages in the development of the actor network for the Web-GIS Tourismus TUM.

- *Interessement.* Once the decision to develop the Web-GIS had been made, the initial actors sought to convince other regional stakeholders to participate in the project. One feature of this phase was lively discussion about regional development and Web-GIS issues. This resulted in a focusing of objectives (strongly influenced by the initial partners) and the generation of new visions about regional development, GIS use etc.
- *Delineation and Coordination.* This phase was characterized by the formation of actor groups and the emergence of leading representatives from each subset. A particularly interesting feature of this stage was that although the system development process had minimal top-down influence, existing administrative structures re-emerged as already established spokespersons of specific organized groups (e.g., chair of the regional tourist board, chief administrative officer of a rural district etc.) took over the leading roles. This became especially obvious at the very end of the project. At the suggestion of the initial project actors, issues concerning future work and the financing of system implementation were discussed in depth with the group representatives and in meetings at institutions such as the Bavarian Parliament. Slightly ironically, therefore, decisions regarding the future of the PPGIS initiative were taken without direct public involvement.

The formation of a network of aligned interests is explained by ANT as a process where one or more initial actors succeed in convincing other potential participants that joining will also benefit their interests. Such an exertion of power and 'enrollment' of other actors (i.e., regional stakeholders) was certainly a feature of the Web-GIS case study.

ANT also views documents, technologies etc. as 'actants' that are part of a network and capable of influencing other actors. In this case, regional development plans, GIS and Web-GIS technology were all instances of such non-human actors. Discussion of development aims led to a recognition of the need for stronger regional collaboration. Similarly, while the content of the Web-GIS was strongly influenced by the people involved in the system development process, the use of the GIS technology itself augmented the knowledge of the partners involved, increased their willingness to think about other regional IT applications and even promoted broader agendas (e.g., a greater use of public transportation). This was apparent in both the development of the Web-GIS Tourismus TUM and, to a greater extent, in the impact of the operational *info-bgl* in Berchtesgaden. The operational state of the latter enhanced the actant role of the Web-GIS which, in turn, contributed to other positive outcomes in terms of public services.

According to ANT, action always takes place within networks of aligned interests. All participating actors are simultaneously part of multiple different networks. As a consequence, the outcome of one network of aligned interest can possibly influence other networks as well. This helps explain the 'diffusion' of ideas, technologies, etc. In the Web-GIS study, the increased GIS awareness and the voluntary involvement of some actors who were not invited to the initial workshop are examples of such diffusion.

The above discussion suggests that the use of ANT concepts is indeed useful in explaining many of the observations made during the two case studies. These include the general manner in which the network of interests developed (e.g., the emergence of actor subgroups) in the Tourismus TUM project and the impacts that occurred through use of the two Web-GIS services (e.g., on technical awareness and the use of public transportation)

15.5 CONCLUSIONS

The social implications identified by analyzing both system development and operational use suggest that a user-centered regional Web-GIS can contribute to:

- Activating a certain degree of moderated public participation;
- Building of regional networks of interest;
- Strengthening regional coherence;
- Stimulating discussions about overall regional development visions and strategies;
- Developing a technology-friendly attitude, resulting in an increased willingness to use innovative technologies;
- Fostering behavior that is in accordance with the aims underpinning a Web-GIS.

Development of the Web-GIS Tourismus TUM took place in a form of moderated participation process activated by a core group of actors (i.e., those involved in system development). This does not mean, however, that the process was a purely 'top-down' one. Instead, diverse regional actors voluntarily enrolled themselves in the process – forming the network of aligned interest – and actively influenced both the conceptual approach and operating concept of the Web-GIS. However, this 'bottom-up' element became more marginalized as the project progressed. Individually diverse actors became less influential as already established power structures re-emerged and resulted in a shift back to a more 'top-down' process in the final stage of the project.

A related point concerns the nature of the actors that participated in the project. These were almost exclusively people already involved in regional development, nature conservation or other NGOs and so must be regarded as self-selected to some degree. Such an outcome has been noted in many other participation processes[2,20,21] and again raises the question of whether the development of the Web-GIS really contributed to an empowerment of regional stakeholders or just cemented already established administrative and political 'top-down' power structures.

The observations and interpretation presented in this study suggest that the use of ANT concepts can provide insights into the interactions and implications associated with such a technical activity. In particular, they highlight the effects that use of the Web-GIS had on other actors. This, in turn, raises issues concerning the ethics of PPGIS projects[22], particularly given the scope for misuse of such technology[23]. At present, Web-GIS is still perceived as a mere tool by many spatial planners, GIS experts and administrators when it really needs to be more widely understood as a social technology.

15.6 ACKNOWLEDGMENTS

This study was conducted as part of the 'High-Tech-Offensive' Bavaria Initiative (HTO 33-5) funded by the Government of Lower Bavaria and the Bavarian Ministry of Agriculture and Forestry.

15.7 REFERENCES

1. Craig, W.J., Harris, T.M., and Weiner, D., Eds., *Community Participation and Geographic Information Systems*, Taylor & Francis, London, 2002.

2. Carver, S. J., The future of participatory approaches using geographic information: developing a research agenda for the 21st Century, *Journal of the Urban and Regional Information Systems Association*, 15, Access and Participatory Approaches I (Electronic only), 61-71, 2003.

3. Hansen, S.H. and Prosperi, D.C., Citizen participation and the Internet – some recent advances, *Computers, Environment and Urban Systems*, 29, 617-629, 2005.

4. ARGE Waldwildnis, http://www.waldwildnis.de, 2005.

5. Neumeier S., Implementation of a web-based GIS for tourists in the Bavarian Forest National Park region, in *Proceedings of GeoTec 2003*, Vancouver, B.C, 16-18 March, 2003.

6. Hogan, M. and Stajan, D., A web-based GIS application to help prompt tourism for the city of Chilliwack, http://giswww1.bcit.ca/giscentre/projects/projects2002/prj_13_final_report.pdf , 2002.

7. Peng, Z-R. and Tsou,M-H., *Internet GIS*, Wiley, Chichester, 2003.

8. Flick, U., *Qualitative Forschung. Theorie, Methoden, Anwendungen in Psychologie und Sozialwissenschaften* (Qualitative Research. Theory, Methods, Applications in Psychology and Social Sciences), Rowohlt Taschenbuch Verlag, Hamburg, 1996.

9. Reccius, A., Konzeptionelles Datenmodell eines Touristischen Geoinformationssystems für die Landkreise Regen und Freyung-Grafenau (Conceptual data model of a tourism GIS for the administrative districts Regen and Freyung-Grafenau), Unpublished diploma thesis, Department of Geography, Technische Universität München, 2000.

10. Semmler, M., Entwicklungen und Potenziale von Geoinformationssystemen im Tourismus, dargestellt am Beispiel Touristik-Online-Service Chiemsee (Development and potential of GIS in tourism, a study

of the Tourism-Online Service Chiemsee), Unpublished diploma thesis, Department of Geography, Technische Universität München, 2000.

11. Latour, B., Materials of power. Technology is society made durable, in *A Sociology of Monsters: Essays on Power, Technology and Domination*, Law, J., Ed., Routledge, London, 1991, 103-131.

12. Law, J., After ANT: complexity, naming and topology, in *Actor Network Theory and After*, Law, J. and Hassard, J., Eds., Blackwell, Oxford, 1999, 1-14.

13. Schulz-Schaeffer, I., Akteur-Netzwerk-Theorie. Zur koevolution von gesellschaft, natur und technik, in *Soziale Netzwerke. Konzepte und Methoden der Sozialwissen-Schaftlichen Netzwerkforschung*, Weyer, J., Ed., R. Oldenbourg Verlag, Munich, 2000, 187-209.

14. Harvey, F., Constructing GIS: actor networks of collaboration, *Journal of the Urban and Regional Information Systems Association*, 13, 29-37, 2001.

15. Tatnall, A. and Gilding, A., Actor-Network Theory and information systems research, in *Proceedings of the 10th Australian Conference on Information Systems*, Wellington, New Zealand, 1-3 December, 1999.

16. Sidorova, A. and Sarker, S., Unearthing some causes of BPR failure: an Actor-Network Theory perspective, in *Proceedings of the American Conference on Information Systems (AMCIS 2000)*, Long Beach, California, August 10-13th, 2000.

17. Walsham, G. and Sahay, S., GIS for district-level administration in India: problems and opportunities, *MIS Quarterly*, 23, 39-66, 1999.

18. Underwood, J., Not another methodology. What ANT tells us about system development, in *Proceedings of the 6th International Conference on Information System Methodologies*, British Computer Society, Salford, UK, 3-4 September, 1998.

19. Latour, B., *Science in Action. How to Follow Scientists and Engineers Through Society*, Harvard University Press, Cambridge, Massachusetts, 1987.

20. Young, I. M., *Justice and Politics of Difference*, Princeton University Press, Princeton, 1990.

21. Leitner, H., McMaster, R.B., Elwood, S., McMaster, S., and Sheppard, E., Models for making GIS available to community organisations: dimensions of difference and appropriateness, in *Community Participation and Geographic Information Systems*, Craig, W.J., Harris, T.M., and Weiner, D., Eds., Taylor & Francis, London, 2002, 37-52.

22. Elwood, S., Critical issues in participatory GIS: deconstructions, reconstructions and new research directions, *Transactions in GIS*, 10, 693-708, 2006.

23. Towers, G., GIS versus the community. Siting power in southern West Virginia, *Applied Geography*, 17, 111-125, 1997.

Subject Index

O

P

Q

R

S

T

U

V

W

lems. The computation of a vector of criteria based on a system of complex differential equations took approximately 3 minutes. In the identification phase of this problem, we identified 16 parameters of the mathematical model using 65 adequacy criteria and defined to what extent the model corresponds to the real system. In the initial problem statement, 4,096 tests were conducted. Only seven solutions met the criteria constraints and thus entered the feasible solution set. After analyzing the obtained results, new boundaries of the design variables were defined. The same number of tests (4,096) was conducted. As a result, 11 more feasible solutions were obtained. On the basis of solution of the identification problem and the definition of the feasible solution set, the problem of optimization by 20 performance criteria was solved next. These criteria were divided into the following groups: (1) comfort, (2) durability, (3) load preservation, and (4) safety. Twenty parameters were varied. Optimization was aimed to improve the prototype. 4,096 tests were conducted and 21 solutions satisfied all constraints and entered the feasible solution set. The Pareto set consisted of 20 solutions.

In addition, we provide below three more relevant examples:

4. In the problems of improving the rear axle housing for a truck (5 performance criteria, 12 pseudo-criteria, and 7 design variables, Section 5.3) and improving the truck frame prototype (4 performance criteria, 3 pseudo-criteria, and 21 design variables, Section 5.4) 256 and 300 trials are performed correspondingly.
5. In the problem of vector identification of parameters of a spindle unit (Section 6.3), we considered dynamic and thermal mathematical models with 23 and 13 criteria adequacy correspondingly. 512 trials are carried out in each problem. In the problems of optimization of design variables of a spindle unit, 6 performance criteria were minimized and 256 trials were performed in each problem. (For the problems 4 and 5, calculation of one criterion vector requires up to 4 minutes.)
6. Chapter 9 discusses the application of the PSI method for the design of the L_1 flight control system implemented on the two turbine-powered dynamically scaled Generic Transport Model (GTM), which is part of the Airborne Subscale Transport Aircraft Research aircraft at the NASA Langley Research Center. The problem statement has been corrected twice. In the first iteration 14 criteria and pseudo-criteria

were investigated, 1,024 trails were performed, and it took approximately 1 minute to compute one vector of criteria. In the second iteration 16 criteria and pseudo-criteria were investigated, 512 trails were performed, and it took approximately 10 minutes to compute one vector of criteria.

In all cases, the results surpassing prototypes in all criteria or in the most important ones were obtained. It is important to note that in order to obtain these results the statements of the problems, as a rule, have been repeatedly corrected.

References

[1] Statnikov, R. B., et al., *MOVI 1.4 (Multicriteria Optimization and Vector Identification) Software Package,* Certificate of Registration, Registere of Copyright, U.S.A., Registration Number TXU 1-698-418, Date of Registration: May 22, 2010.

[2] Statnikov, R. B., and J. B. Matusov, *Multicriteria Analysis in Engineering,* Dordrecht/Boston/London: Kluwer Academic Publishers, 2002.

[3] Egorov, I. N., et al., "Multicriteria Optimization of Complex Engineering Systems from the Design to Control," *Journal of Machinery Manufacture and Reliability,* Russian Academy of Sciences, No. 2, 1998, pp. 10–20.

[4] Beknev, V. S., I. N. Egorov, and V. S. Talyzina, "Multicriteria Design Optimization of Multistage Axial Flow Compressor," *Proceedings of the 36th International Gas Turbine Congress,* Budapest, Hungary, 1991.

[5] Egorov, I. N., and G. V. Kretinin, "Search for Compromise Solutions of the Multistage Axial Flow Compessor's Stochastic Optimization Problem," *Proceedings of the 3rd ISAIF (International Symposium on Experimental and Computational Aerothermodynamics of Internal Flows III),* Beijing, China, 1996, pp. 112–120.

[6] Egorov, I. N., et al., "Problems of Design and Multicriteria Control for Adjustable Technical Systems," *Doklady Physics* (Russian Academy of Sciences), Vol. 43, No. 3, 1998, pp. 181–184, translated from *Doklady Akademii Nauk,* Vol. 359, No. 3, 1998, pp. 330–333.

[7] Statnikov, R. B., et al., "Multicriteria Analysis Tools in Real-Life Problems," *Journal of Computers & Mathematics with Applications,* Vol. 52, 2006, pp. 1–32.

[8] Wasserman, P. D., *Advanced Methods in Neural Computing,* New York: Van Nostrand Reinhold, 1993.

[9] Anderson, T. W., *An Introduction to Multivariate Statistical Analysis,* 3rd ed., New York: Wiley-Interscience, 2003.

[10] Chatterjee, S., and A. S. Hadi, "Influential Observations, High Leverage Points, and Outliers in Linear Regression," *Statistical Science,* Vol. 1, No. 3, 1986, pp. 379–416.

[11] Vapnik, V., *Statistical Learning Theory,* New York: John Wiley & Sons, 1998.

[12] Statnikov, R. B., et al., "Visualization Approaches for the Prototype Improvement Problem," *Journal of Multi-Criteria Decision Analysis,* No. 15, 2008, pp. 45–61.

[13] Bondarenko, M. I., et al., "Construction of Consistent Solutions in Multicriteria Problems of Optimization of Large Systems," *Physics-Doklady,* Vol. 39, No. 4, 1994, pp. 274–279, translated from *Doklady Rossiiskoi Akademii Nauk,* Vol. 335, No. 6, pp. 719–724.

8

Adopting the PSI Method for Database Search

8.1 Introduction

Database search engines are used in essentially all areas of modern life. There are two major types of information retrieval problems that are being solved by search engines. In the first type, a search engine matches a query of the client to a database and reports back results. For example, the client uses PubMed service (http://www.ncbi.nlm.nih.gov/pubmed/) to identify articles that talk about the BRCA1 gene in breast cancer using the query "BRCA1 breast cancer." Another example is when the client uses the Google search engine (http://www.google.com/) to identify a poem from the phrase "to whose immortal eyes." Usually, this type of problem can be solved efficiently and automatically (i.e., without interaction with the client) given that the database contains a sought solution.

In the second type of problem, the client uses a search engine to retrieve solutions (alternatives) from the database that are described by a set of contradictory characteristics. For example, such problems are widespread when searching for an airline ticket, a matching partner, or real estate on the Web. In a typical scenario, the database contains a very large number of alternatives described by a set of characteristics. Because of the large number of alternatives, the search engine allows the client to impose constraints on characteristics such that the resulting filtered set of alternatives satisfies all requests of the client. If these constraints are defined correctly, they determine so-called the feasible solution set where the most preferable solution should be then sought by the client. However, in the majority of cases, the client defines constraints intuitively and thus often incorrectly. As such, many interesting solutions that, under certain assumptions, could have been feasible will be lost. This results in the poor or even empty feasible solution set. This is especially the case when the client deals with many contradictory characteristics and many alternatives[1]. Current search engines do not provide tools for systematic construction of the feasible solution set. This is a major shortcoming of existing database search engines, because a search for the most preferable solution depends, to a large extent, on the definition of the constraints, and this insurmountable task is shifted to the client's shoulders.

The problem of database search involves many contradictory characteristics and a large number of alternatives and necessitates decision-making [1]. The ultimate results also critically depend on the quality of database [2], the type of constraints on characteristics [3], and other factors. An important feature of the database search problem is that filtering of the number of solutions via the construction of the feasible solution set should precede decision-making.

That is why main thesis of this chapter is that systematic construction of the feasible solution set has a fundamental value for the client. In this chapter, we modify the PSI method to allow the construction of the feasible solution set specifically for the problems of database search. We propose a modified PSI method called DBS-PSI (DBS is an acronym for "database search") that allows us to significantly improve the quality of the search results by explicitly constructing and analyzing the feasible solution set and identifying the most preferable solutions.

8.1.1 Characteristics of Alternatives: Criteria and Pseudo-Criteria

We define a criterion as the characteristic of the alternative[2] that is related to the alternative's quality by monotonous dependence. In other words, all other things being equal, the more (or the less) the value of the criterion, the better the alternative. Contrary to the criterion, a pseudo-criterion is not related to an alternative's quality by monotonous dependence [4]. Pseudo-criteria do not need to be optimized; only their constraints have to be satisfied. These constraints are determined either by known and generally accepted standards or by the client's preferences. In most cases, constraints on both criteria and pseudo-criteria are not rigidly set. An algorithm allowing us to determine these constraints is provided in [5]. Examples of criteria and pseudo-criteria are given next.

1. In rare cases, when there are a few alternatives and characteristics, this task may be realistic for the client.

2. Examples of alternatives in database search engines are airline tickets (travel search Web sites), women/men (online dating services), houses (real estate search Web sites), cars (online services to buy cars), and so on.

Let us consider four popular uses of database search engines and determine several criteria and pseudo-criteria to illustrate these definitions:

- Using travel search Web sites to buy airline tickets, for example, Kayak travel search engine (http://www.kayak.com). The criteria are cost, number of stops, and duration. The pseudo-criteria are takeoff time, landing time, airline, airport (see Figure 8.1). For example, the client most often specifies a range of constraints on takeoff time and landing time. However, if he or she wanted to depart/arrive as early/late as possible, takeoff time and landing time should be considered as criteria. The choice of destination airport may also be a criterion if the client has to hurry from the airport to work, and as a result he or she chooses the airport closest to his or her office.
- Using online dating services to look for a matching partner, for example, Yahoo! Personals (http://personals.yahoo.com/). The criteria are educa-

Figure 8.1 Kayak travel search engine. (Source: Available at http://www.kayak.com.) Fragments.

tion level and income level. The pseudo-criteria are appearance and lifestyle. Appearance contains the subsets of more specific pseudo-criteria such as ethnicity, height, body type, eye color, and hair color. Lifestyle contains the subset of more specific pseudo-criteria such as marital status, profession, religion, and interests. In general, when we talk about one's height, weight, and age, it generally makes no sense to say that the higher (lower) the values of these characteristics, the better the alternative. These characteristics must lie within certain boundaries.

- Using real estate search websites to buy a house, for example, Realtor.com (http://www.realtor.com/). The criteria are cost, distance to work, and school district (for school-age children). The pseudo-criteria are ceiling height and number of floors. If the client is handicapped and uses a wheelchair, the number of floors becomes a criterion.
- Using online services to buy cars, for example, Cars.com (http://www.cars.com/). The criteria are cost, fuel consumption, and operating cost. The pseudo-criteria are wear resistance, durability, and the operating life of car parts.

8.1.2 General Statement of the Problem and Solution Approach

Assume that a database contains N alternatives (solutions), each of which is described by k characteristics[3] (pseudo-criteria and criteria) Φ_v^j, for $v = 1,\ldots, k$ and $j = 1, \ldots, N$. These characteristics can be contradictory, so improving some characteristics leads to the deterioration of the others. Also assume that the database contains so many alternatives that the client experiences significant difficulties while analyzing them and choosing the most preferable alternative. The problem is how to filter the set of alternatives without losing any alternatives that are acceptable to the client. This can be accomplished by constructing the feasible and Pareto optimal sets.

We are given a vector of characteristics $\Phi = (\Phi_1, \Phi_2, \ldots, \Phi_k)$, where $\Phi_1, \Phi_2, \ldots, \Phi_c$ are criteria that we want to minimize and $\Phi_{c+1}, \Phi_{c+2}, \ldots, \Phi_k$ are pseudo-criteria. Alternatives Φ^j that: (1) satisfy all criteria constraints, that is, Φ_v^j (v = 1, …, c) such that $\Phi_v^j \le \Phi_v^{**}$ (or $\Phi_v^j \ge \Phi_v^{**}$ in the case of maximization) and (2) satisfy constraints on pseudo-criteria, that is, Φ_v^j (v = c + 1, …, k) such that $\Phi_v^* \le \Phi_v^j \le \Phi_v^{**}$, constitute the feasible solution set D. Here Φ_v^{**} is the worst value of criteria which the client considers as acceptable; constraints

3. It is worth noting that often client can a priori state "rigid" requirements to certain characteristics that cannot be changed under any circumstances. In such cases, when searching for the most preferable solution, it is necessary to exclude alternatives that did not satisfy those requirements and then search for feasible solutions as described in this section.

on pseudo-criteria Φ_v^* and Φ_v^{**} correspond to the best and worst values, respectively.

We solve the general problem by first constructing the feasible solution set D by imposing constraints on these criteria and pseudo-criteria mentioned via dialogues of the client with the computer. Next, we define the Pareto optimal set $P \subseteq D$ using only criteria $\Phi_1, \Phi_2, \ldots, \Phi_c$. Recall that an alternative $\Phi^{j_0} \in D$ is called a Pareto optimal if there exists no alternative $\Phi^j \in D$ such that $\Phi_v^j \leq \Phi_v^{j_0}$ for all $v = 1, \ldots, c$ and $\Phi_{v_0}^j < \Phi_{v_0}^{j_0}$ for at least one $v_0 \in \{1, \ldots, c\}$. In other words, alternative Φ^{j_0} cannot be improved by all criteria $\Phi_1, \Phi_2, \ldots, \Phi_c$ simultaneously. A set $P \subseteq D$ is called a Pareto optimal set if it consists of only Pareto optimal alternatives. Finally, the client determines the alternative $\Phi^{j_0} \in P$, which is the most preferable among the alternatives belonging to the set P.

8.1.3 Motivation of the Problem Statement

Concessions Φ_v^* and Φ_v^{**} are something like a market where we have the opportunity to trade until we get what we need. Of course, not all concessions are possible, but that is outside the scope of our work. By investigating the set of Pareto solutions, the clients see what can and what cannot be achieved. They are able to choose the most preferable alternative in this set, $\Phi^{j_0} \in P$. At the same time, they know that there exist no better alternatives in the database. The Pareto optimal set plays an important role in vector optimization problems because it can be analyzed more easily than the feasible solution set and because the optimal alternative always belongs to the Pareto optimal set, irrespective of the system of preferences used by the client for comparing alternatives belonging to the feasible solution set.

8.1.3.1 Problem Considered by This Work

We primarily consider problems with the following features:

- Characteristics are contradictory.
- The dimensionality of the characteristics vector may reach dozens.
- The number of alternatives may reach many thousands.
- The constraints on these characteristics are often stringent.
- The constraints on characteristics Φ_v^* and Φ_v^{**} can be defined in an interactive mode, in the process of dialogues of the client with the computer.

8.1.3.2 What Do the Client and Search Engine Developers Need to Know?

Search engine developers must provide all necessary characteristics (criteria and pseudo-criteria). The client must understand which constraints on characteristics are acceptable and which are not, without differentiating between criteria and pseudo-criteria when constructing the feasible set. After determining the feasible set (feasible alternatives), the client must decide personally which one of these characteristics is a criterion and which one is a pseudo-criterion. This is because the same characteristics may be a criterion for one client and pseudo-criteria for another client. Next come the analysis of the obtained alternatives and the selection of the most preferable one.

8.2 DBS-PSI Method

8.2.1 DBS-PSI Method as a New Paradigm of a Database Search

We mentioned earlier difficulties in constructing the feasible solution set correctly. To do this, the client must see what concessions he or she should make on characteristics and what he or she will get in return[4]. We want to show the process of generating concessions to the client and to interactively engage him or her in it.

Recall that the PSI method was designed to help the client determine constraints on characteristics (criteria and pseudo-criteria) and thus to construct the feasible solution set. After determining the feasible solution set, the Pareto optimal set is constructed only with consideration of the criteria. Then the client will search for the most preferable solution on the Pareto optimal set. The new DBS-PSI method [5] is designed to be used for a database search, and it consists of three main stages, presented next.

Stage 1: Compilation of Analysis Tables Via Computer

In this stage, the client transforms the database into analysis tables[5] as follows. For each characteristic an analysis table is compiled so that the values of $\Phi_\nu^1,\ldots,\Phi_\nu^N$ are arranged in increasing order; that is,

$$\Phi_\nu^{i_1} \le \Phi_\nu^{i_2} \le \ldots \le \Phi_\nu^{i_N}, \; \nu = 1,\ldots,k \tag{8.1}$$

4. We note that in search engines it is possible to revise constraints. However, the client is expected to do this at a very intuitive level.

5. The foundation of the PSI method is test tables. The principal difference between test tables and analysis tables is that in the former case, solutions are obtained with the use of generators (e.g., uniformly distributed sequences), and in the latter case solutions are obtained from a database. Also, the parameters are absent in the search engines that are discussed in this section.

where $i_1, i_2, \ldots, i_N$ are the numbers of alternatives (a separate set for each v). Taken together, the k tables form complete analysis tables. In analysis tables, the list of nonnumerical criteria is arranged in order of the client's preferences.

Stage 2: Selection of Characteristic Constraints

This stage includes the dialogue of a client with the computer. Let $\min\Phi_v$ and $\max\Phi_v$ be unfeasible solutions for the vth characteristic (pseudo-criterion and/or criterion), and let $\mathbf{\Phi}_v^*$ and/or $\mathbf{\Phi}_v^{**}$, respectively, be the minimum possible concessions from $\min\Phi_v$ and/or $\max\Phi_v$, ($\min\mathbf{\Phi}_v^* \le \mathbf{\Phi}_v^j \le \mathbf{\Phi}_v^{**} \le \max\Phi_v$), where $[\min\Phi_v; \mathbf{\Phi}_v^*)$ and $(\mathbf{\Phi}_v^{**}; \max\Phi_v]$ are unfeasible solution domains. Then the remainder of table $[\mathbf{\Phi}_v^*; \mathbf{\Phi}_v^{**}]$ is the maximum search interval for feasible solutions. After examining the characteristics arranged in each of the analysis tables in increasing (decreasing) order, starting with $\min\Phi_v$ and/or $\max\Phi_v$, the client analyzes each value. The first feasible value found will be $\mathbf{\Phi}_v^*$ ($\mathbf{\Phi}_v^{**}$). If $\min\Phi_v$ and/or $\max\Phi_v$ are feasible values, there is no need to make the corresponding concessions. In general, we will denote the feasible solution search interval as $[\mathbf{\Phi}_v^*; \mathbf{\Phi}_v^{**}]$. If the selected values of $[\mathbf{\Phi}_v^*; \mathbf{\Phi}_v^{**}]$ are not a maximum, then many interesting solutions may be lost, since some of the characteristics are contradictory.

The client has to consider one characteristic (analysis table) at a time and specify the respective constraints $\mathbf{\Phi}_v^*$ and $\mathbf{\Phi}_v^{**}$. Then the client proceeds to the next analysis table, and so on. Note that the revision of the characteristic constraints does not cause any difficulties to the client.

It is important to emphasize that a human-computer dialogue is very convenient for the client: he or she does not have to change the values of some characteristics at the expense of others. He or she sees one analysis table and sets the appropriate constraints; then he or she repeats the same process with another table, and so on. After the client defines the constraints, the computer searches for the feasible solutions. For example, if the client has defined the constraints on the first and second characteristics, he or she immediately obtains information on feasible solutions based on these characteristics. After defining the constraints on the third characteristic, the client obtains feasible solutions based on three characteristics, and so on. In other words, the client sees what price he or she pays for making concessions on various characteristics (i.e., what the client loses and gains). Such an analysis gives the client valuable information on the advisability of revising various characteristic constraints with the aim of improving the basic characteristics. Thus, in this method, the client imposes constraints on each characteristic in succession and follows the construction of the feasible solution set step by step. Moreover, the client can go back at any time and reexamine the constraints depending on the feasible solution set obtained.

Stage 3: Verification of the Solvability of a Problem Via Computer

Let the client fix a characteristic (e.g., Φ_{v_1}), and consider the corresponding analysis table (8.1). Let S_1 be the number of values in the table satisfying the selected characteristic constraints:

$$\Phi^{*}_{v_1} \leq \Phi^{i_1}_{v_1} \leq \ldots \leq \Phi^{i_{S_1}}_{v_1} \leq \Phi^{**}_{v_1} \tag{8.2}$$

Then characteristic Φ_{v_2} is selected by analogy with Φ_{v_1} and the values of $\Phi^{i_1}_{v_2},\ldots,\Phi^{i_{S_2}}_{v_2}$ in the analysis table are considered. Let the table contain $S_2 \leq S_1$ values such that $\Phi^{*}_{v_2} \leq \Phi^{i_j}_{v_2} \leq \Phi^{**}_{v_2}$, where $1 \leq j \leq S_2$. Similar procedures are carried out for each characteristic. If at least one alternative can be found for which all characteristic constraints are valid simultaneously, then the feasible solution set is nonempty and our problem is solvable. Otherwise, the client should return to Stage 2 and make certain concessions on characteristic constraints Φ^{*}_{v} and Φ^{**}_{v}. The procedure is iterated until the feasible solution set is nonempty. Now the Pareto optimal set is constructed. This is done by removing those feasible alternatives that can be improved with respect to all characteristics simultaneously.

As a rule, the client makes more concessions on the less important characteristics. These concessions may allow him or her to obtain a substantial gain in the most significant characteristics.

Notice that the number of characteristics must be no less than necessary. The client always wishes to optimize not one, but all of the most important characteristics, many of which are antagonistic. The greater the number of characteristics taken into account, the richer the information obtained about alternatives. The DBS-PSI method allows us to consider as many characteristics as necessary.

8.3 Searching for a Matching Partner

The purpose of this example is to give readers an idea of how to construct feasible and Pareto optimal solutions on the basis of analysis tables. Earlier we mentioned the example of using online dating services to look for a matching partner. Let us show how a male client will search for a matching female partner using the DBS-PSI method. Suppose the search characteristics are Height (Φ_1), Weight (Φ_2), Age (Φ_3), Income (Φ_4) and Education (Φ_5). Let the client have access to a database with alternatives. According to Stage 1 of the DBS-PSI method, the client should construct analysis tables (see Figure 8.2). From Figure 8.2 it can be seen that the database contains a large number of alternatives[6] and the values of the characteristics vary over fairly large ranges: from a

6. For example, #12,089, #34,006, #95,175, #30,617, #88,945, and so on. Each of the women in the analysis tables is designated as an alternative #.

#	Height, Φ_1	#	Weight, Φ_2	#	Age, Φ_3	#	Income, Φ_4	#	Education, Φ_5
284	4'11"	5,298	99.2 lbs	971	25 years	1,021	$25,000	12,907	Some high school
…	…	…	…	…	…	…	…	…	…
689	5'2"	88	125.7 lbs	472	30	894	$25,000	9,705	High school graduate
…	…	…	…	…	…	…	…	…	…
27	**5'3"**	…	…	…	…	…	…	12,096	High school graduate
63	5'3"	**7,632**	**127.9 lbs**	…	…	…	…	63	Some college
…	…	97	130.1 lbs	…	…	14,007	$35,000	…	…
12,089	5'3"	…	…	888	34	7	$50,000	…	…
92	5'4"	894	130.1 lbs	**7,632**	**34**	…	…	…	…
…	…	…	…	…	…	**27**	**$50,000**	**7,632**	**Some college**
894	5'4"	10	132.3 lbs	9	35	80,092	$50,000	**27**	**College graduate**
…	…	…	…	…	…	21	$75,000	…	…
1,964	5'4"	1,093	138.9 lbs	23	36	…	…	…	…
38	5'5"	…	…	894	36	…	…	…	…
7,632	**5'5"**	…	…	…	…	…	…	…	…
30,617	5'5"	**27**	**143.3 lbs**	…	…	**7,632**	**$75,000**	…	…
…	…	…	…	**27**	**38**	245	$100,000	…	…
994	5'6"	…	…	…	…	…	…	…	…
…	…	30,617	150 lbs	300	40	6,005	$100,000	88,945	College graduate
…	…	…	…	…	…	711	$150,000	992	Postgraduate
…	…	…	…	…	…	…	…	…	…
76	6'11"	66	183 lbs	34,006	65	90,661	$150,000	95,175	Postgraduate

Figure 8.2 Fragment of analysis tables.

height of 4'11" (alternative #284) to 6'11" (alternative #76), from a weight of 99.2 lbs (alternative #5,298) to 183 lbs (alternative #66), from 25 years of Age (alternative #971) to 65 years of Age (alternative #34,006), from an Income of $25,000 (alternative #1,021) to $150,000 (alternative #90,661), and from an Education of Some High School (alternative #12,907) to Postgraduate (alternative #95,175). The more alternatives there are in the analysis tables, the greater the client's choice.

According to Stage 1, the client should construct search intervals for feasible solutions. After studying the analysis table for the first characteristic (Height), the client has determined the search interval for feasible solutions $[\Phi_1^*;\Phi_1^{**}]$ = [5'2"; 5'6"] (see Figure 8.2).

Alternatives #689, …, #27, …, #994 correspond to this interval. The client then proceeds to the second analysis table, to Weight, and determines the search interval for feasible solutions $[\Phi_2^*;\Phi_2^{**}]$ = [125.7 lbs; 150 lbs]. Alternatives #88, …, #7,632, …, #30,617 correspond to this interval. When switching from one analysis table to another (from Height to Weight), the computer records the total number of alternatives that satisfy the constraints on these characteristics simultaneously. As it turns out, there are four of these alternatives, #27, #894, #7,632 and #30,167. If there were no feasible alternatives, client should revise the constraints on the first and/or second characteristics. The client then proceeds to the third analysis table (Age) and constructs the interval $[\Phi_3^*;\Phi_3^{**}]$ = [30 years; 40 years] that is satisfied by alternatives #472, …, #888, …, #300. The total number of alternatives that satisfies the constraints on three characteristics simultaneously is reduced to three, #27, #894 and #7,632. Then the client examines Income (the fourth analysis table). Not a single alternative has satisfied the constraints $150,000, and $100,000. After setting the constraint to $75,000, the alternative #7,632 is found that satisfies all constraints on four characteristics. Finally, the client reduces the constraint to $50,000. As a result, the total number of alternatives that satisfy constraints on four characteristics is two, namely, #27 and #7,632 (see Figure 8.2).

In the last analysis table (the "Education" characteristic), not a single alternative satisfied the "Postgraduate" constraint. After setting the constraint to "Some College," two feasible alternatives (matching partners) #27 and #7,632 are obtained (shown with bold in the Figure 8.2). These alternatives satisfy all constraints on five characteristics. The values of characteristics of matching partners #27 and #7,632 also are given in Table 8.1. All constraints and matching partners were determined in an interactive mode.

In Figure 8.2, the areas of search for feasible alternatives are highlighted in tan (pseudo-criteria) and in light green (criteria) accordingly. Since our example is concerned with the construction of feasible solutions #27 and #7,632 only, some alternatives and their corresponding values of criteria have been replaced with "…" in the analysis table.

8.3.1 Conclusions of the Example

Constructing and studying analysis tables allowed the client to obtain two feasible alternatives. After that the client may wish to determine the optimal alternative set with consideration of all or the most important criteria. For example, if the client had searched for the optimal solutions according to the criterion

Table 8.1
Input Data and Some Search Results

Characteristics	Height, Φ_1	Weight, lbs, Φ_2	Age, years, Φ_3	Income, $, Φ_4	Education, Φ_5
Range of characteristic values	4′11″–6′11″	99.2–183	25-65	25,000–150,000	Some high school, some college, college graduate, postgraduate
Intervals of search for feasible solutions	5′2″–5′6″	125.7–150	30–40	50,000–150,000	Some college, college graduate, postgraduate
Feasible Alternatives (matching partners)					
#27	5′3″	143.3	38	50,000	College graduate
#7,632	5′5″	127.9	34	75,000	Some college

Income, the search engine would have given him alternative #7,632; alternative #27 would be better for the criterion Education. If two criteria, Income and Education, are considered simultaneously, the Pareto optimal solution set will contain the two specified solutions.

However, according to Yahoo! Personals (http://personals.yahoo.com/), upon starting a search, the client should impose constraints on age and geographical area (see Figure 8.3). In our opinion, these recommendations are incorrect, since the client can lose compromise solutions that are acceptable to him. These constraints should be defined in the process of studying the analysis tables; after finding the feasible alternatives (matching partners) in an interactive mode, the remaining solutions will automatically become unfeasible.

8.4 Summary

It is often the case that database search involves many contradictory criteria and many alternatives, and it is necessary to search for compromise solutions. This

Figure 8.3 Yahoo! Personals search engine. (Notes: available at http://personals.yahoo.com/.) Fragment.

can be efficiently accomplished by construction and analysis of the feasible solution set. However, current search engines do not systematically construct the feasible solution set, thus "hiding" many potentially interesting solutions from the client. To address this problem, we introduced the DBS-PSI method, which allows the client to:

1. Construct analysis tables.
2. Determine the feasible solution set in the process of dialogues with the computer.
3. Find the Pareto optimal set.
4. Identify the most preferable solution.

The process of searching for the feasible solution set based on analysis tables is client-friendly: clients see what they lose and what they gain for every concession. The DBS-PSI method as a new paradigm of database search can be used in all areas of modern life and promises to increase efficiency of modern search engines. The relevancy of the proposed methodology is apparent given ongoing expansion of the World Wide Web and search engines.

References

[1] Saaty, T., "Decision Making with the Analytic Hierarchy Process," *International Journal of Services Sciences,* Vol. 1, No. 1, 2008, pp. 83–98.

[2] Li, X., et al., "Problems and Systematic Solutions in Data Quality," *International Journal of Services Sciences,* Vol. 2, No. 1, 2009, pp. 53–69.

[3] Stefansen, C., and S. E. Borch, "Using Soft Constraints to Guide Users in Flexible Business Process Management Systems," *International Journal of Business Process Integration and Management,* Vol. 3, No. 1 2008, pp. 26–35.

[4] Sobol', I. M., and R. B. Statnikov, *Selecting Optimal Parameters in Multicriteria Problems,* 2nd ed., (in Russian), Moscow: Drofa, 2006.

[5] Statnikov, R. B., et al., "DBS-PSI: A New Paradigm of Database Search," *International Journal of Services Sciences,* Vol. 4, No. 1, 2011, pp. 1–13.

9

Multicriteria Analysis of L_1 Adaptive Flight Control System

9.1 Objective of the Research

Inner-loop adaptive flight control systems[1] are seen as an appealing technology [1] to improve aircraft performance with reduced pilot compensation in adverse flight conditions or in the event of control surface failures and vehicle damage. Under these conditions, which are characterized by a high degree of uncertainty with respect to a nominal aircraft model, the achievable levels of performance and flying qualities (FQ) that a nonadaptive flight control system (FCS) can provide might be limited. However, in applications with stringent performance and robustness specifications [2], several limitations of conventional adaptive control architectures have been identified that render the design of these conventional adaptive controllers overly challenging. Among these limitations [2, 3], are of special relevance: (1) the lack of transient characterization of the closed-loop response, (2) the limited analysis framework for robustness and performance guarantees for closed-loop adaptive systems, and (3) the lack of systematic design guidelines to solve the trade-off between adaptation, performance, and robustness.

The L_1 adaptive control theory [4] appeared recently as a method for the design of robust adaptive control architectures using fast estimation schemes, with the potential to overcome the limitations mentioned above. The key feature of L_1 adaptive control is the decoupling of adaptation and robustness. In fact, in L_1 adaptive control architectures, the speed of adaptation is limited

1. Chapter 9 is written in collaboration with Enrick Xargay (University of Illinois at Urbana-Champaign) and Vladimir Dobrokhodov (Naval Postgraduate School).

only by the available hardware (computational power and high-frequency sensor noise), while the trade-off between performance and robustness can be addressed via conventional methods from classical and robust control. Fast adaptation, which allows for the compensation of the undesirable effects of rapidly varying uncertainties and significant changes in system dynamics, is critical towards achieving guaranteed transient performance without enforcing persistency of excitation, applying gain scheduling of the control parameters, or resorting to control reconfiguration or high-gain feedback. Moreover, the systematic design procedures of the L_1 adaptive control theory significantly reduce the tuning effort required to achieve a desired closed-loop performance, which translates into a reduction in both the design cycle time and the development costs. With these features, the L_1 adaptive control architectures provide a suitable framework for the development of advanced flight critical systems.

In this chapter, we present the preliminary results of the application of the PSI method for the design optimization of the L_1 flight control system implemented on the GTM AirSTAR [5, 6] aircraft's current primary flight test vehicle, the GTM tail number T2. The T2 is a twin-engine jet-powered and dynamically-scaled (5.5%) civil transport aircraft, designed and instrumented to perform control law evaluation, experiment design and modeling research, in-flight failure emulation, and flight in upset conditions. The research control law developed for the NASA AirSTAR flight test vehicle has a primary objective of achieving tracking for a variety of tasks with guaranteed stability and robustness in the presence of uncertain dynamics, such as changes due to rapidly varying flight conditions during standard maneuvers, and unexpected failures. Ideally, the flight control system should provide level 1 flying qualities under nominal as well as adverse flight conditions.

The L_1 flight control system (L_1 FCS) used for this application is a three-axis angle of attack (α), roll rate (p), and sideslip angle (β) all-adaptive flight control system. The FCS consists of two decoupled L_1 controllers, one for the longitudinal channel and another one for the control of the lateral-directional dynamics. On one hand, the longitudinal L_1 controller is implemented as a single-input single-output controller and uses feedback in angle of attack and pitch rate to generate an elevator control signal in order to track angle of attack reference signals; on the other hand, the lateral/directional L_1 controller is a multiple-input multiple-output architecture and uses feedback in sideslip angle, roll rate, and yaw rate to generate aileron and rudder commands in order to track sideslip-angle and roll-rate reference signals with reduced coupling. In the current L_1 FCS, the pilot directly adjusts the thrust level using the throttle lever. The reader is referred to [7–9] for a more detailed explanation of the L_1 FCS implemented on the NASA AirSTAR flight test vehicle.

The main challenge for the design of the L_1 FCS is the optimal tuning of its elements to provide desired flying qualities with satisfactory robustness margins. While the theory of L_1 adaptive control provides partial systematic design guidelines to address the trade-off between performance and robustness, the optimization of the design of the L_1 adaptive controller is still largely open and hard to address. The main difficulty is the nonconvex and nonsmooth nature of the underlying optimization problem that involves the L_1 norm of cascaded linear systems. Randomized algorithms have been proven to be useful in control-related nonconvex optimization problems, and therefore they appear as appealing methods for the optimal design of L_1 adaptive controllers [4, 10].

The optimization design methodology proposed in this chapter unfolds in three sequential steps. First, we take advantage of the systematic design guidelines of the L_1 adaptive control theory to find a nominal prototype solution satisfying a given set of control specifications. Then, taking the prototype solution as a reference design, the PSI method is used for the construction of the feasible solution set and for determining an initial direction of improvement for the design of the flight control system [11–13]. Finally, the PSI method is again used to determine an optimal design that satisfies performance and robustness constraints and improves the prototype design. In particular, this design methodology demonstrates the suitability of the PSI method as a tool for formulating and solving multicriteria optimization problems for the design of adaptive FCS and also illustrates that the consistent application of the systematic design guidelines of L_1 adaptive control becomes particularly beneficial for the construction of the feasible solution set.

Following this approach, we next present preliminary results on the application of the PSI method and the MOVI software [12] to the design optimization of the L_1 FCS for the NASA GTM. We notice that, although both the PSI and the software package MOVI were specially developed to address problems with design variable and the criteria spaces of high dimensionality, for the sake of clarity in the presentation and discussion of the results, we keep the design problem within a reasonable complexity, and the design procedure is applied to the design of the longitudinal channel only. Finally, we also notice that the results included in this study are obtained by the MOVI software package combined with the MATLAB environment and are based on the full nonlinear simulation of the two-engine-powered, dynamically-scaled GTM AirSTAR system tail number T2, which was released by NASA in December 2009 [14].

The formulation of the design optimization of the L_1 FCS of the NASA AirSTAR flight test vehicle is addressed in Section 9.2. In particular, this section provides a detailed discussion of the workflow of the optimization process, including the identification of a nominal prototype design, the construction of the feasible solution set, and the improvement of the prototype. Finally, Section 9.3 provides solution of the optimization problem and summarizes the key results and presents the main conclusions.

9.2 Prototype: Criteria and Design Variables

The design of the longitudinal L_1 FCS is based on the linearized short-period dynamics of the GTM at the reference flight condition (80 kt, 1,000 ft). The L_1 FCS with its main elements and the design variables (*DV*) is represented in Figure 9.1 with a following corresponding explanation of the control architecture. The motivated choice of the *DV*s will be given in the following section. Since the airplane is level 1 FQ at this flight condition[2], the desired dynamics of the state predictor are chosen to be close to those of the actual aircraft.

For the nominal prototype design, the natural frequency of the poles of the system is reduced from $6\frac{\text{rad}}{\text{sec}}$ to $5.5\frac{\text{rad}}{\text{sec}}$, while the damping ratio is increased from 0.47 to 0.85. A first-order lowpass filter with DC gain 1 and a bandwidth of $20\frac{\text{rad}}{\text{sec}}$ was used in the matched contribution to the elevator command, while two cascaded first-order lowpass filters were used in the unmatched channel, both having DC gain equal to 1 and bandwidths of $5\frac{\text{rad}}{\text{sec}}$ and $7\frac{\text{rad}}{\text{sec}}$, respectively. Finally, the adaptation sampling time was set to $\frac{1}{600}$ sec, which corresponds to the minimum integration step allowed in the AirSTAR flight control computer. A first-order prefilter with $20\frac{\text{rad}}{\text{sec}}$ of bandwidth was added to shape the pilot command. This prototype design of the state predictor,

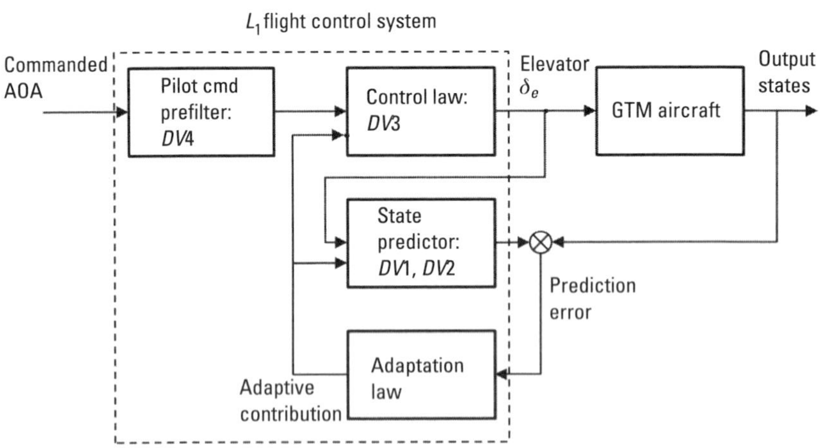

Figure 9.1 Longitudinal channel of the L_1 flight control architecture.

2. This FQ rating is based on offset-landing tasks performed in the AirSTAR real-time simulator at NASA LARC in the absence of atmospheric turbulence.

the lowpass filters, the adaptation sampling rate, and the prefilter delivers the angle of attack α (AOA) response similar to the desired one; see Figure 9.2 where α_{des} designates the desired angle-of-attack response.

It is worth noting here that this nominal design ensures a time-delay margin of the inner loop of approximately 85 ms and a gain margin of 7.2 dB, in wings-level flight at the flight condition of 80 knots and 1,000 ft. At this flight condition, the FQs are predicted to be level 1 and the FCS design has no predicted pilot induced oscillation (PIO) tendencies (for an acquisition time of 1.5 seconds). Naturally, these metrics are the performance criteria, and their initial values provided above for the nominal design will serve as the guidelines for the motivated definition of the criteria constraints.

We next present preliminary results of the application of the PSI method and the MOVI software to the design optimization of the L_1 FCS for the NASA GTM airplane [11–13]. In particular, the objective of the optimization task is to minimize the difference between the desired and actual AOA responses, while ensuring satisfactory FQs and not overloading the elevator actuators. For this purpose, a number of criteria reflecting the differences were formulated. The corresponding discussion is presented in the following sections.

In what follows, we provide a description of the design variables alone with the performance and robustness metrics used for this study and obtain a prototype design that achieves desired flying qualities with satisfactory robustness margins. Then we utilize the capability of the PSI method and the software package MOVI to improve the prototype design.

9.2.1 Design Variables

Next we define the set of design parameters of the flight control architecture considered for optimization. Since the primary objective is to improve the flying qualities of the prototype design while guaranteeing satisfactory robustness margins, we include the natural frequency and the damping ratio of the poles of the state-predictor dynamics (which can speed up or slow down the response of the augmented aircraft), and the bandwidth of the lowpass filter in the matched channel (which can be used to adjust the time-delay margin of the inner loop) as optimization parameters; see Figure 9.1 illustrating their place in the L_1 flight control architecture. In addition, we also consider the optimization of the bandwidth of the prefilter, which can be used to shape the pilot command as to prevent elevator rate limiting and avoid structural mode interaction. The following list summarizes the set of optimization parameters that define the design variable space:

- *DV1:* Natural frequency of the state-predictor poles (rad/sec);
- *DV2:* Damping ratio of the state-predictor poles;

- *DV3:* Bandwidth of the "matched" lowpass filter (rad/sec);
- *DV4:* Bandwidth of the pilot-command prefilter (rad/sec).

Therefore, the design variable vector is $DV = \{DV1, DV2, DV3, DV4\}$. We notice that, consistent with the theory of L_1 adaptive control, the adaptation sampling time is set to match the minimum integration step allowed in the AirSTAR flight control computer, and therefore it is not included as an optimization parameter.

9.2.2 List of Criteria and Pseudo-Criteria

The set of design criteria considered in this study is chosen to evaluate performance and robustness properties of the GTM aircraft augmented with the L_1 FCS. To provide an adequate assessment of the performance characteristics and flying qualities of the L_1-augmented aircraft both pilot-off-the-loop and pilot-in-the-loop performance metrics are included in the design procedure. The metrics considered can thus be classified in three categories:

1. Pilot-off-the-loop performance metrics;
2. Robustness metrics;
3. Flying qualities and PIO metrics.

Because the present material addresses only the design of the longitudinal channel of the L_1 FCS, the set of metrics used in this study are mainly based on the (time-domain) longitudinal response of the GTM with the L_1 FCS closing the inner loop. Further, we provide a detailed description of the specific metrics used for the design improvement of the adaptive control system. We note that some of the metrics used in this study were also proposed in [15] for the evaluation of aircraft augmented with an adaptive FCS.

9.2.2.1 Pilot-off-the-Loop Performance Metrics

This first set of metrics evaluates the pilot-off-the-loop performance of the augmented aircraft by characterizing its response to step inputs. In particular, the pilot-off-the-loop performance metrics are based on the time-domain response of the augmented aircraft to a step command of 3° held for 4 seconds in the AOA (see Figure 9.2), starting from a wings-level flight condition. The metrics capture the deviation of the actual response of the aircraft from a given desired response, which is defined to provide satisfactory flying qualities without reaching the physical limits of the platform, as well as different measures of control activity, load factor, and cross-coupling.

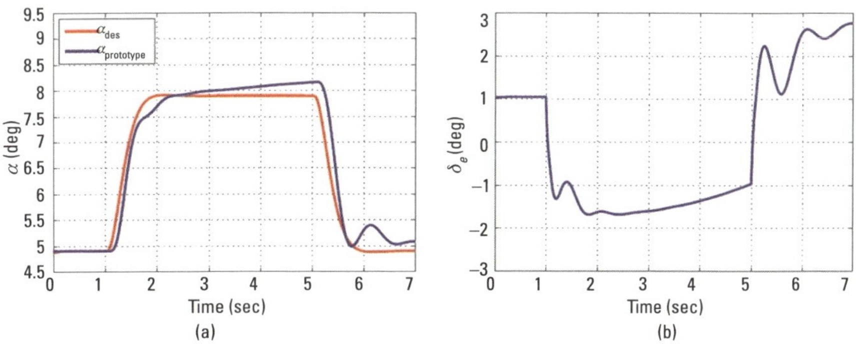

Figure 9.2 Prototype design, 3°-AOA step response (a) Angle of attack, α; (b) elevator deflection, δ_e.

Next we provide a description of the metrics included in the study. First, however, we need to introduce some key notation to facilitate the definition of these metrics. Below, alone with the previously defined α (AOA) and α_{des}, the α_{cmd} is the angle-of-attack pilot command; β is the angle of sideslip; β_{des} is the sideslip-angle desired response; p is the roll rate; p_{des} is the roll-rate desired response; A_z is the vertical acceleration; δ_e is the elevator deflection command; and $\tilde{x}(t)$ is the L_1 state predictor error.

Let t_0 be the time instant at which the step command is applied, and define t_f as the final time instant considered for the performance evaluation ($t_f = t_0 + 4$ seconds). With these notations, the metrics are defined as follows:

- *P1, final deviation:* This metric captures the final deviation of the actual AOA response from the desired AOA response at 4 seconds after the application of the step command. This metric is set to zero if the actual response reaches the AOA reference command before the end of the 4-second step:

$$P1 = \begin{cases} \left|\alpha\left(t_f\right) - \alpha_{des}\left(t_f\right)\right| & \text{if } \alpha(t) < \alpha_{cmd}, \forall t \in \left[t_0, t_f\right] \\ 0 & \text{otherwise} \end{cases}$$

  This metric can be used to penalize or exclude sluggish responses. In this study this metric is normalized to the amplitude of the step command (3°).

- *P2, maximum deviation from desired AOA response:* This metric captures the maximum deviation (in absolute value) of the actual AOA response from the desired AOA response:

$$P2 = \max_{t \in [t_0, t_f]} |\alpha(t) - \alpha_{des}(t)|$$

This metric can be redefined in terms of normalized AOA responses.

- *P3, integral deviation from desired AOA response:* This metric is defined as the (truncated) L_2 norm of the deviation of the actual AOA response from the desired AOA response:

$$P3 = \int_{t_0}^{t_f} |\alpha(t) - \alpha_{des}(t)| \, dt$$

Similar to *P*2, this metric can be redefined in terms of normalized AOA responses.

- *P4, overshoot in AOA response:* This metric captures possible overshoots and low-damping characteristics in the AOA response:

$$P4 = \begin{cases} \max_{t \in [t_0, t_f]} |\alpha(t)| \text{ if } \alpha(t) > \alpha_{cmd}, & \text{for some } t \in [t_0, t_f] \\ \alpha_{cmd} & \text{otherwise} \end{cases}$$

Similar to *P*1, this metric can be normalized to the amplitude of the step command (3°).

- *P5, maximum deviation from desired AOA rate response:* This metric captures the maximum rate deviation (in absolute value) of the actual AOA response from the desired AOA response:

$$P5 = \max_{t \in [t_0, t_f]} |\dot{\alpha}(t) - \dot{\alpha}_{des}(t)|$$

Similar to *P*2 and *P*3, this metric can be redefined in terms of normalized AOA responses.

- *P6, integral deviation from desired AOA rate response:* This metric is defined as the (truncated) L_2 norm of the rate deviation of the actual AOA response from the desired AOA response:

$$P6 = \int_{t_0}^{t_f} |\dot{\alpha}(t) - \dot{\alpha}_{des}(t)| \, dt$$

Similar to *P2*, *P3*, and *P5*, this metric can be redefined in terms of normalized AOA responses.

The metrics *P1* to *P6* provide a good first characterization of the transient response of the augmented aircraft when compared to a given desired response. Next, we present a set of metrics that can be extracted from the same step response experiment utilizing different flight dynamics characteristics which complement the AOA-based metrics defined above.

- *P7, maximum vertical acceleration:* Load factor (and passenger comfort) requirements can be captured by the maximum vertical acceleration during the step response:

$$P7 = \max_{t \in [t_0, t_f]} |A_z(t)|$$

  This metric can also be normalized to the amplitude of the step command (3°).
- *P8, control effort:* This metric is defined as the (truncated) L_2 norm of the elevator deflection command:

$$P8 = \int_{t_0}^{t_f} |\delta_e(t)| \, dt$$

  Similar to *P1*, *P4*, and *P7*, this metric can be normalized to the amplitude of the step command (3°). This metric can be used to penalize flight control designs that require a high control activity to achieve a desired control objective. It is important to note, however, that a high control effort might just be the result of a faster AOA response, and therefore a large P8 might not always be an undesirable response characteristic.
- *P9, maximum elevator rate:* Excessive control rate can be identified by the following metric:

$$P9 = \max_{t \in [t_0, t_f]} |\dot{\delta}_e(t)|$$

  This metric, which can also be normalized to the amplitude of the step command (3°), can be used to penalize designs with high elevator rates in order to prevent undesirable effects from rate limiting.

- *P10, maximum elevator acceleration:* High-order derivatives of the control commands are coupled to the flexible modes of the aircraft. The following metric, based on the second derivative of the elevator command, can be used to capture excessive accelerations and oscillations in the control command that could potentially lead to unwanted structural mode interactions:

$$P10 = \max_{t \in [t_0, t_f]} \left| \ddot{\delta}_e(t) \right|$$

This metric can also be normalized to the amplitude of the step command (3°).

- *P11, maximum of L_1 prediction error:* This metric captures the maximum error between the actual system state and the state of the L_1 state predictor, usually denoted by $\tilde{x}(t)$:

$$P11 = \max_{t \in [t_0, t_f]} \left\| \tilde{x}(t) \right\|_\infty$$

In L_1 adaptive control architectures, the accurate estimation of system uncertainties and the performance guarantees rely on the (small) "size" of the prediction error $\tilde{x}(t)$. This metric can thus be used to monitor the correct functioning of the L_1 adaptive controller.

- *P12, maximum deviation in cross-coupling dynamics:* This metric captures the lateral-directional coupling induced by a command in the longitudinal channel:

$$P12 = \max_{t \in [t_0, t_f]} \left| \delta_e(t) \right| \left(\left(\beta(t) - \beta_{des}(t) \right)^2 + \left(p(t) - p_{des}(t) \right)^2 \right)$$

We notice that this metric, which can also be normalized to the amplitude of the step command (3°), provides valuable information for the design of the lateral-directional FCS, rather than the longitudinal FCS. In fact, for the design of the longitudinal L_1 FCS, this metric provides little information and it would be more convenient to analyze the coupling in the AOA response induced by a command either in roll rate or sideslip angle. The analysis of the response of the system to commands in the lateral-directional channel and the design of the lateral-directional L_1 FCS are left (deliberately) for future work.

- *P13, integral deviation in cross-coupling dynamics:* This metric is the integral version of the previous cross-coupling metric and is defined as follows:

$$P13 = \int_{t_0}^{t_f} |\delta_e(t)| \left((\beta(t) - \beta_{des}(t))^2 + (p(t) - p_{des}(t))^2 \right) dt$$

Similar to *P*12, this metric would be more adequate for the design of the lateral-directional control system, and it is included in this study only to illustrate a set of additional metrics that can be derived from the response of the augmented aircraft to a command in the longitudinal channel.

9.2.2.2 Robustness Margins

In this preliminary study, the only robustness metric considered for optimization is the *time-delay margin* of the closed-loop adaptive system defined at the input of the aircraft (time delay inserted at the elevator deflection command), and it is also derived from the time-domain response of the augmented aircraft. For a given wings-level flight condition and with the pilot-off-the-loop, a small perturbation in the trim (initial) condition is introduced. The time-delay margin is determined as the minimum time delay that produces sustained oscillations in the AOA response as the L_1 FCS tries to stabilize the aircraft at the given trim condition. In this chapter, this robustness metric will be denoted by *R*1.

We notice that the time delay introduced in the elevator control channel is in addition to any time delay that is already modeled in the AirSTAR simulation environment, which amounts approximately to 25 ms.

9.2.3 Criteria Addressing FQ and PIO Characteristics

Finally, predictions for both flying qualities and PIO tendencies have also been included in order to complement the pilot-off-the-loop performance metrics presented above. For this study, we consider the time-domain Neal-Smith (TDNS) flying qualities and PIO criteria, which were specifically developed for nonlinear aircraft dynamics and nonlinear FCS. The TDNS criterion is the counterpart in the time domain of the frequency-domain Neal-Smith criterion, and it is based on a step-tracking task with different acquisition-time requirements. For a detailed description of this criterion, the reader is referred to [16]. The reader can also find in [17] a study on the prediction of flying qualities and adverse pilot interactions in the GTM augmented with the L_1 FCS.

We use four different metrics, extracting all of them from the TDNS criterion for an acquisition time of 1.5 seconds, to characterize the FQ and PIO tendencies of the augmented aircraft:

- *FQ1, tracking performance:* In the TDNS criterion, the root-mean-squared tracking error is used to evaluate the closed-loop performance with the pilot in the loop. A value of zero means that the pilot is able to perfectly track (with zero error) the reference command after the specified acquisition time.
- *FQ2, pilot workload:* In the TDNS criterion, the pilot workload is given by the pilot compensation phase angle (in degrees), which is derived from the optimal pilot model obtained from the criterion. A value of zero means that there is no need for either pilot lead compensation or lag compensation.
- *FQ3, FQ level:* The two metrics *FQ*1 and *FQ*2 are used to determine the predicted *FQ* level based on the *FQ* boundaries proposed in the criterion. *FQ*3 is a discrete metric, and it only admits the values 1, 2, and 3, which correspond to level 1, level 2, and level 3 flying qualities, respectively.
- *FQ4, PIO tendency:* The TDNS criterion also provides a prediction for the susceptibility of the augmented aircraft to PIO. This PIO-susceptibility metric is used to complement the flying qualities metrics discussed above. According to the TDNS criterion, a value above 100 implies that the augmented aircraft is PIO-prone, whereas a value below 100 indicates a PIO-immune configuration.

The set of FQ metrics is used at the second step when improving a prototype design of the longitudinal channel of the L_1 FCS. For the first step of the prototype design and the exploration of the feasible set, only a subset of these metrics is utilized. The full set of metrics is used in the last stage to optimize the design of the adaptive control system.

Based on the objectives of the task and previous flight control design expertise the following vectors of criteria {*P*1, *P*2, *P*3, *P*4, *P*5, *P*6, *FQ*1, *FQ*2, *FQ*3, *FQ*4, *R*1} and pseudo-criteria {*P*7, *P*8, *P*9, *P*10, *P*11, *P*12, *P*13} were defined.

9.2.4 Criteria Constraints

Based on the metrics defined, the final design of the L_1 FCS should ideally verify the set of control objectives at the reference flight condition of 80 knots of (equivalent) airspeed and 1,000 ft of altitude. Corresponding to this flight conditions a set of three criteria constraints were defined a priori:

$$P1 \leq 0.1, FQ3 = 1 \text{ and } P4 \leq 1.2$$

The first and second conditions address directly the control specifications, namely, the final value of the step response within 10% of the desired, and the predicted level 1 FQ. The third inequality imposes a 20% constraint on the overshoot in the step response, establishing thus a (loose) bound on the acceptable transient performance characteristics of the actual AOA response. Due to the significant difficulty of defining all criteria constraints consistent with the feasibility of the solution, the rest of the constraints were identified interactively while analyzing the test tables.

9.3 Solutions and Analysis

At this point the design optimization task and the nominal prototype solution were completely defined. This section presents last two steps of iterative application of the PSI method to the design improvement of the longitudinal channel of the L_1 FCS. It is critical to mention that the objective of the first iteration is achieved utilizing a limited number of criteria consistent with the control metrics used by the L_1 synthesis procedure. Numerical implementation of this first step is relatively efficient with the "computational price" of one solution measured in minutes. At the second step, while utilizing an extended set of criteria involving complex numerical procedures, the efficiency of numerical implementation becomes critical; the "computational price" of one solution at this step is measured in tenth of minutes.

9.3.1 First Iteration

For the purpose of constructing the feasible solution set, we consider the following initial intervals of design variables, which have been defined to include the initial prototype vector (see Table 9.1).

This first step is used to find an initial direction of improvement for the design of the adaptive FCS. More precisely, we aim here at determining tight intervals for the design variables characterizing the state predictor (*DV*1 and *DV*2) that would provide level 1 *FQ* and would not deviate significantly from the desired response defined previously. To this end, the design is to be minimized with respect to the following reduced number of criteria {*P*1, *P*2, *P*3, *P*4, *P*5, *P*6, *FQ*1, *FQ*2}, which is a subset of the metrics described earlier.

The robustness metric *R*1 and the PIO metric *FQ*4 are not included in this first step because their evaluation is computationally expensive; these

Table 9.1
Initial Intervals of Design Variables

Design Variable	Prototype	Initial Intervals of Variation of Design Variables	
		Min	Max
DV1	5.50E+00	4.00E+00	8.00E+00
DV2	8.50E-01	5.00E-01	1.10E+00
DV3	2.00E+01	5.00E+00	3.00E+01
DV4	2.00E+01	1.00E+01	5.00E+01

metrics will be considered in the next step of the optimization process when the domain of the design variables becomes significantly refined.

The metrics *P7* through *P10* are not included in the set of criteria to be minimized because improved flying qualities may require "high" values of these metrics. Nevertheless, including these metrics in the optimization process as pseudo-criteria may provide useful information regarding the dynamics of the augmented aircraft. They also provide additional performance characteristics of the set of Pareto optimal solutions. In particular, we note that the metric *P9*, which is subject to "soft" control specifications, is included as a pseudo-criterion (see [11, 13] for a detailed justification of this choice). Similarly, the metric *P11*, which can be used to monitor the correct operation of the L_1 adaptive controller, does not need to be minimized as long as it remains a couple of orders of magnitude below the system state (truncated) L_∞ norm. Finally, as explained previously, the metrics *P12* and *P13* are included for the sake of completeness and should be considered only for the design of the lateral-directional control system. Therefore, they are not included as criteria in the current task.

Next we present the results obtained in this first iteration of the optimization process, which are based on 1,024 tests [11–13]. Out of these 1,024 tests, 427 vectors did not satisfy the a priori given criteria constraints. The solutions that did not satisfy the constraints entered the table of criteria failures [12]. In particular, 160 design variable vectors failed the 20% overshoot criterion constraint on *P4*, and 267 constraints failed the level 1 *FQ* criterion constraint *FQ3*, while the 10% overshoot constraint on *P1* was not effective, therefore not rejecting any solutions. Analysis of the design variable vectors contained in the table of criteria failures indicates that: (1) solutions characterized by low natural frequencies and high damping ratios of the state-predictor poles ($DV1 < 5$ and $0.95 < DV2$) fail the level 1 *FQ* criterion constraint *FQ3*; and (2) solutions with very high natural frequencies of the state-predictor poles ($7.6 < DV1$) or solutions with high natural frequencies and low damping ratios of the state-predictor poles ($7 < DV1 < 7.6$ and $DV2 < 0.6$) fail the overshoot criterion constraint *P4*. The first set of design variable vectors leads to slow responses of the augmented aircraft to pilot commands, which require moderate or considerable pilot compensation and thus result in a predicted level 2 *FQ* that is inadmissible. The second set corresponds to the solutions providing fast responses to step commands with big overshoots, which differ significantly from the desired response for the augmented aircraft.

The remaining 597 vectors that did satisfy a priori given criteria constraints were used to construct the test table. While tightening the criteria constraints in the test table, the following criteria constraints were formulated (see Table 9.2). Note that while analyzing the test table, the constraint of *P4* was significantly tightened to the value of 1.02. It is also worth mentioning that the response on criteria *P1* is not presented in the table because all solutions of the test table provided identical response, *P1* = 0.

Only 20 solutions were found to be feasible according to these criteria constraints, all of them contributing to the Pareto optimal solutions. A fragment of the criteria table is given in Table 9.3.

Analysis of the criteria table shows that solution #993 is the most preferable one. This solution is equivalent to others with respect to criterion *P4*; it is superior to others over a set of five criteria {*P2*, *P3*, *P6*, *FQ1*, *FQ2*} and is weaker than the prototype only with respect to the criterion P5. Furthermore, the remaining 19 solutions are better than the prototype with respect to four criteria {*P2*, *P3*, *FQ1*, *FQ2*}. However, none of the solutions are more superior than the prototype with respect to criterion P5. In particular, this observation implies that if the prototype design vector was sampled by the system, then it would belong to the Pareto set.

The analysis of the histograms demonstrates the filtering effect of the criteria constraints (see Figure 9.3). The feasible solutions for the design variables *DV1* and *DV2* are located in the middle of the interval, with no feasible solutions in the left and right ends of the intervals, and slightly shifted from the prototype design (marked with a yellow triangle ∇). These histograms clearly identify feasible intervals for the design variables *DV1* and *DV2* characterizing

Table 9.2
Criteria Constraints

$P2 \le 0.2$	(min)	$P9 \le 15$	(pseudo)
$P3 \le 0.2$	(min)	$P10 \le 300$	(pseudo)
$P4 \le 1.02$	(min)	$P11 \le 0.25$	(pseudo)
$P5 \le 1$	(min)	$P12 \le 0.01$	(pseudo)
$P6 \le 0.3$	(min)	$P13 \le 0.01$	(pseudo)
$P7 \le 0.25$	(pseudo)	$FQ1 \le 0.1$	(min)
$P8 \le 5$	(pseudo)	$FQ2 \le 45$	(min)

Table 9.3
Fragment of Criteria Table

Criteria		Prototype	Pareto Optimal Solutions						
			#241	#281	#329	#409	#649	#825	#993
P2	(min)	1.30E-01	1.04E-01	8.40E-02	9.14E-02	8.97E-02	6.03E-02	7.44E-02	5.39E-02
P3	(min)	1.54E-01	1.16E-01	1.03E-01	1.06E-01	1.04E-01	8.72E-02	9.51E-02	8.84E-02
P4	(min)	1.0E+00	1.00E+00	1.00E+00	1.00E+00	1.00E+00	1.01E+00	1.00E+00	1.00E+00
P5	(min)	3.15E-01	5.37E-01	9.36E-01	9.68E-01	6.29E-01	8.63E-01	8.58E-01	6.89E-01
P6	(min)	1.49E-01	1.97E-01	2.21E-01	2.58E-01	2.03E-01	1.74E-01	1.94E-01	1.32E-01
P7	(pseudo)	1.51E-01	1.65E-01	1.74E-01	1.84E-01	1.72E-01	1.83E-01	1.78E-01	1.75E-01
P8	(pseudo)	3.24E+00	3.3E+00	3.29E+00	3.31E+00	3.30E+00	3.31E+00	3.30E+00	3.31E+00
P9	(pseudo)	5.96E+00	1.17E+01	7.55E+00	9.1E+00	1.09E+01	1.11E+01	9.36E+00	1.16E+01
P10	(pseudo)	1.07E+02	2.42E+02	1.34E+02	1.72E+02	2.16E+02	2.24E+02	1.76E+02	2.44E+02
P11	(pseudo)	7.45E-02	7.79E-02	6.02E-02	6.82E-02	8.10E-02	7.72E-02	7.22E-02	7.77E-02
P12	(pseudo)	1.01E-04	1.84E-04	1.87E-04	2.08E-04	2.06E-04	2.29E-04	2.14E-04	2.09E-04
P13	(pseudo)	3.16E-05	6.18E-05	6.87E-05	8.01E-05	7.00E-05	8.18E-05	7.58E-05	7.03E-05
FQ1	(min)	1.23E-02	6.73E-02	9.29E-02	9.74E-02	6.86E-02	9.05E-02	7.80E-02	8.85E-02
FQ2	(min)	5.36E+01	4.42E+01	4.35E+01	4.03E+01	4.22E+01	3.79E+01	4.10E+01	4.00E+01

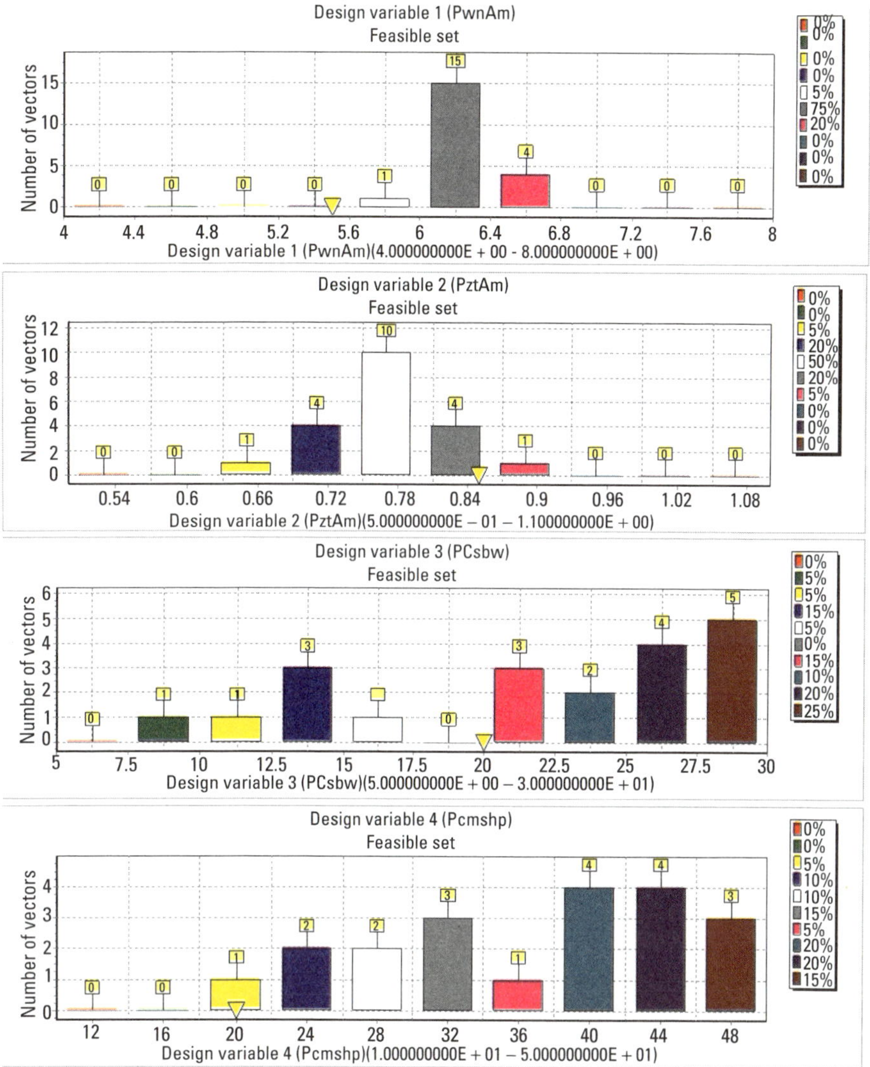

Figure 9.3 PSI iteration 1, distribution of feasible solutions; yellow triangle shows the prototype design.

the state-predictor dynamics. On the other hand, the feasible solutions for the design variables *DV3* and *DV4* are located at the right boundaries of the intervals. These observations will be used in the next step of the optimization process to modify the initial intervals in the direction of higher density of feasible solutions to improve richness of the feasible solution set.

It is also important to analyze the influence of the design variables on criteria and pseudo-criteria and to evaluate the degree of improvement (or degradation) of the Pareto optimal solutions with respect to the prototype design. Figures 9.4 to 9.8 show some of these dependencies, which provide valuable insight into the effect of the design variables on the augmented aircraft dynamics. Preliminary conclusions that can be drawn from this analysis are detailed next.

Figure 9.4 shows a typical dependency of criteria *P2* (*P3* and *P6* have the same trend) on the design variable *DV1*. First, we notice that there are no feasible solutions in the range 4 < *DV1* < 4.6, which indicates that such design variable vectors did not satisfy the constraints. Second, it can be seen that the deviations of the actual response (*P2*, *P3*, and *P6*) of the augmented aircraft from the desired model become large for big values of the design variable *DV1*. Third, all three dependencies show that the set of Pareto optimal solutions, when compared to the prototype design, provide step responses with reduced deviations in terms of both (truncated) L_∞ and L_2 norms. The rate deviations for the Pareto optimal solutions are, however, larger than the rate deviations for the prototype design.

Figure 9.5 presents the dependencies of the flying qualities criteria *FQ2* (*FQ1* has the same trend) on the design variable *DV1*, which indicate that faster responses of the augmented aircraft (large *DV1*) result in improved (predicted) flying qualities. This trend is particularly evident in Figure 9.5, in which the reduced pilot (lead) compensation is limited to large values of the design variable *DV1*. Also, when compared to the prototype design, the set of optimal solutions improve the criterion *FQ1* by 20% to 50%, and the criterion *FQ2* by 15% to 30%.

As expected, a faster response of the augmented aircraft (large *DV1*) leads to an increase in the control effort (pseudo-criteria P8), as well as an increase in both the maximum elevator rate and the maximum elevator acceleration (pseudo-criteria *P9* and *P10*).

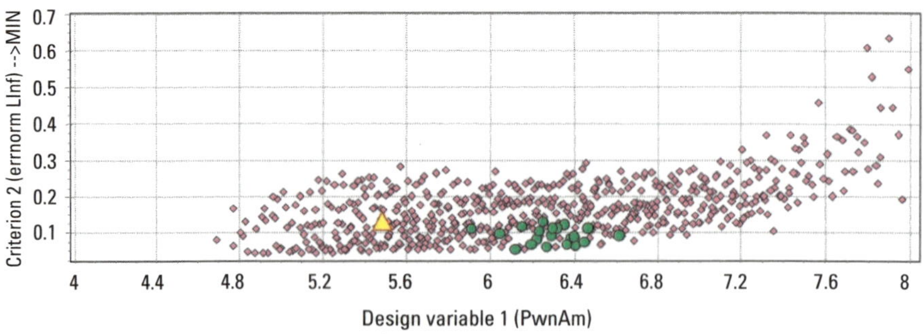

Figure 9.4 PSI iteration 1; dependencies of the criterion *P2* (maximum AOA deviation) on the design variable *DV1*.

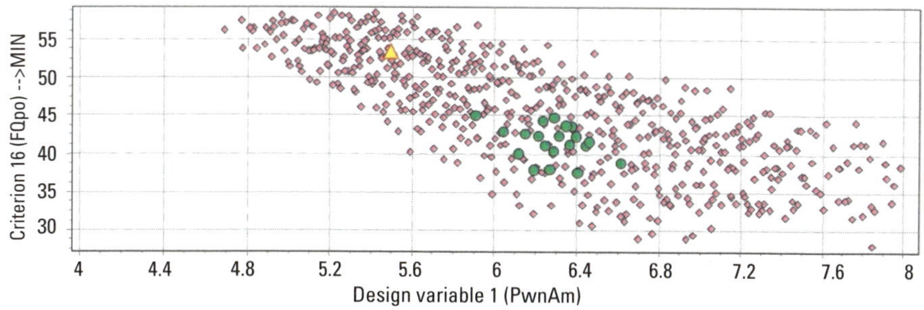

Figure 9.5 PSI iteration 1; dependencies of the criterion *FQ2* (pilot workload) on the design variable *DV1*.

Figure 9.6 illustrates the dependencies of criteria *FQ*1 on the design variable *DV*2. (*FQ*2 has the same trend.) While the dependency of *FQ*2 on *DV*2 is not obvious, small values of *DV*2 seem to limit the achievable tracking performance with the pilot in the loop (*FQ*1). This would imply that the augmented aircraft with low-damping characteristics result in degraded (predicted) flying qualities.

Finally, the dependencies of pseudo-criteria *P*9 and *P*10 on the design variable *DV*4 show that the Pareto optimal solutions are located along a "straight" line with positive slope where high values of the criteria correspond to high values of the design variable *DV*4. This implies that the bandwidth of the command prefilter in the L_1 FCS (*DV*4) can be set to limit the maximum elevator rate and the maximum elevator acceleration of the Pareto optimal solutions.

Dependencies between criteria provide useful information about the solutions in the Pareto set and the trade-offs between criteria. The analysis of trade-offs becomes especially helpful in the final stages of the optimization process, when a decision about the most preferable solution has to be made. These

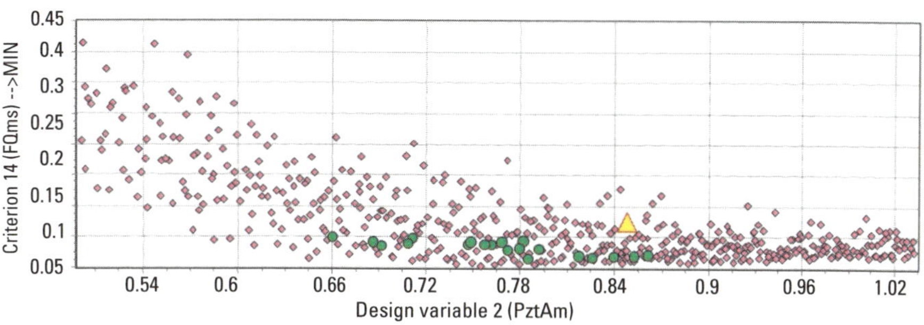

Figure 9.6 PSI iteration 1; dependencies of the criterion *FQ1* (tracking performance) on the design variable *DV2*.

dependencies are useful to explore the trade-offs in the design of the controller and to identify directions in the design variable space that may lead to improved L_1 FCS design over the prototype design.

Next we present and discuss a set of results showing dependencies between criteria illustrating these trade-offs and determining possible directions of improvement of the prototype solution:

Figure 9.7 shows the dependency between *P*2 and *P*3 (maximum and integral deviations from the desired response) as well as the localization of the Pareto optimal solutions with respect to the prototype design in the *P*2-*P*3 plane. It can be seen that the solutions of the Pareto set are located on a "straight" line passing through the prototype design, and all of them improve the prototype in terms of the metrics *P*2 and *P*3.

Figure 9.8 shows the location of the prototype as well as the Pareto optimal solutions in the FQ plane of the TDNS criteria, thus characterizing the (predicted) flying qualities of the solutions. The improvement with respect to the prototype design in terms of predicted FQ is evident in this *FQ*2-*FQ*1

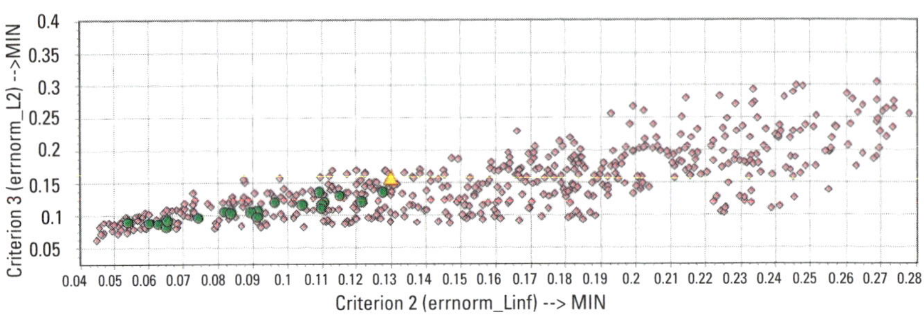

Figure 9.7 PSI iteration 1; dependencies between criteria *P*2 and *P*3 (maximum and integral deviation of AOA).

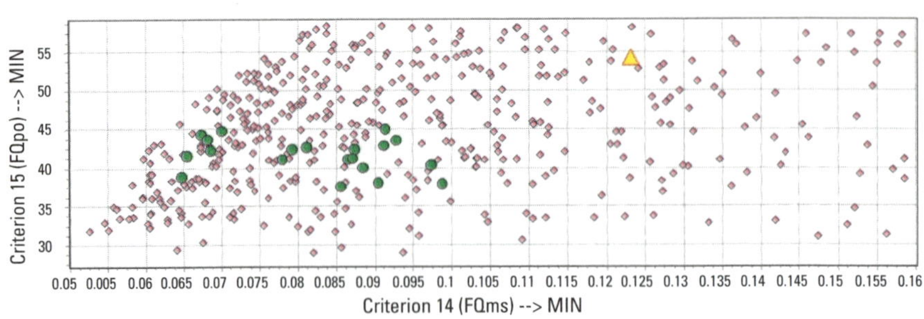

Figure 9.8 PSI iteration 1; dependencies between criteria *FQ*1 and *FQ*2 (tracking performance and pilot workload).

graph. Also, the boundary for level 1 FQ of the TDNS criterion can be easily recognized in this figure (see [16]).

Finally, Figure 9.9 shows the dependency between the FQ criterion *FQ*2 and the control effort *P*8. This figure shows that reduced pilot (lead) compensation is only possible for increased control effort. In fact, all of the solutions in the Pareto set present higher control efforts than the prototype solution.

The analysis of the obtained solutions allows us to define the direction of further search. In particular, the results have provided tight intervals for the design variables *DV*1 and *DV*2 characterizing the state-predictor dynamics, and have exposed the necessity of extending the initial intervals of initial intervals of variation of the design variables *DV*3 and *DV*4. It is worth emphasizing that this significant improvement of the prototype solution has been achieved utilizing a reduced number of criteria, {*P*2, *P*3, *P*4, *P*5, *P*6, *FQ*1, *FQ*2}. Based on these results and the conclusions drawn from them, a new experiment is carried out to: (1) improve the feasible solution set, and (2) determine an optimal solution of the L_1 FCS design that improves the prototype with respect to extended set of criteria {*P*2, *P*3, *P*4, *P*5, *P*6, *FQ*1, *FQ*2, *FQ*4, *R*1}.

9.3.2 Second Iteration

The improvement of the feasible solution set is based on the analysis of the histograms in Figure 9.3 and Table 9.3. The analysis of the histograms results in adjusting the initial problem statement by changing the intervals of variation of the design variables and is given in Table 9.4.

The criteria constraints remain unchanged, whereas the design is now to be optimized with respect to the following new set of criteria {*P*2, *P*3, *P*4, *P*5, *P*6, *FQ*1, *FQ*2, *FQ*4, *R*1}. All these criteria are to be minimized except for *R*1, which is to be maximized.

Next we present the results obtained in this second iteration of the optimization process, which are based on 512 tests. As mentioned previously, adding

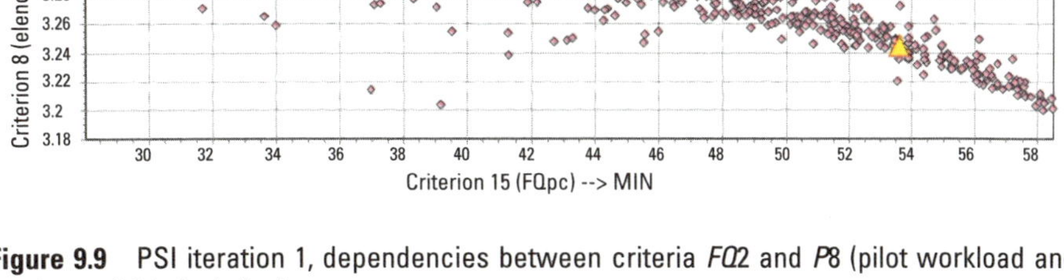

Figure 9.9 PSI iteration 1, dependencies between criteria *FQ*2 and *P*8 (pilot workload and maximum AOA deviation).

Table 9.4
Refined Intervals of Design Variables

Design Variable	Prototype	Initial Intervals of Variation of Design Variables	
		Min	**Max**
DV1	5.50E+00	5.50E+00	7.00E+00
DV2	8.50E-01	6.50E-01	0.90E+00
DV3	2.00E+01	9.80E+00	4.00E+01
DV4	2.00E+01	1.80E+01	6.50E+01

criteria *R*1 and *FQ*4 in the optimization process results in a significant increase in the computational time. Approximately 124 vectors satisfied previously assigned criteria constraints (see Table 9.2). All these solutions are Pareto optimal. As might be expected, the coefficient of selection efficiency γ increased by more than 12 times. The histograms in the second iteration have better distributions (higher concentration) of the feasible solutions than in the first iteration. The top of Figure 9.10 represents the distribution of 124 solutions for *DV1*.

After the analysis of the test table, stronger criteria and pseudo-criteria constraints were considered as presented in Table 9.5. According to these new constraints, only six solutions were found to be feasible, and all of them are

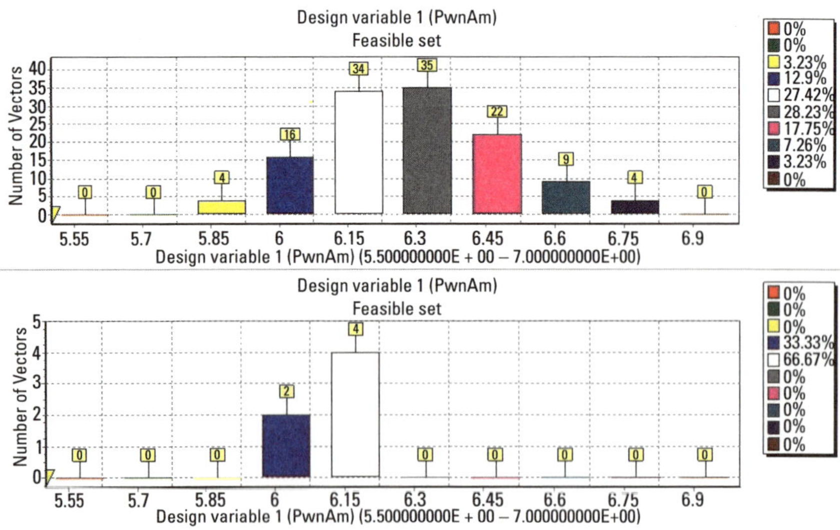

Figure 9.10 PSI iteration 2, distribution of feasible solutions of *DV1* with the original (top) and with tightened criteria constraints (bottom).

Table 9.5
Second Iteration, Refined Criteria Constraints

$P2 \leq 0.1$	(min)	$P10 \leq 200$	(pseudo)
$P3 \leq 0.15$	(min)	$P11 \leq 0.1$	(pseudo)
$P4 \leq 1.02$	(min)	$P12 \leq 0.01$	(pseudo)
$P5 \leq 1$	(min)	$P13 \leq 0.01$	(pseudo)
$P6 \leq 0.25$	(min)	$FQ1 \leq 0.1$	(min)
$P7 \leq 0.2$	(pseudo)	$FQ2 \leq 45$	(min)
$P8 \leq 5$	(pseudo)	$FQ4 \leq 5$	(min)
$P9 \leq 10$	(pseudo)	$R1 \geq 80$	(max)

Pareto optimal. The values of design variables and criteria of the Pareto optimal solutions are given in Tables 9.6 and 9.7. The new distribution of the feasible solutions for these criteria and pseudo-criteria constraints is significantly tighter as expected and is shown in the bottom of Figure 9.10. These new histograms clearly identify tight intervals for all of the design variables in which the optimal solutions lie.

Furthermore, from the analysis of Table 9.6, it follows that all solutions of second iteration as well as the #993 from the first iteration belong to the very tight intervals of the first and second design variables. The first three parameters (*DV*1–*DV*3) of #993 and #106 are almost identical. Observe that #993, while providing good response of many criteria, does not satisfy new constraints on criteria *P*9 and *P*10 (the elevator workload). Moreover, the analysis of the #993 also shows that it fails to satisfy constraint of the flying qualities criteria *FQ*3.

The analysis of test tables, dependencies of criteria on design variables, and dependencies between criteria allows us to determine the most preferable solutions. In particular, Figure 9.11 shows the influence of the bandwidth of the "matched" lowpass filter (*DV*3) on the (pilot-off-the-loop) trade-off between performance criterion *P*2 (*P*3 shows the same trend) and robustness (*R*1) of the augmented aircraft. From this observation we can conclude that criteria *P*2 (*P*3) and *R*1 are contradictory with respect to the design variable *DV*3. This means that improvement of the tracking performance requires an increase in the bandwidth of the lowpass filter, which in turn results in degradation of the time delay margin of the augmented aircraft, as predicted by theory.

Figure 9.12 shows the dependencies of the flying qualities criterion *FQ*1 (*FQ*2 shows a similar trend) on the design variable *DV*2. While in the first PSI iteration the dependency of the criterion *FQ*2 on the design variable *DV*2 was not obvious, now it becomes more apparent that a smaller damping ratio seems to result in reduced (lead) pilot compensation.

Table 9.6
Second Iteration: Table of Design Variables

Design variable	Prototype	#993, First Iteration	Pareto Optimal Solutions					
			#106	**#202**	**#254**	**#318**	**#358**	**#462**
DV1	5.50E+00	6.12E+00	6.00E+00	5.99E+00	6.24E+00	6.23E+00	6.10E+00	6.18E+00
DV2	8.50E+00	7.09E-01	7.34E-01	7.49E-01	7.76E-01	7.33E-01	7.81E-01	7.18E-01
DV3	2.0E+01	2.70E+01	2.52E+01	1.67E+01	1.81E+01	2.18E+01	1.69E+01	1.58E+01
DV4	2.0E+01	4.93E+01	3.16E+01	3.20E+01	2.10E+01	2.72E+01	2.57E+01	3.11E+01

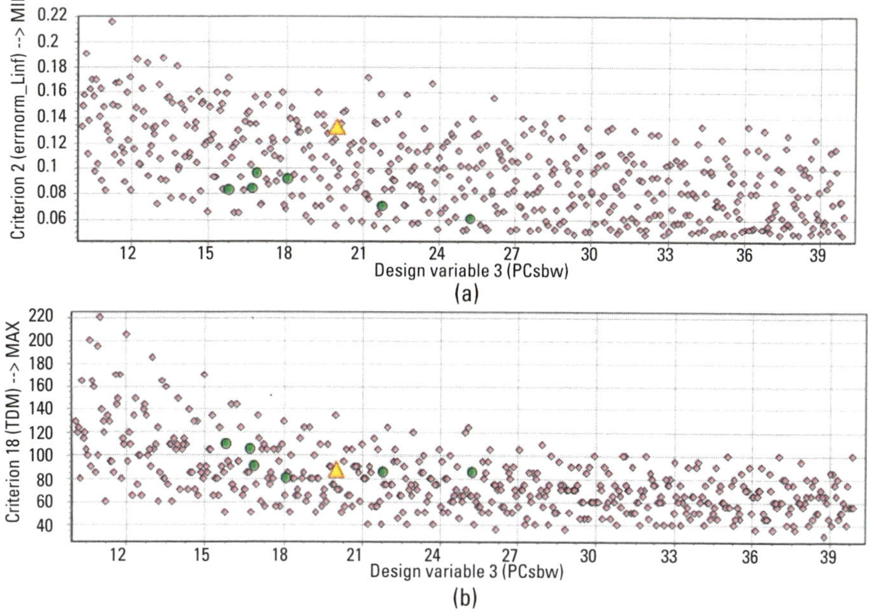

Figure 9.11 PSI iteration 2, dependencies of criteria *P*2 and *R*1 on the design variable *DV*3. (a) Criterion *P*2 (max AOA deviation) versus design variable *DV*3. (b) Criterion *R*1 (time delay margin) versus design variable *DV*3.

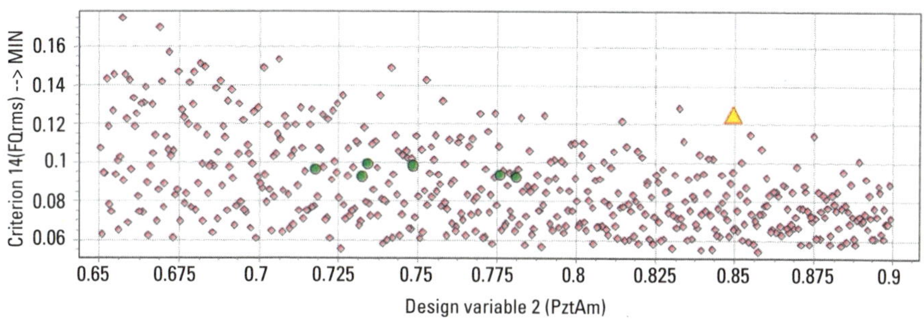

Figure 9.12 PSI iteration 2, dependencies of criteria *FQ*1 (tracking performance) on the design variable *DV*2.

Dependency between *P*2 and *P*3 is analogous to those obtained in first iteration; therefore, the corresponding figure is not presented. All of the Pareto optimal solutions improve the prototype design in terms of peak and integral deviations from the desired response.

The dependency of flying qualities criteria *FQ*1 and *FQ*2 obtained in this iteration is also similar to those obtained in the first iteration, thus demonstrating

Table 9.7
Second Iteration: Table of Criteria

Criteria		Prototype	Pareto Optimal Solutions #106	#202	#254	#318	#358	#462
P2	(min)	1.30E-001	6.01E-01	8.45E-02	9.17E-02	7.04E-02	9.63E-02	8.28E-02
P3	(min)	1.54E-01	9.60E-02	1.13E-01	1.11E-01	9.66E-02	1.18E-01	1.04E-01
P4	(min)	1.0E+00	1.00E+00	1.00E+00	1.00E+00	1.00E+00	1.01E+00	1.00E+00
P5	(min)	3.15E-01	6.13E-01	6.81E-01	8.85E-01	8.43E-01	7.71E-01	8.63E-01
P6	(min)	1.49E-01	1.28E-01	1.67E-01	2.16E-01	1.78E-01	1.94E-01	2.11E-01
P7	(pseudo)	1.51E-01	1.68E-01	1.67E-01	1.72E-01	1.76E-01	1.68E-01	1.78E-01
P8	(pseudo)	3.24E+00	3.29E+00	3.29E+00	3.29E+00	3.30E+00	3.29E+00	3.31E+00
P9	(pseudo)	5.96E+00	9.09E+00	9.1E+00	7.9E+00	9.05E+00	8.43E+00	9.54E+00
P10	(pseudo)	1.07E+02	1.77E+02	1.78E+02	1.43E+02	1.72E+02	1.59E+02	1.85E+02
P11	(pseudo)	7.45E-02	6.62E-02	6.60E-02	6.09E-02	6.74E-02	6.31E-02	6.95E-02
P12	(pseudo)	1.01E-04	1.81E-04	1.70E-04	1.80E-04	2.01E-04	1.69E-04	1.98E-04
P13	(pseudo)	3.16E-05	6.02E-05	5.89E-05	6.58E-05	7.13E-05	6.01E-05	7.27E-05
FQ1	(min)	1.23E-02	9.93E-02	9.81E-02	9.34E-02	9.26E-02	9.20E-02	9.64E-02
FQ2	(min)	5.36E+01	4.33E+01	4.41E+01	4.38E+01	4.14E+01	4.46E+01	4.10E+01
FQ4	(min)	4.68E+00	4.08E+00	4.19E+00	3.87E+00	3.88E+00	3.84E+00	3.97E+00
R1	(max)	8.50E+01	8.50E+01	1.05E+02	8.00E+01	8.50E+01	9.00E+01	1.00E+02

significant improvement of predicted flying qualities over the prototype design but now in the extended criteria space.

Finally, Figure 9.13 shows the dependency between criteria *P*3 and *R*1, which illustrates the trade-off between performance and robustness of the closed-loop adaptive system with the pilot-off-the-loop. While all of the optimal solutions reduce the deviations from the desired response with respect to the prototype design, only three of these solutions exhibit a better time-delay margin than the prototype design (#202, #462), and two exhibit a similar margin (#106, #318)[3].

As a result of the iterative two-step correction of initial constraints, the six Pareto optimal solutions have been found. Analyzing the criteria table shows that there are no solutions better than the prototype by all criteria simultaneously. This analysis revealed that four solutions, #106, #202, #358, and #462, surpassed the prototype by six criteria simultaneously.

Although all six solutions are practically equivalent, some advantage is given to the design vector #202 since it provides a better trade-off between the (predicted) flying qualities (*FQ*1, *FQ*2) and the time-delay margin (*R*1), while minimizing the difference with the desired response (see Figure 9.14). The comparative analysis of the AOA and the elevator workload demonstrates that the optimal solution provides faster response (rise time) to the commanded AOA command while minimally increasing the elevator workload that is naturally expected.

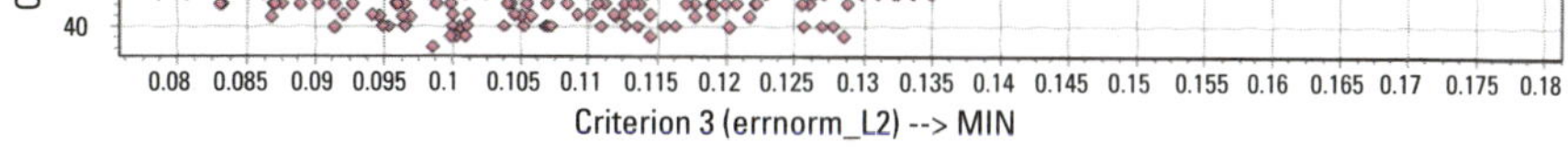

Figure 9.13 PSI iteration 2; dependencies between criteria *R*1 (time delay margin) and *P*3 (integral deviation of AOA).

3. Notice that for the determination of the time-delay margin, metric *R*1, we insert incremental time delays of 5 ms in the elevator command channel. This choice was made for consistency with the determination of the time-delay margin in the AirSTAR real-time simulator at NASA LARC. As a consequence, however, we cannot identify the time-delay margin with higher accuracy.

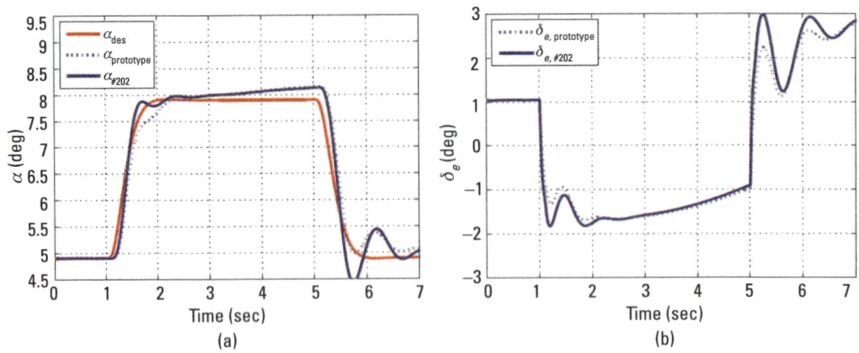

Figure 9.14 Optimal design #202, 3° AOA step response. (a) Angle of attack, α; (b) elevator deflection, δ_e.

9.3.3 Conclusion

The construction of the feasible solutions of the L_1 adaptive controller satisfying the desired performance and robustness specifications is a critical step for the optimal design of the L_1 FCS. In particular, the eigenstructure of the state-predictor state matrix and the bandwidth of the lowpass filters are the key elements that characterize the performance of the L_1 FCS. To optimize the design of these elements, a 16-criteria problem of improving a prototype solution was formulated and solved. The results presented demonstrate the application of PSI method to the multicriteria design optimization of the L_1 FCS implemented on the GTM AirSTAR aircraft. The study has addressed both the construction of the feasible solution set and the improvement of a nominal prototype design initially synthesized by the L_1 theory.

On one hand, the results have demonstrated that the consistent application of the systematic design guidelines of L_1 adaptive control becomes particularly beneficial for the construction of the feasible solution set. Moreover, the results of this study are consistent with the theoretical claims of the theory of L_1 adaptive control in terms of robustness and performance. On the other hand, the developed procedure and the obtained results confirm the suitability of the PSI method and the MOVI software package for the multicriteria optimization of an adaptive FCS subject to desired control specifications.

The optimal design of the L_1 FCS significantly improved understanding of the design trade-offs between adaptation and robustness. Finally, the design guidelines learned during the interaction with GTM model utilizing MOVI software contributed to the successful flight verification and validation of the designed all-adaptive control law at NASA LARC.

References

[1] Jacklin, S. A., et al., "Verification, Validation, and Certification Challenges for Adaptive Flight-Critical Control System Software," *AIAA Guidance, Navigation and Control Conference,* AIAA-2004-5258, Providence, RI, August 2004.

[2] Wise, K. A., E. Lavretsky, and N. Hovakimyan, "Adaptive Control in Flight: Theory, Application, and Open Problems," *American Control Conference,* Minneapolis, MN, June 2006, pp. 5966–5971.

[3] Jacklin, S. A., "Closing Certification Gaps in Adaptive Flight Control Software," *AIAA Guidance, Navigation and Control Conference,* AIAA-2008-6988, Honolulu, HI, August 2008.

[4] Hovakimyan, N., and C. Cao, *L_1 Adaptive Control Theory,* Philadelphia, PA: Society for Industrial and Applied Mathematics, 2010.

[5] Jordan, T. L., W. M. Langford, and J. S. Hill, "Airborne Subscale Transport Aircraft Research Testbed-Aircraft Model Development," *AIAA Guidance, Navigation and Control Conference,* AIAA-2005-6432, San Francisco, CA, August 2005.

[6] Jordan, T. L., et al., "AirSTAR: A UAV Platform for Flight Dynamics and Control System Testing," *AIAA Aerodynamic Measurement Technology and Ground Testing Conference,* AIAA-2006-3307, San Francisco, CA, June 2006.

[7] Xargay, E., N. Hovakimyan, and C. Cao, "L_1 Adaptive Controller for Multi-Input Multi-Output Systems in the Presence of Nonlinear Unmatched Uncertainties," *American Control Conference,* Baltimore, MD, June–July 2010.

[8] Gregory, I. M., et al., "L_1 Adaptive Control Design for NASA AirSTAR Flight Test Vehicle," *AIAA Guidance, Navigation and Control Conference, AIAA-2009-5738,* Chicago, IL, August 2009.

[9] Gregory, I. M., et al., "Flight Test of an L_1 Adaptive Controller on the NASA AirSTAR Flight Test Vehicle," *AIAA Guidance, Navigation and Control Conference,* Toronto, Canada, August 2010.

[10] Kim, K. K., and N. Hovakimyan, "Development of Verification and Validation Approaches for L_1 Adaptive Control: Multi-Criteria Optimization for Filter Design," *AIAA Guidance, Navigation and Control Conference,* Toronto, Canada, August 2010.

[11] Statnikov, R. B., and J. B. Matusov, *Multicriteria Analysis in Engineering,* Dordrecht/Boston/London: Kluwer Academic Publishers, 2002.

[12] Statnikov, R. B., et al., *MOVI 1.4 (Multicriteria Optimization and Vector Identification) Software Package,* Certificate of Registration, Registere of Copyright, U.S.A., Registration Number TXU 1-698-418, Date of Registration: May 22, 2010.

[13] Sobol, I. M., and R. B. Statnikov, *Selecting Optimal Parameters in Multicriteria Problems,* 2nd ed., (in Russian), Moscow: Drofa, 2006.

[14] Xargay, E., et al., "L_1 Adaptive Flight Control System: Systematic Design and V&V of Control Metrics," *AIAA Guidance, Navigation, and Control Conference,* Toronto, Ontario, Canada, August 2–5, 2010.

[15] Stepanyan, V., et al., "Stability and Performance Metrics for Adaptive Flight Control," *AIAA Guidance, Navigation and Control Conference,* Chicago, IL, AIAA-2009-5965, August 2009.

[16] Bailey, R. E. and T. J. Bidlack, *Unified Pilot-Induced Oscillation Theory, Volume IV: Time-Domain Neal-Smith Criterion,* Tech. Rep. WL-TR-96-3031, Air Force Wright Laboratory, December 1995.

[17] Choe, R., et al., "L_1 Adaptive Control Under Anomaly: Flying Qualities and Adverse Pilot Interaction," *AIAA Guidance, Navigation and Control Conference,* Toronto, Canada, August 2010.

Conclusions

All real-life optimization problems, such as the design, identification, design of controlled systems, and operational development of prototypes, are essentially multicriteria. Constraints on the design variables, functional dependences, and criteria determine the feasible solution set. This set is the domain where one should be looking for the optimal solutions.

A statement of real-life optimization problem primarily involves justification of the feasible set. It is usually assumed that the statement of the optimization problem is the prerogative and skill of the expert. However, there are few who understand that the expert is practically unable to define the feasible set in the best way. The skill of the expert is a necessary, but by no means sufficient condition for the correct statement of the real-life problem. Since criteria are contradictory, the definition of the feasible solution set represents significant, sometimes insurmountable, difficulties. As a result, the expert solves the incorrectly stated optimization problem. If the problem has not been stated correctly, the application of optimization methods, however good they may be, is often ineffective. This situation is common today when solving real-life problems. Unfortunately, the expert will not be able to find the instructions on how to correctly determine the feasible solution set, and hence to state the optimization problem. All the numerous works devoted to real-life optimization discuss how to solve the problem but not how to state it correctly.

In order to construct the feasible solution set, a method called the parameter space investigation (PSI) has been created and successfully integrated into various fields of industry, science, and technology.

On the basis of the PSI method and MOVI software, we have: (1) demonstrated how to help an expert state and solve optimization problems and thus answered the fundamental question of where to search for optimal solutions; (2) explained the necessity of multicriteria analysis for constructing a feasible solution set; (3) shown tools for multicriteria analysis; and (4) presented the statement and solution of real-life optimization problems.

The correct determination of the feasible solution set is a major challenge in real-life optimization problems. Multicriteria analysis of the feasible solution set always provides the valuable information for decision-making. This analysis demonstrates the work of all constraints; the cost of making concessions in various constraints (i.e., what are the losses and the gains, expediency of modification of constraints, and resources for improving the object in all criteria).

The definition of the feasible solution set on the basis of the PSI method/ MOVI software has, for the first time, allowed the correct statement and solution of real-life problems.

Appendix
Examples of Calculation of the Approximate Compromise Curves

This appendix considers two simple problems where compromise curves can be found by analytical procedure; the calculated approximate compromise curves are compared to the exact ones [1].

Analytical Approach

If two functions $\Phi_1(\alpha)$ and $\Phi_2(\alpha)$ are differentiable, then one can try to find the locus of contact of level surfaces $\Phi_1(\alpha) = b_1$ and $\Phi_2(\alpha) = b_2$, see [1, 2]. In these points

$$\mathbf{grad}\ \mathbf{\Phi}_1 = -\lambda\ \mathbf{grad}\ \mathbf{\Phi}_2$$

This vector equation is equivalent to a set of n scalar algebraic equations

$$\partial\Phi_1/\partial a_j = -\lambda\partial\Phi_2/\partial\alpha_j \quad (j = 1, 2, \ldots, n)$$

which, generally speaking, defines the following curve in the design variable space:

$$\alpha_1 = \varphi_1(\lambda), \ldots, \alpha_n = \varphi_n(\lambda)$$

If a segment of this curve, on which $\lambda \geq 0$, belongs to the feasible set D_α, then it belongs to the Pareto optimal set E_α as well. For example, Figure A.1, which corresponds to Example A.1, shows arc AA', which consists of Pareto optimal points, and arc $A'A''$, which consists of points of contact, which are not Pareto optimal, although both arcs are described by the same equation:

$$\alpha_2 = 4\alpha_1/(3\alpha_1 - 1)$$

In this case the segment of the compromise curve is determined by the following parametric equations:

$$\Phi_1 = \Phi_1(\varphi_1(\lambda), \ldots, \varphi_n(\lambda)),$$
$$\Phi_2 = \Phi_2(\varphi_2(\lambda), \ldots, \varphi_n(\lambda)), \quad (\lambda \geq 0)$$

Along this curve

$$d\Phi_1 = \sum_{j=1}^{n} (\partial\Phi_1/\partial a_j) d\varphi_j = -\lambda \sum_{j=1}^{n} (\partial\Phi_2/\partial a_j) d\varphi_j = -\lambda d\Phi_2$$

which gives the following expression for the inclination of the compromise curve:

$$d\Phi_2/d\Phi_1 = -1/\lambda$$

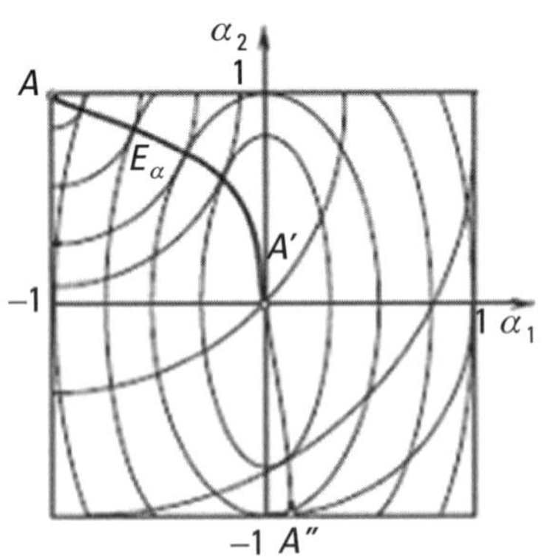

Figure A.1 Level curves Φ_1 = const and Φ_2 = const, set of points of contact $AA'A''$, Pareto optimal points—arc AA'.

Example A.1

The following two criteria are defined over square

$$D_\alpha = \{-1 \le \alpha_1 \le 1, -1 \le \alpha_2 \le 1\}:$$
$$\Phi_1 = 4\alpha_1^2 + \alpha_2^2 \text{ and } \Phi_2 = (\alpha_1 + 1)^2 + (\alpha_2 - 1)^2$$

these criteria need to be minimized.

Absolute minimums of the functions Φ_1 and Φ_2 are reached, respectively, at points (0, 0) and (–1, 1), which belong to D_α. Therefore, line E_α should link these points up. The above-mentioned analytical procedure gives a segment AA' of the hyperbola $a_2 = 4\alpha_1 (3\alpha_1 - 1)^{-1}$ shown in Figure A.2(a).

Part of D_α, located to the right of the line $AA'A''$, maps to the curvilinear figure $BB'B''B'''B_R$ at the criteria plane, while part of D_α located to the left of line $AA'A''$ maps to $BB'B''B_L$. Therefore, points located in the middle of D_Φ have two prototypes in D_α. Line BB' is the compromise curve [Figure A.2(b)].

The results of this problem's calculation are shown in Figures A.3 and A.4. Figure A.3 shows the exact compromise curve $\Phi_2 = u(\Phi_1)$ and the approximate line $\Phi_2 = u_N(\Phi_1)$. For $N = 511$, exact and approximate curves coincide within 0.01. Figure A.4 shows all trial points in D_α and in D_Φ. Notice the increased density of these points in the middle part of D_Φ.

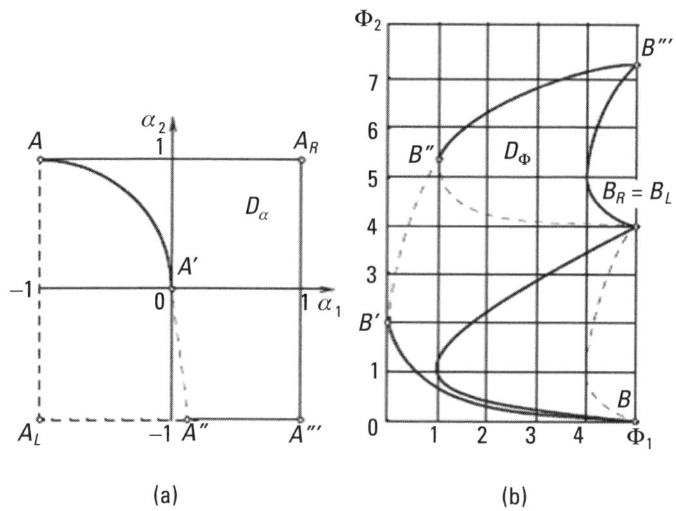

Figure A.2 (a) Feasible solution set D_α in the design variable space and (b) feasible solution set D_Φ in the criteria space. Arc AA' represents set E_α, while arc BB' represents set E_Φ.

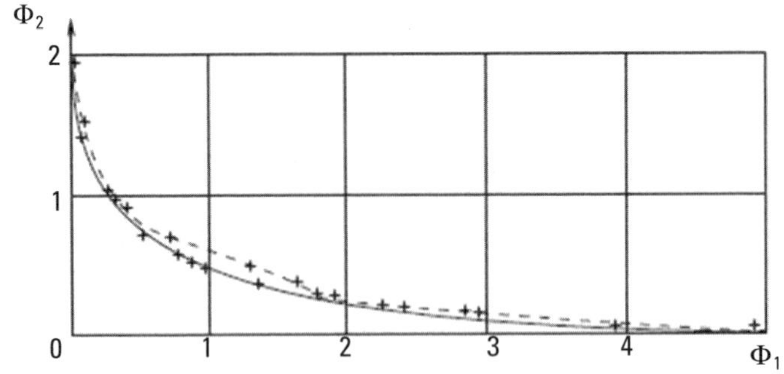

Figure A.3 Exact (solid line) and approximate (dotted line) compromise curve for $N = 63$. Crosses represent approximately Pareto optimal points for $N = 255$.

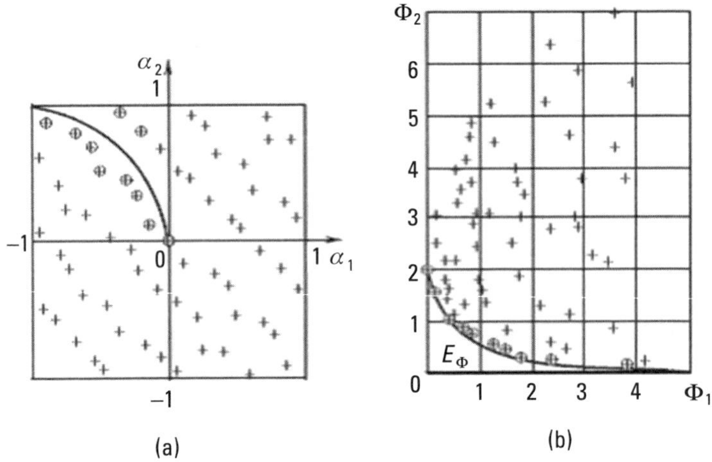

Figure A.4 Trial points (a) in D_α in design variable space and (b) in D_Φ in criteria space for $N = 63$. Approximate Pareto optimal points are circled.

Example A.2

The following two criteria are defined over square $\Pi = (-0.5 \leq \alpha_1 \leq 0.5, 0 \leq \alpha_2 \leq 1)$:

$$\Phi_1 = \alpha_1^2 + 4\alpha_2^2 \text{ and } \Phi_2 = (\alpha_1 + 1)^2 + (\alpha_2 - 1)^2$$

these criteria need to be minimized taking into account functional constraint

$$|\alpha_2 - \alpha_1 - 0.375| \geq 0.125$$

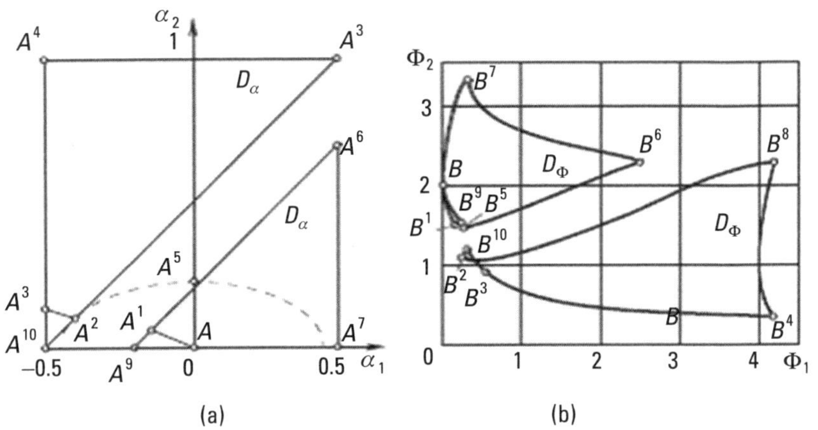

3

Figure A.5 Feasible solutions set (a) D_α in design variable space and (b) D_Φ in criteria space. Set E_α consists of two segments AA^1A^5 and $A^2A^3A^4$, while set E_Φ consists of segments BB^1B^5 and $B^2B^3B^4$.

In this problem set D_α is represented by square Π with a strip-shaped cutout (Figure A.5). Set E_α consists of two segments AA^1 and A^2A^3 of the hyperbola $\alpha_2 = -\alpha_1(3\alpha_1 + 4)^{-1}$, segment A^3A^4 of the border line $\alpha_1 = -0.5$, and segment A^1A^5 of the border line $\alpha_2 - \alpha_1 = 0.25$. Figure A.5 also shows set D_Φ on the criteria plane. However, those parts of D_Φ that consist of points that have two prototypes in D_α are much smaller here than in Example A.1.

The exact compromise curve is also shown in Figure A.6. Figure A.7 shows trial points in D_Φ. Results obtained for $N = 63$ are not sufficient to conclude that set D_Φ consists of two separate parts; however, results for $N = 255$ make that certain. Here q is number of points that satisfy the functional constraint $|\alpha_2 - \alpha_1 - 0.375| \geq 0.125$.

The computation above was performed using the LP_τ generator. Below we also show Example A.2 for 64, 256, 512, and 1,024 trials[1] (Figures A.8 through A.11) performed using a random number generator. In these figures the Pareto optimal set is marked in red in the criteria space.

1. This example has been performed by Douglas S. Parten.

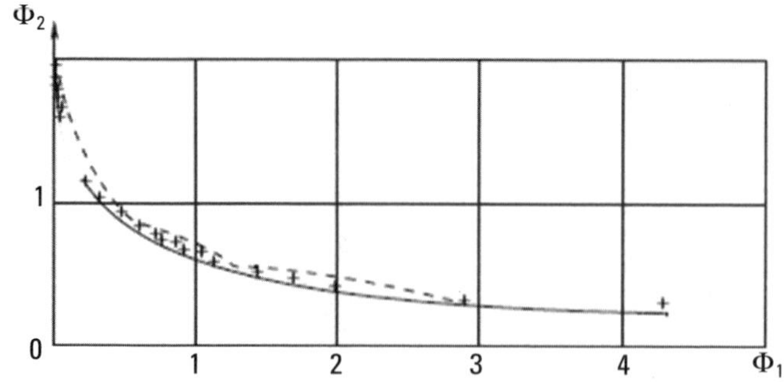

Figure A.6 Exact (solid line) and approximate (dotted line) compromise curves for $N = 63$. Crosses represent approximate Pareto optimal points for $N = 255$.

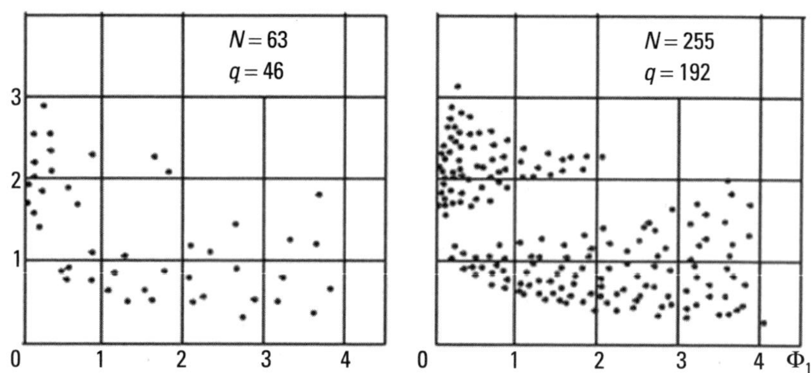

Figure A.7 All q trial points in the criteria space for $N = 63$ and for $N = 255$.

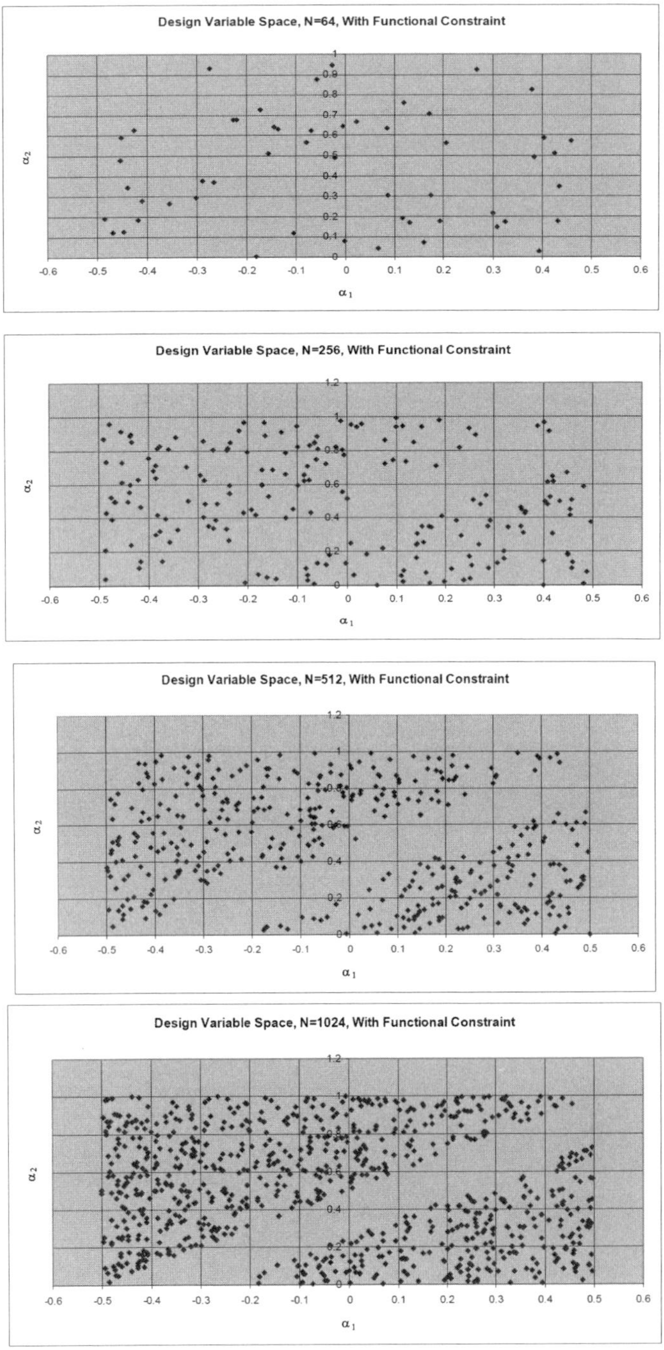

Figure A.8 Investigations with a functional constraint. The left column shows the trial points in the design variable space. The right column shows the trial points in the criteria space.

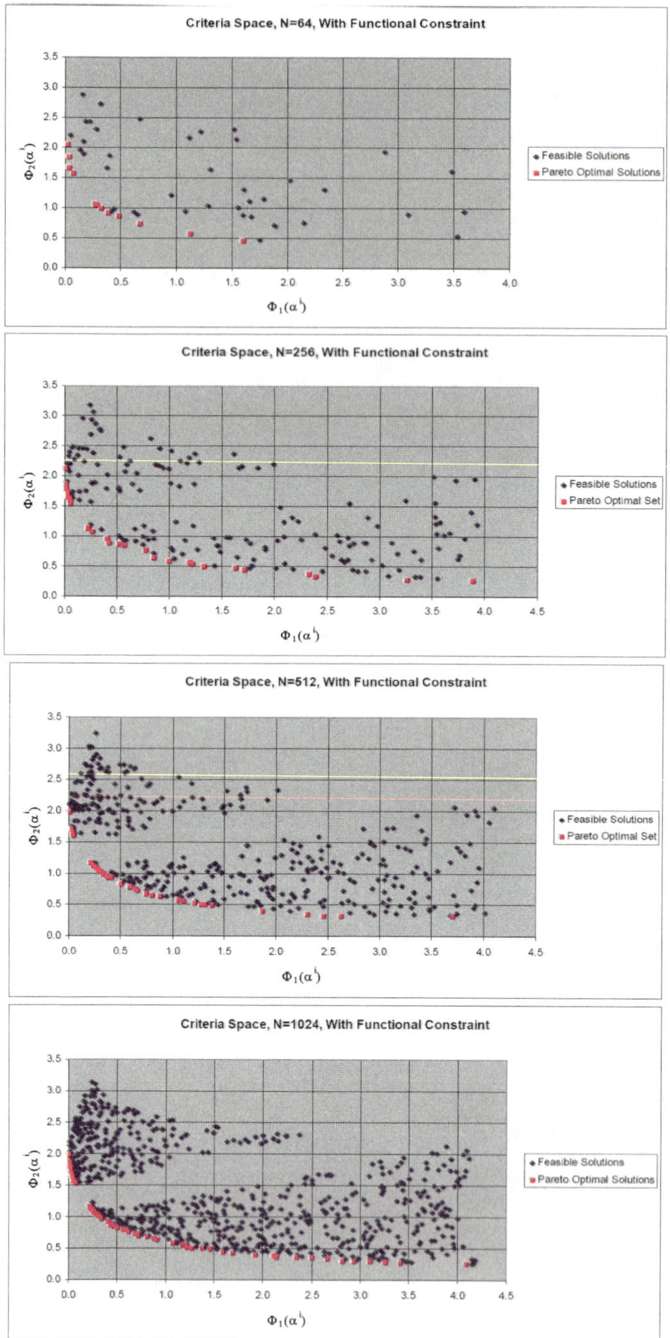

Figure A.8 (continued).

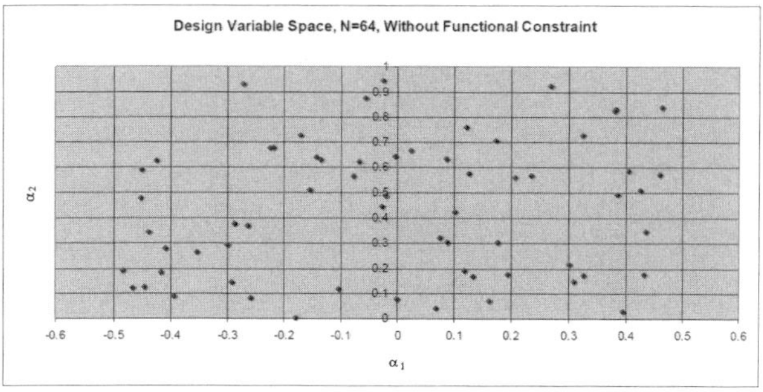

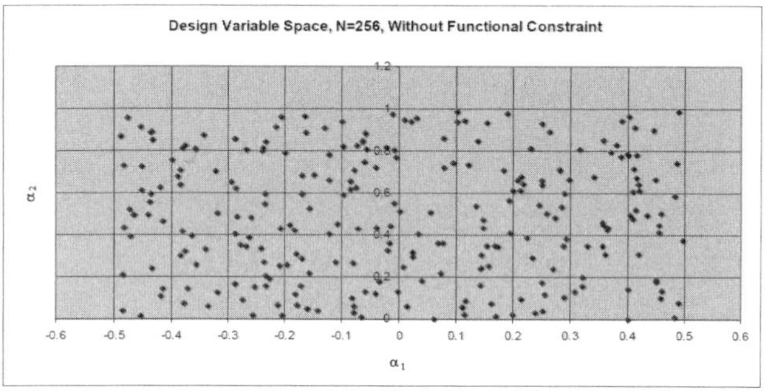

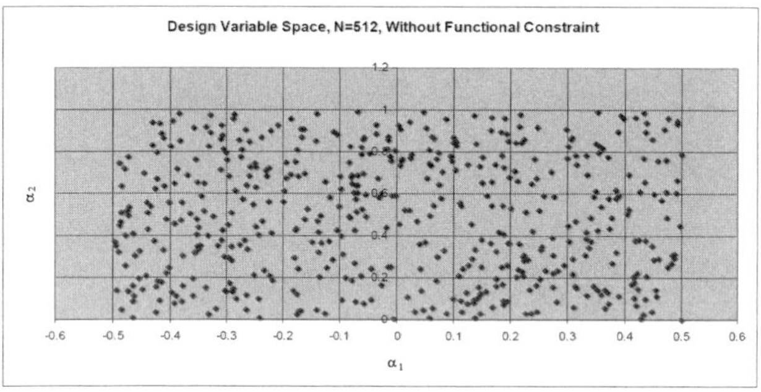

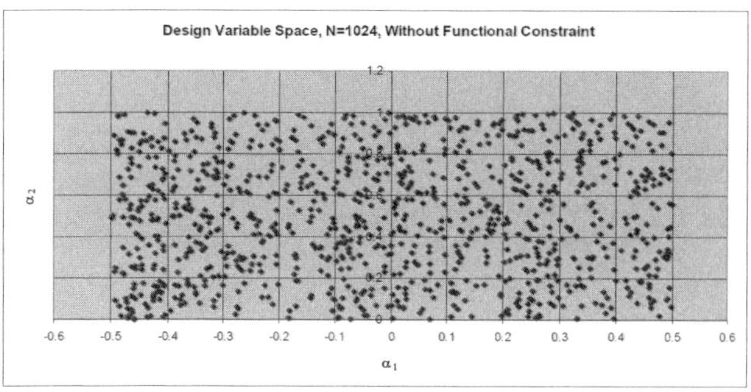

Figure A.9 Investigations without a functional constraint. The left column shows the trial points in the design variable space. The right column shows the trial points in the criteria space.

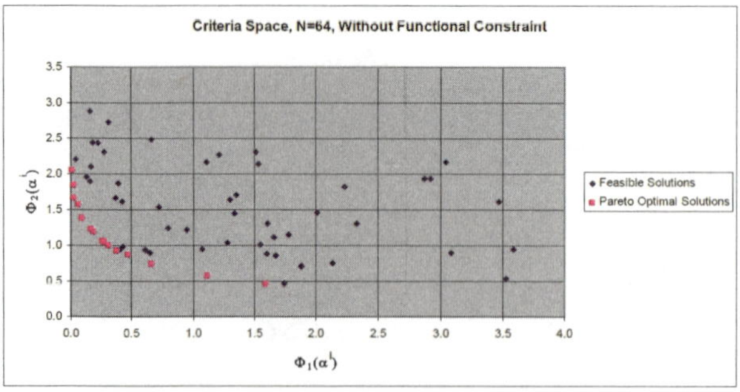

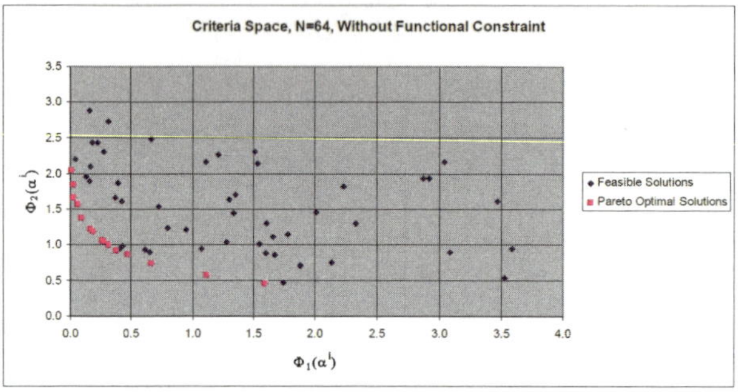

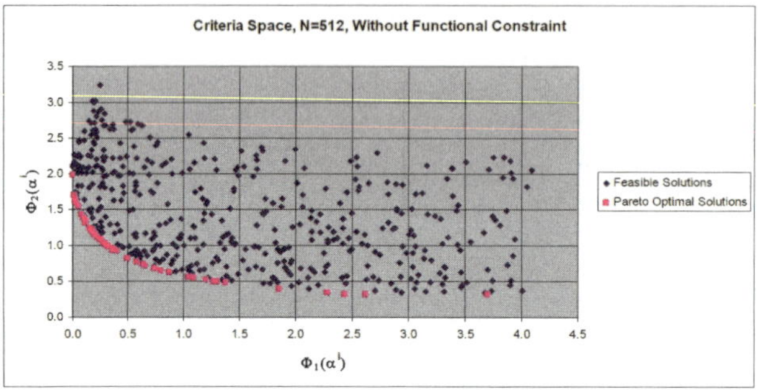

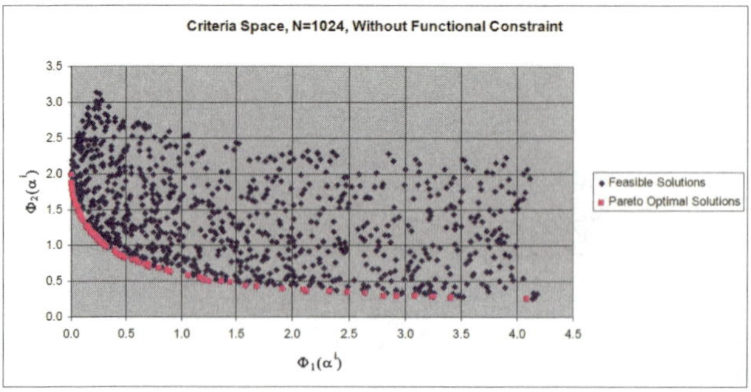

Figure A.9 (continued).

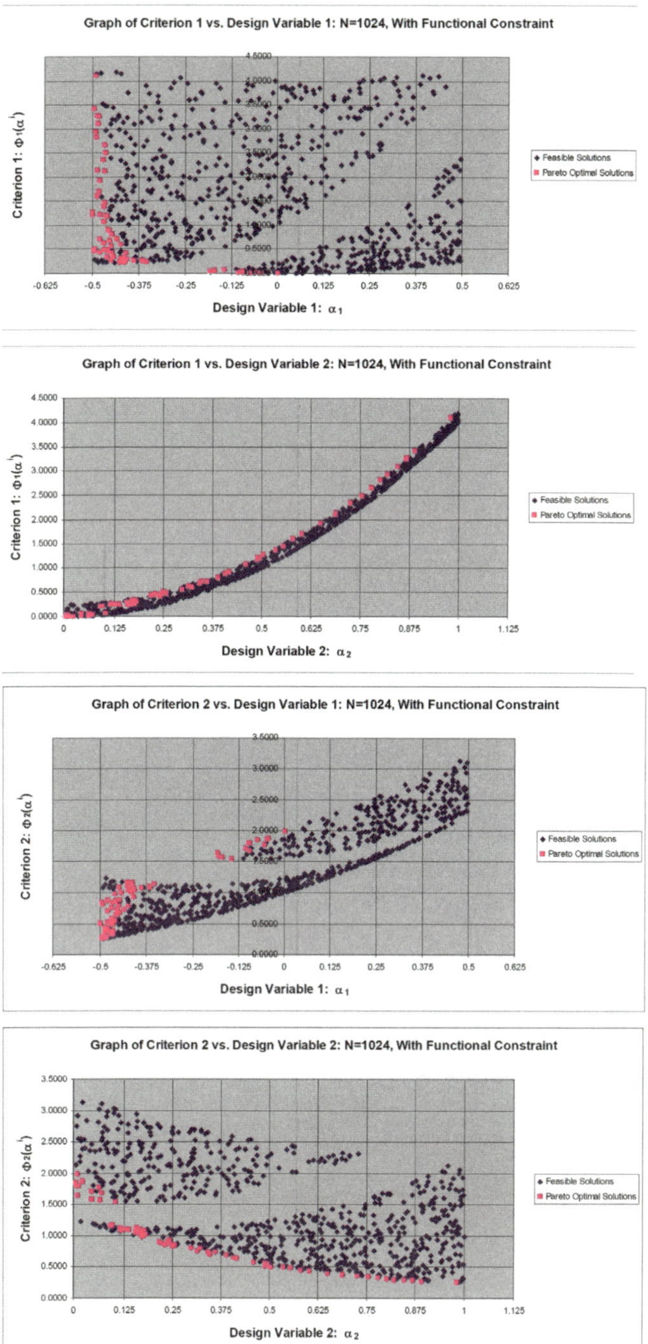

Figure A.10 Graphs of criteria versus design variables for $N = 1{,}024$ trials, with a functional constraint.

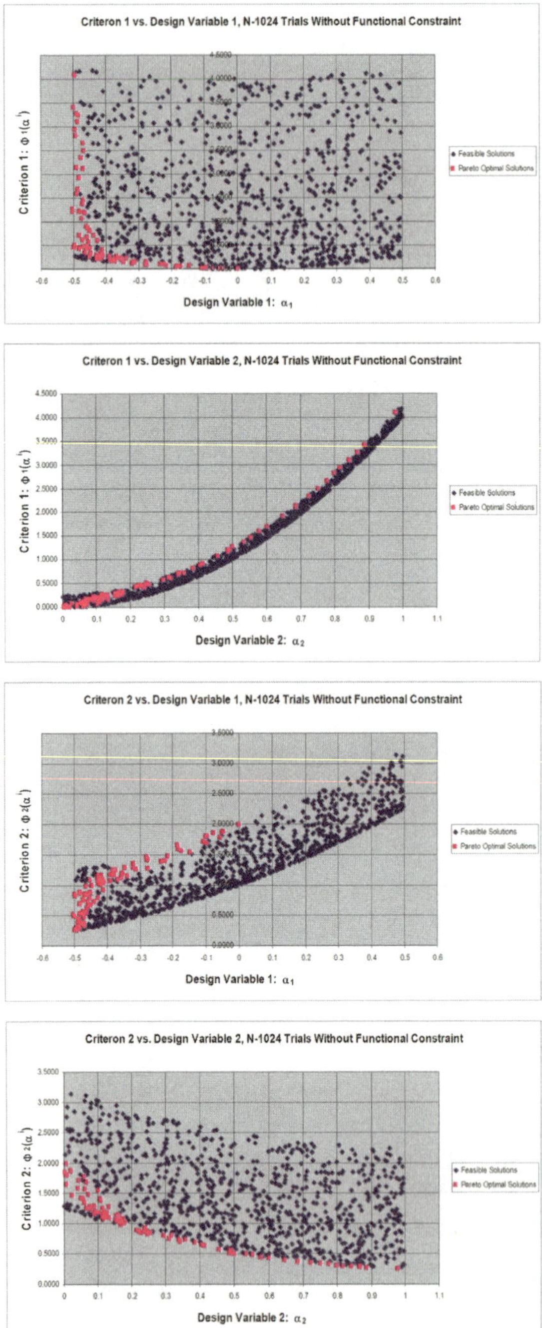

Figure A.11 Graphs of criteria versus design variables for N= 1,024 trials, without a functional constraint.

References

[1] Sobol', I. M., and R. B. Statnikov, *Selecting Optimal Parameters in Multicriteria Problems,* 2nd ed., (in Russian), Moscow: Drofa, 2006.

[2] Bartel D.L., and R. W. Marks, "The optimal design of mechanical systems with competing design objectives," *Journal of Engineering for Industry, Trans. ASME,* Vol. 96, No. 1, 1974, pp.171–178.

About the Authors

Roman Statnikov is a professor and principal research scientist in the Russian Academy of Sciences, Mechanical Engineering Institute, Russian Academy of Sciences, and since 2002 he has been a senior researcher and instructor at the Naval Postgraduate School in Monterey, California. Professor Statnikov has developed the parameter space investigation (PSI) method for the correct statement and solution of multicriteria problems jointly with Professor I. M. Sobol'. The PSI method has been widely integrated into various fields of industry, science, and technology. The PSI method has been used in design of the space shuttle, nuclear reactors, missiles, automobiles, ships, aircraft, metal tools, and bridges. Professor Statnikov has developed an interdisciplinary course "Multicriteria Analysis in Engineering" and taught it online since 2004 at the Naval Postgraduate School. He is also a consultant for many companies.

Alexander Statnikov is an assistant professor in the Center for Health Informatics and Bioinformatics and Department of Medicine, New York University School of Medicine, New York, New York. He is also the director of the Computational Causal Discovery Laboratory.

Index

meeh